Electron Correlations in Solids, Molecules, and Atoms

NATO ADVANCED STUDY INSTITUTES SERIES

A series of edited volumes comprising multifaceted studies of contemporary scientific issues by some of the best scientific minds in the world, assembled in cooperation with NATO Scientific Affairs Division.

Series B: Physics

Recent Volumes in this Series

Volume 78 – Excitations in Disordered Systems
edited by M. F. Thorpe

Volume 79 – Artificial Particle Beams in Space Plasma Studies
edited by Bjørn Grandal

Volume 80 – Quantum Electrodynamics of Strong Fields
edited by W. Greiner

Volume 81 – Electron Correlations in Solids, Molecules, and Atoms
edited by Jozef T. Devreese and Fons Brosens

Volume 82 – Structural Elements in Particle Physics and Statistical Mechanics
edited by J. Honerkamp, K. Pohlmeyer, and H. Römer

Volume 83 – Physics on Ion–Ion and Electron–Ion Collisions
edited by F. Brouillard

Volume 84 – Physical Processes in Laser–Materials Interactions
edited by M. Bertolotti

Volume 85 – Fundamental Interactions: *Cargèse 1981*
edited by Maurice Lévy, Jean-Louis Basdevant, David Speiser, Jacques Weyers, Maurice Jacob, and Raymond Gastmans

Volume 86 – Surface Mobilities on Solid Materials: Fundamental Concepts and Applications
edited by Vu Thien Binh

This series is published by an international board of publishers in conjunction with NATO Scientific Affairs Division

A Life Sciences B Physics	Plenum Publishing Corporation London and New York
C Mathematical and Physical Sciences	D. Reidel Publishing Company Dordrecht, The Netherlands and Hingham, Massachusetts, USA
D Behavioral and Social Sciences E Applied Sciences	Martinus Nijhoff Publishers The Hague, The Netherlands

Electron Correlations in Solids, Molecules, and Atoms

Edited by

Jozef T. Devreese and Fons Brosens

University of Antwerpen (RUCA and UIA)
Antwerpen, Belgium

PLENUM PRESS • NEW YORK AND LONDON

Published in cooperation with NATO Scientific Affairs Division

Library of Congress Cataloging in Publication Data

NATO Advanced Study Institute on Electron Correlations in Solids, Molecules, and Atoms (1981: Turnhout, Belgium)
Electron correlations in solids, molecules, and atoms.

(NATO advanced study institutes series. Series B, Physics; v. 81)
"Proceedings of a NATO Advanced Study Institute on Electron Correlations in Solids, Molecules, and Atoms, held at Corsendonk Conference Center, between July 20-31, 1981, in Turnhout, Belgium"—Verso t.p.
Includes bibliographies and indexes.
Contents: Determination of S (q,ω) by inelastic electron and X-ray scattering/S. Schnatterly—Charge density wave phenomena in potassium/A. Overhauser—Electron-hole liquid/K. S. Singwi—[etc.]
1. Electron configuration—Congresses. 2. Solid state physics—Congresses. 3. Molecules—Congresses. 4. Atoms—Congresses. I. Devreese, J. T. (Jozef T.) II. Brosens, Fons. III. North Atlantic Treaty Organization. IV. Title. V. Series.
QC176.8.E4N335 1981 530.4 82-10160

ISBN-13: 978-1-4613-3499-6 e-ISBN-13: 978-1-4613-3497-2
DOI: 10.1007/978-1-4613-3497-2

Proceedings of a NATO Advanced Study Institute on Electron Correlations in Solids, Molecules, and Atoms, held at Corsendonk Conference Center, between July 20-31, 1981, in Turnhout, Belgium

Softcover reprint of the hardcover 1st edition 1983

A Division of Plenum Publishing Corporation
233 Spring Street, New York, N.Y. 10013

PREFACE

From July 20 till 31, 1981, the Advanced Study Institute on "Electron Correlations in Solids, Molecules and Atoms", sponsored by NATO, was held at the University of Antwerpen (U.I.A.), in the Conference Center Corsendonk.

In the last few years, the problem of many-electron correlations has gained renewed attention, due to recent experimental and theoretical developments.

From the theoretical point of view, more sophisticated treatments of the homogeneous electron gas model evolved, including dynamical aspects of the electron correlation in the dielectric response. Furthermore, the homogeneous electron gas, which served as a model for simple metals, was extended to include spin- and charge-density waves and phasons. The concept of elementary excitations too was introduced not only in perfectly ordered metallic crystals, but also in magnetic alloys, in liquid metals and alloys, in semiconductors, and even in molecules and atoms. Fairly accurate quantitative calculations of these effects recently became possible, ranging from plasmon frequencies in atoms, over dielectric response of semiconductors and resistivity in magnetic alloys to electron-hole liquids and their phase separation.

The recent technological evolution allowed for more accurate measurements in previously unaccessible domains, e.g. X-ray scattering and fast electron energy loss at large wavevector. Moreover, these new developments opened new perspectives in physics, accompanying or even introducing the new concepts which also evolved in the theory.

The aim of the Advanced Study Institute was to collect these new insights and developments in the vast domain of electron correlations. For this purpose, adequate discussion periods formed an important ingredient of the Institute, where lecturers and participants had active interactions and gained a deeper total picture. The formal lectures, presented in a coherent and didactic way, provided the basis for these discussions.

We would like to thank the lecturers for their appreciated contributions, and the participants for their active participation at workshops and informal discussions. We also thank Dr. P. Van Camp for his important contribution in the preparation of the scientific aspect of the meeting, and Mr. Van Den Brempt and his staff at the Conference Center for their assistance in the organisational aspect. Furthermore, we thank Drs. J. Van Royen and Drs. M. Marien for their help in preparing the author and subject index and Mrs. H. Evans for her contribution in the preparation of the manuscript. It is our pleasure to express special thanks to the Vlaamse Wetenschappelijke Stichting and its president for providing support for the publication of these proceedings.

Finally, financial support by NATO, and the co-sponsorship of Agfa-Gevaert N.V., Bell Telephone Mfg. Co. N.V., Compagnie Maritime Belge, EBES, Esso Belgium N.V., IBM Belgium N.V., Kredietbank, Metallurgie Hoboken Overpelt, Siemens N.V., Sociale Dienst V.E.V., Hessenatie and the Vlaamse Wetenschappelijke Stichting, are gratefully acknowledged.

CONTENTS

DETERMINATION OF S(q,ω) BY INELASTIC ELECTRON AND X-RAY SCATTERING

S. Schnatterly

Physics Department
University of Virginia, Charlottesville
Virginia 22901, USA

One of the major efforts involved in studying many particle systems is the discovery of the effective force between the particles. In atomic and condensed matter physics the underlying force is the Coulomb potential. Due to the presence of many particles simultaneously interacting the net potential between any pair of particles is never just the pure $1/r$ potential discovered by Coulomb, but is strongly modified, usually being reduced. This phenomenon is called screening. Let us begin with a classical example which is easily described. We shall see that even here there are surprises.

Consider an ionic solution with equal numbers of positive and negative ions diffusing through the solvent. At any given instant the solution might be pictured as consisting of pairs of nearest neighbor ions of opposite charge. What is the force between pairs? At first sight it appears that there should be a $1/r^3$ dipole-dipole interaction between pairs. This is not however the correct result.

Debye and Huckel obtained an approximate solution to this problem using mean field theory [1]. If we wish to know the force between positive ions, they proposed averaging over the positions of the negative ions. This results in a spherically symmetric negative charge cloud surrounding each positive charge. The net interaction between two such clouds can easily be evaluated and the result is a Yukawa potential : $V(r) = \dfrac{e^2 \exp(-\alpha r)}{r}$.

It is not at all obvious that the mean field solution

to this problem should be an accurate one since all the charges are point-like and the diffusional motion is slow compared with speed of light which is the rate at which the electric fields are transmitted between particles.

A few years ago a rigorous solution to this problem was obtained by applying methods developed recently in constructive quantum field theory to classical statistical mechanics [2] . The result is that the mean field solution obtained by Debye and Hückel is indeed correct : The Coulomb potential is exponentially screened. Somehow the ions arrange themselves cleverly with respect to their neighbors so that the sum of all the multipole terms in an expansion of the potential adds up to form a Yukawa potential.

The subtelty of this classical problem should serve as a warning. It is convenient that such a simple model is accurate. The complexity comes in trying to understand why this should be so.

In considering quantum systems, two extremes are a system of particles whose wave functions extend nearly uniformly throughout the entire sample volume, such as a metal, and a system with highly localized particles such as a molecular solid. We shall be concerned here with the first type of system.

Consider first a simple metal in which it is a reasonable approximation to smear out the positively charged ion cores into a uniform background charge. Then we have no periodic field and the only interactions to be dealt with are those between the electrons themselves. One way of describing the properties of such a many-particle system is by defining a complex dielectric response function $\varepsilon(q,\omega)$ [3] . In addition to providing a common language for the description of different experiments, many basic properties of the many-particle system such as the ground state energy and compressibility can be expressed in terms of $\varepsilon(q,\omega)$.

Suppose an external potential ϕ (r,t) is applied to the system. The electrons will respond to ϕ_e producing an induced potential $\phi_i(r,t)$. Then the total potential is $\phi_T(r,t) = \phi_e + \phi_i$. This is the potential which a given electron experiences as it moves through the system. We shall almost always be dealing with the Fourier Transform functions $\phi(q,\omega)$ instead of $\phi(r,t)$.

We wish to calculate the induced density $\rho_i(q,\omega)$. We shall assume that ρ_i is linearly related to ϕ_T :

$$\rho_i(q,\omega) = \alpha(q,\omega)\ \phi_T(q,\omega) \ .$$

It is important that the total potential ϕ_T appears in this relation rather than the external potential ϕ_e.

Since only electrostatic forces are important here, we can relate ϕ_i to ρ_i using Poisson's equation :

$$\nabla^2\phi_i(r,t) = -4\pi\rho_i(r,t) .$$

So that

$$\phi_i(q,\omega) = \frac{4\pi}{q^2}\rho_i(q,\omega)$$

Now we have :

$$\phi_i(q,\omega) = \frac{4\pi}{q^2}\alpha(q,\omega)\phi_T(q,\omega)$$

The dielectric response function is defined by :

$$\frac{1}{\varepsilon(q,\omega)} = \frac{\phi_T(q,\omega)}{\phi_e(q,\omega)}$$

This can be related to our function $\alpha(q,\omega)$ using the above results :

$$\frac{1}{\varepsilon(q,\omega)} = \frac{\phi_T(q,\omega)}{\phi_T(q,\omega) - \phi_i(q,\omega)} = \frac{1}{1 - \dfrac{\phi_i(q,\omega)}{\phi_T(q,\omega)}} = \frac{1}{1 - \dfrac{4\pi}{q^2}\alpha(q,\omega)}$$

Or,

$$\varepsilon(q,\omega) = 1 - \frac{4\pi}{q^2}\alpha(q,\omega)$$

Analytical Properties of $\varepsilon(q,\omega)$:

The real and imaginary parts of $\varepsilon(q,\omega)$ are not independent functions. The requirement that the system response cannot preceed the applied potential (i.e. that the response be causal) gives rise to an integral relationship between the real and imaginary parts of $\varepsilon(q,\omega)$. Let us write $\varepsilon(q,\omega) = \varepsilon_1(q,\omega) + i\varepsilon_2(q,\omega)$ with ε_1 and ε_2 both real. Then the requirement of causality gives rise to the Kramers-Kronig relations :

$$\varepsilon_1(q,\omega) = 1 - \frac{2}{\pi} P \int_0^\infty \frac{x\varepsilon_2(q,x)}{x^2 - \omega^2} dx$$

$$\varepsilon_2(q,\omega) = - \frac{2\omega}{\pi} P \int_0^\infty \frac{\varepsilon_1(q,x) - 1}{x^2 - \omega^2} dx$$

So if an experiment can determine either of these functions, the other can be derived from it. Unfortunately no experiment is known which accomplishes this directly. The quantities which are directly measurable are related to functional combinations of ε_1 and ε_2. For example in analyzing inelastic scattering measurements, a more useful version of the above relations is :

$$\mathrm{Re}\left(\frac{1}{\varepsilon(q,\omega)} - 1\right) = \frac{2}{\pi} P \int_0^\infty \mathrm{Im}\left(\frac{1}{\varepsilon(q,x)}\right) \frac{x}{x^2 - \omega^2}\, dx$$

$$\mathrm{Im}\left(\frac{-1}{\varepsilon(q,\omega)}\right) = \frac{2\omega}{\pi} P \int_0^\infty \mathrm{Re}\left(\frac{1}{\varepsilon(q,x)} - 1\right) \frac{dx}{x^2 - \omega^2}$$

Similarly in analyzing optical reflectivity measurements what is needed is a relation between the measured reflectance $R(\omega)$ and the corresponding phase $\theta(\omega)$. The result is

$$\theta(\omega) = \frac{\omega}{\pi} \int_0^\infty \frac{\ell n(R(x)/R(\omega))}{\omega^2 - x^2}\, dx$$

Then the dielectric response function $\varepsilon(\omega) = (n + ik)^2$ can be obtained from :

$$\sqrt{R}\, e^{i\theta} = \frac{n - 1 + ki}{n + 1 + ki}$$

These Kramers-Kronig relations require integrating a measured quantity over the range $0 \leqslant \omega \leqslant \infty$. Since no experiment covers this entire range, end-point extrapolations become necessary. Generally speaking, the wider the range that can be covered, the more accurate will be the result.

A very useful relation is the oscillator strength sum rule :

$$\int_0^\infty \omega\, \mathrm{Im}\left(\frac{-1}{\varepsilon(q,\omega)}\right) d\omega = \int \omega\, \mathrm{Im}\, \varepsilon(q,\omega)\, d\omega = \frac{\pi}{2} \omega_p^2$$

This relation can be used as a constraint either in data reduction or in devising models for $\varepsilon(q,\omega)$. For example in analyzing electron scattering data if the sample thickness is unknown, there is an unknown overall scale factor relating the number of counts observed at a given energy to $\varepsilon(q,\omega)$. The oscillator strength sum rule can be used to obtain that scale factor.

THE RANDOM PHASE APPROXIMATION

A widely used simple model for the dielectric response of a metallic system is called the Random Phase Approximation (RPA) [3,4]. It consists of making the approximation :

$$\alpha(q,\omega) = e^2 Q(q,\omega)$$

where

$$Q(q,\omega) = \sum_k \frac{n(k) - n(k+q)}{\omega + \hbar\left(\frac{k^2}{2m} - \frac{(k+q)^2}{2m}\right) + i\delta}$$

Q is the response function of a non-interacting electron gas. It is a sum over all possible transitions between filled states n(k) and empty states n(k+q) with momentum larger by q and energy higher by $\hbar\omega$. Energy and momentum conservation are the only restrictions on the sum. Since no interactions between the electrons are included n(k) and n(k+q) are simply the Fermi occupation numbers familiar from statistical mechanics.

Since we are using Q for α, how can we say that the RPA is any better than a simple non-interacting electron gas model? The answer has to do with the way Q is used. We defined α to give us the induced charge density in response to the total potential ϕ_T which is the sum of the applied and induced potentials. The induced potential is an aspect of the electron-electron interaction. In fact the induced potential is just the average of the particle interaction, averaged over the electrons' orbits. Fluctuations about the average are ignored in this approximation.

As a simple example let us evaluate the induced potential caused by an external static potential ϕ_e in the limit as $q \to 0$ according to the RPA. We have :

$$\phi_i = \phi_T - \phi_e$$

$$= \frac{q^2}{4\pi\alpha}\phi_i - \phi_e$$

$$\phi_i\left(1 - \frac{q^2}{4\pi\alpha}\right) = -\phi_e$$

$$\phi_i = -\frac{\phi_e}{1 - \frac{q^2}{4\pi\alpha}} = -\frac{\phi_e}{1 - \frac{q^2}{4\pi e^2}\frac{1}{Q}}$$

In the limit $q \to 0$, Q remains finite so we have $\phi_i = -\phi_e$. The electrons respond so as to completely cancel the applied potential. The result is familiar from electrostatics and was first demonstrated by Michael Faraday with the "ice pail" experiment. So we see that the induced potential is by no means a small effect to be thought of as a refinement.

The dielectric function $\varepsilon(q,\omega) = 1 - \frac{4\pi}{q^2} Q(q,\omega)$ is called the Lindhard dielectric function [5] .

STATIC SCREENING

Suppose a static potential $\phi_e(q,o)$ is applied to the system. This can be done by dissolving impurities in the metal which have a different valence than the atoms of the host metal. This problem is similar to the ionic case described earlier. The potential in the neighborhood of the impurity is given by the Fourier transform of :

$$\phi_T(q,o) = \frac{\phi_e(q,o)}{\varepsilon(q,o)}$$

where $\phi_e(q,o)$ is the fourier transform of the impurity potential. For example if the impurity potential is e^2/r then $\phi_e(q,o) = 4\pi e^2/q^2$. The static limit of the Lindhard function is :

$$\varepsilon(q,o) = 1 + \frac{k_{FT}^2}{2q^2} [1 + (\frac{4_{kF}^2 - q^2}{4q_{kF}}) \ln |\frac{q + 2k_F}{q - 2k_F}|]$$

where $k_F = (3\pi^2 n)^{1/3}$ is the Fermi wave vector and $k_{FT} = \frac{6\pi ne^2}{E_F}$ is the Fermi-Thomas wave vector. If this function were independent of q, as is approximately the case is an insulator, then the Coulomb field surrounding the impurity would retain its 1/r dependence but be altered in magnitude. The variation with q is significant however in the metallic case. The logarithmic singularity at $q = 2k_F$ gives rise to a striking quantum mechanical effect first noticed by Friedel. The effect is to cause the long range form of the induced charge density to vary as :

$$\rho_i(r) = \frac{A \cos(2k_F r)}{(k_F r)^3}$$

This long range oscillatory behavior is a direct result of the sharpness of the Fermi distribution - the fact that states with $k > k_F$ are unoccupied with those with $k < k_F$ are filled at $T = 0$. Experimental verification of the above formula for $\rho_i(r)$ provides evidence for the sharpness of the occupation number distribution function in real metals.

The best direct experimental observation of Friedel oscillations comes from NMR measurements on magnetic impurities in noble metals. The magnetic impurity scatters spin up conduction electrons differently than spin down electrons. Therefore the

Friedel oscillations are slighlty different for spin up and spin down conduction electrons. In other words there is an induced static spin density wave surrounding the impurity.

The spin density wave contributes to the magnetic field seen by the nuclei of the noble metal atoms near the impurity. NMR measurements on the noble metal host reveal satellite lines at both higher and lower fields than that of the unperturbed host nuclei. These satellites are very weak and difficult to observe. Figure 1 shows an example of some recent measurements on Cu Mn [6] . The solid line is a model calculation which includes the static limit of the Lindhard function. The overall agreement is quite satisfactory.

FINITE FREQUENCY, SMALL WAVE VECTOR RESPONSE

If a long wavelength, finite frequency potential is applied to a degenerate electron system, with frequency less than the classical plasma frequency $\omega_p = (\frac{4\pi ne^2}{m})^{1/2}$, it responds almost as completely as for the $q \to 0$ $\omega \to 0$ limit. For the range $\omega/q \gg v_F$, $q \ll k_F$ the Lindhard function reduces to

$$\varepsilon(q,\omega) = 1 - \frac{\omega_p^2}{\omega^2} (1 + \frac{3}{5} \frac{q^2 v_F^2}{\omega^2})$$

For this entire range of independent variable values, the denominator of $Q(q,\omega)$ does not vanish, so $\varepsilon(q,\omega)$ has no imaginary part. This means that the electron plasma responds to the applied potential without internal dissipation. No energy is lost in the process of causing the electrons to respond to and screen out the applied potential. This also means that the collective modes of the electrons, the plasma oscillations, are undamped. The condition for a collective mode is :

$$\varepsilon(q,\omega) = 0 \quad \text{or} \quad \omega^2 \approx \omega_p^2 + \frac{3}{5} v_F^2 q^2$$

As q increases, this small q expansion eventually becomes inaccurate. More important, at a certain value of q, the denominator of Q develops a zero and the plasma oscillations become damped. This is called Landau damping. In this process the plasma oscillation transfers all of its energy and momentum to a single electron. See figure 2. The value of q for which this begins to be possible is approximately given by :

$$q_c \approx \frac{\omega_p}{v_F}$$

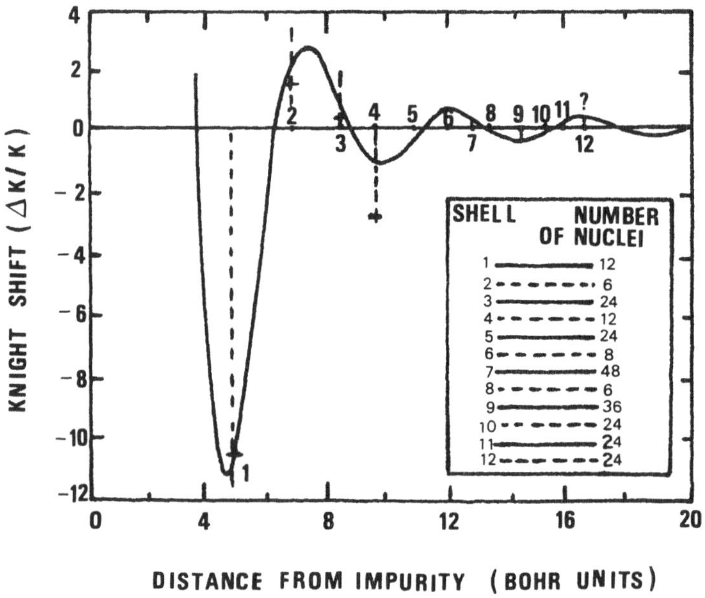

Figure 1. Relative Knight shift of Cu NMR as a function of distance from a Mn impurity. See reference 7.

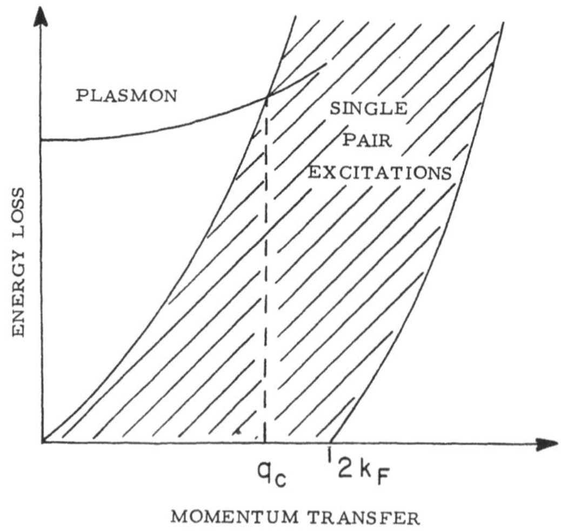

Figure 2. Plasmon dispersion according to the RPA. No damping is possible in this approximation for $q < q_c$.

Within the RPA there is no damping for $q < q_c$. We shall see later that real metals do exhibit damping in this regime.

ULTRAVIOLET OPTICAL PROPERTIES

This was the first method used to obtain quantitative information about the dielectric response function of representative solids in the range of the plasma frequency. Today such measurements are considerably more reliable than they used to be because of the availability of synchrotron light sources and improved vacuum systems and surface preparation methods. Synchrotron sources greatly extend the wavelength range available improving the accuracy of the Kramers-Kronig analysis. In addition reflectivity measurements in the 5-30 eV range can be supplemented with absorption coefficient measurements using thin films at higher energies.

Optical measurements probe the transverse response function since the electron field of the photon is perpendicular to its wave vector. The Lindhard function above is the longitudinal dielectric function for the system. It has been shown that the longitudinal and transverse functions are the same in the limit $q\langle r\rangle \ll 1$ where $\langle r\rangle$ is the mean spacing between electrons in the system. This inequality is well satisfied for optical absorption studies. In this energy range of interest the photon wavelengths are always large compared with mean interparticle spacings. So results of optical measurements can be compared with the $q \to 0$ limit of the Lindhard function.

Shiles et al have recently published a composite analysis of the optical properties of Al [7]. They used various published optical measurements covering the range 0.04 eV to 10,000 eV, and bombined them using self-consistent Kramers-Kronig analysis. By evaluating the oscillator strength sum rule they were able to indicate where the data seemed to be most in need of improvement.

Figure 2 shows their determination of $K(\omega)$ where the index of refraction is given by $n(\omega) + iK(\omega)$. The substantial drop at 15 eV is due to passing through the plasma frequency. The metal is considerably more transparant above $\hbar\omega p$ than below it. At higher energies, core excitations appear : the LII, III threshold at 72 eV (2P $\to$ conduction band) and the K edge at 1500 eV (1S $\to$ conduction band). High resolution measurements of the absorption profiles of these edges reveal information about how the conduction electrons respond to the presence of the suddenly created core hole.

Figure 3 shows the reflectance in the neighborhood of the plasma frequency and corresponding values of the energy loss function $\mathrm{Im}\left(\frac{-1}{\varepsilon(q,\omega)}\right)$. There is considerable variation in the data

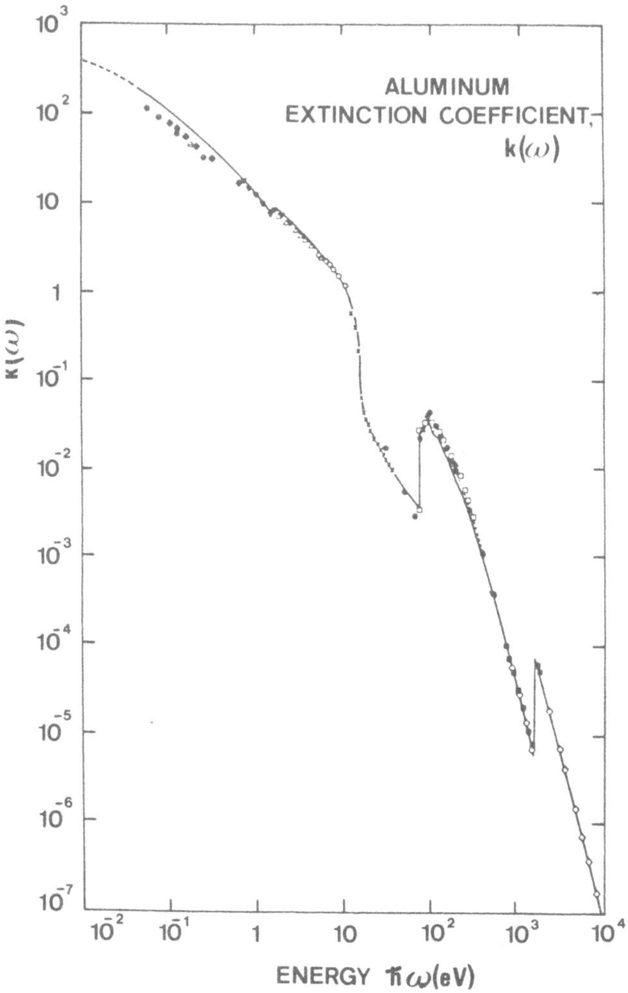

Figure 3. Extinction coefficient of Al as determined in Reference 7.

due largely to the fact that surface contamination and surface roughness can significantly alter the results.

Optical measurements such as these can provide a value for the only free parameter in the Lindhard function, ω_p. Knowledge of this parameter allows the others (k_F, v_F) to be calculated if one is willing to assume an average effective mass for the conduction band.

It is clear from these measurements that damping is present in these $q \approx 0$ experimental results. The Lindhard function provides no account of this damping. We shall inquire later about the physical origin of the damping.

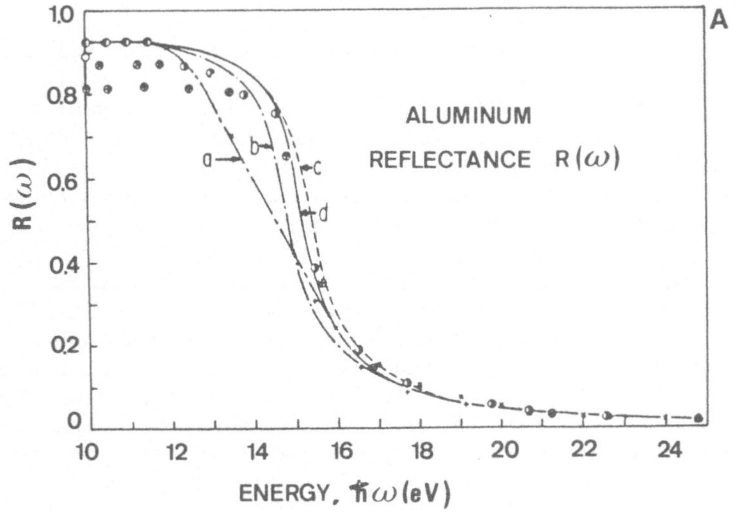

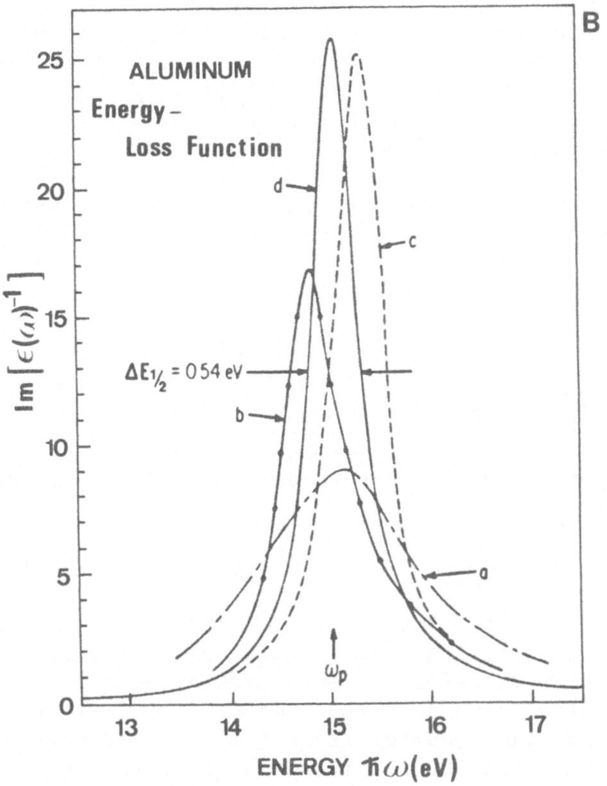

Figure 4. Normal incidence relectance and Im $(\frac{-1}{\varepsilon(o,\omega)})$ in the plasmon energy range for Al as determined in reference 7.

INELASTIC ELECTRON SCATTERING

When a fast electron passes through a thin solid film the electrons in the solid feel a pulse of electric field. That pulse can be fourier analyzed and each fourier component thought of as an "effective photon". IES is therefore capable of playing the role of a wideband photon source extending from the infrared to the x-ray range. For spectroscopic purposes this is roughly equivalent to a synchrotron light source.

There are important differences between optical absorption and IES however. In optical absorption, virtually no momentum is transferred to the sample. An "effective photon", however, which has a longitudinal rather than transverse electric field, can do so. Therefore in forward scattering, involving very little momentum transfer, IES and optical absorption measurements reveal the same information. For example, they both involve the dipole selection rule in the case of localized excitations. Away from the forward direction IES provides new information not obtainable with opticle absorption. In the case of a localized excitation (e.g. core excitations), multipole terms beyond dipole enter into the transition matrix element. In the case of delocalized excitations (e.g. plasmons, valence excitations), the dispersion of the excitation can be measured throughout the Brillouin Zone.

If the electron energy is high energy so multiple scattering is not a problem (or can be removed), first order time dependent perturbation theory (or the Born approximation) leads directly to an expression for the inelastic scattering cross section [8] :

$$\frac{d^2\sigma}{d\omega\, d\Omega} = \frac{4\hbar}{a_o^2 q^4} S(q,\omega)$$

where $S(q,\omega)$, the dynamic structure factor of the system is given by :

$$S(q,\omega) = \frac{1}{2\pi\hbar N} \int dt\, e^{i\omega t} \langle n_q(t)\, n_{-q}(0)\rangle$$

In the above, a_o is the Bohr radius, $\hbar q$ the momentum, and $\hbar\omega$ the energy transferred to the sample by the fast electron. $n_q(t)$ is the space fourier transform of the electron density at time t and N the number of electrons in the sample. $S(q,\omega)$ is proportional to the space and time fourier transform of the density-density correlation function. The importance of this result is that on the left hand side we have something directly measurable by an experimentalist-counts per second in a given energy and solid angle interval. On the right is an important correlation function

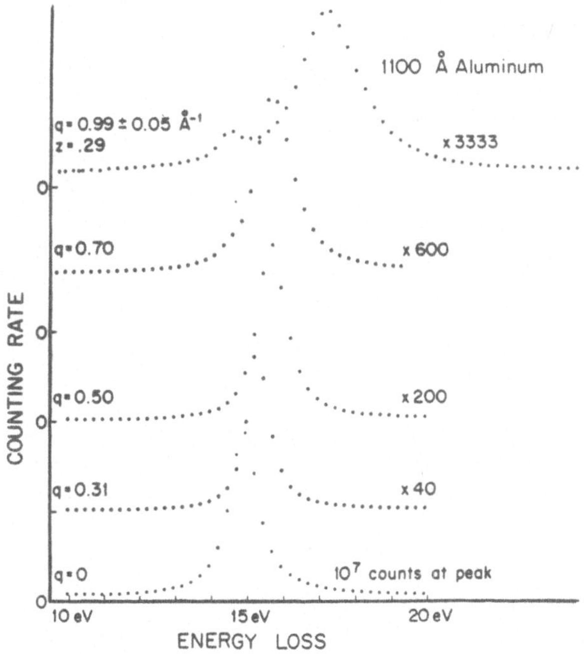

Figure 5. Al IES measurements in the plasmon energy range for valence q values.

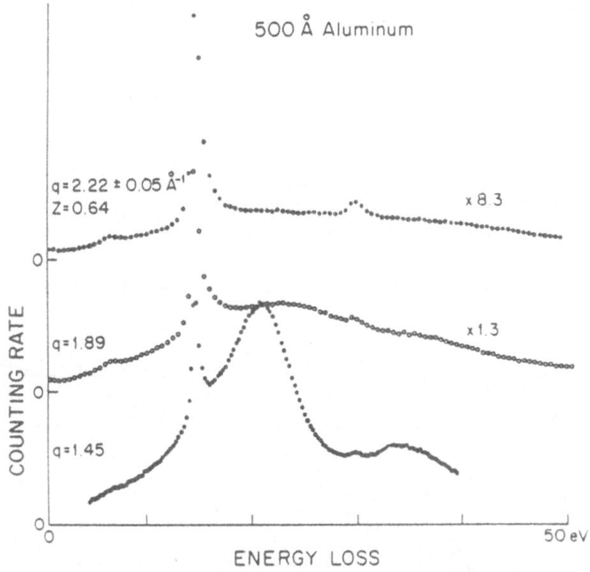

Figure 6. Al IES measurements in the plasmon energy range for various q values.

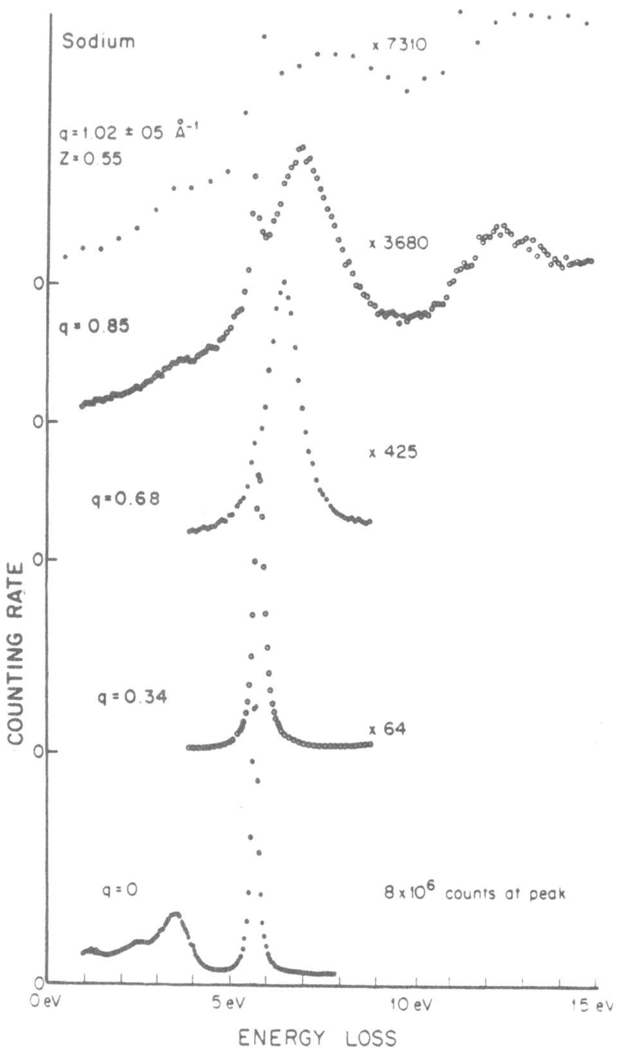

Figure 7. Na IES measurements in the plasmon energy range for various q values.

describing how the electrons in the solid are distributed, a function which is implicit in any complete theory of electronic structure.

Using the fluctuation-dissipation theorem we can relate $S(q,\omega)$ to $\varepsilon(q,\omega)$:

$$S(q,\omega) = \frac{q^2}{4\pi^2 e^2 n} \mathrm{Im} \frac{-1}{\varepsilon(q,\omega)}$$

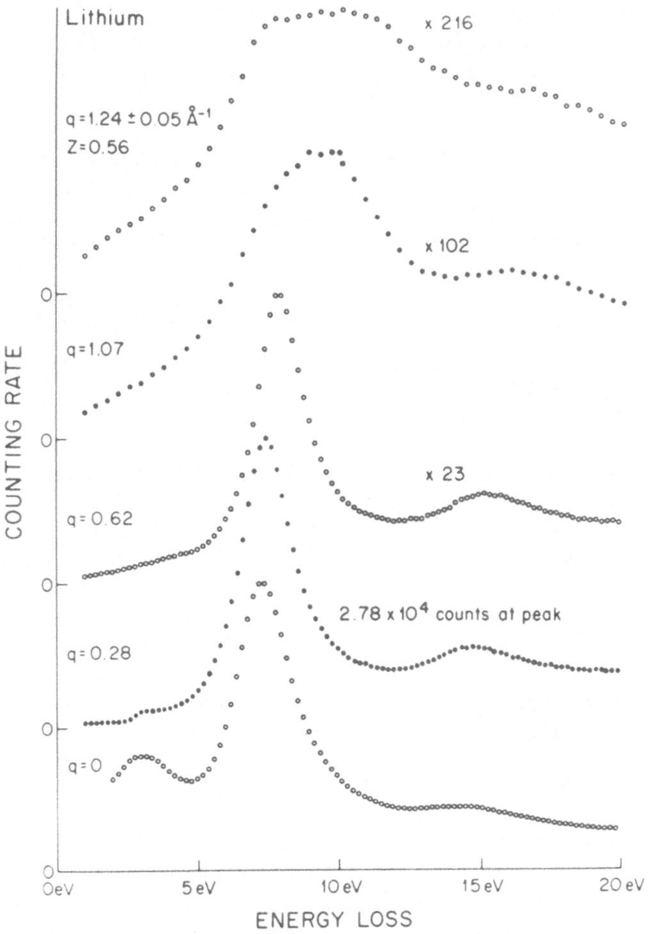

Figure 8. Li IES measurements in the plasmon energy range for various q values.

where n is the electron density in the sample. Therefore we have :

$$\frac{d^2\sigma}{d\omega\ d\Omega} = \frac{h}{\pi^2 e^2 a_o^2 n} \frac{1}{q^2} \text{Im} \frac{-1}{\varepsilon(q,\omega)}$$

So measurements of the IES cross section give direct information about $\varepsilon(q,\omega)$. The advantage of IES over the other techniques mentioned above is that both q and ω can be varied (quasi-independently). The capability of IES to study electronic excitations is similar to that of neutron scattering to study lattice vibrational excitations.

Figures 9-12 show IES measurements of Al, Na and Li in the plasmon energy range for various momentum transfers q [9]. The dispersion of the plasmon (i.e. its change in energy with q) is clearly evident. Also evident is the fact that the width of the peak is non-zero and varies with q starting from q = 0. The RPA result is that the width of the peak should remain zero up to q_c at which point it sharply increases. Figures 9-12 show the measured plasmon energies and widths plotted versus q^2. It is usually stated that both quantities increase linearly with q^2 for small q. This is often the case but exceptions are evident in the data shown here. The Al plasmon dispersion appears to change slope for q^2 below about 0.5 $Å^{-2}$. This is a relatively small effect but is quite repeatable and has been noticed by others. The Al plasmon width increases linearly with q^2 up to $q \simeq q_c$ where the slope increases by about a factor of 6.

The Na dispersion appears linear within experimental error and the width piecewice linear as in Al. The Li results are more complex. The error bars are larger for Li because the plasmon width is greater. Even with the larger errors it appears that the dispersion shows non linear behavior over the range shown. The width is more anomalous, showing a decrease with q before increasing, with a minimum around 0.45 k_F.

IES spectroscopy is capable of producing more than just the position and width of the plasma line : The shape of the spectrum over a wide energy range can be accurately determined. The largest problem in making such measurements is multiple scattering. Two kinds of multiple scattering are most important : elastic-inelastic and inelastic-inelastic.

In the first case the fast electron scatters twice in passing through the sample, once elastically and once inelastically. Since elastic scattering is nearly q-independent while the inelastic cross section falls as Y_q-2, the importance of this type of double scattering increases with q. The effect of this double scattering can be seen clearly in figure 6. The sharp peak at 15 eV is caused by fast electrons that created a $q \approx 0$ plasmon and scattered elastically with momentum equal to that set by the spectrometer. This reproduces the $q \simeq 0$ spectrum on top of the high q background. This kind of double scattering which is usually referred to as thermal diffuse scattering can usually be removed from the data since the q = 0 spectral shape is known.

Double inelastic scattering is more complex. The problem is that the shape of $S(q,\omega)$ depends on q, and all that is known about the two q values involved is that they add to give the value set by the spectrometer. Various approximations have been made to handle this problem. Since the double inelastic spectrum is generally broad and featureless it is often adequate to assume

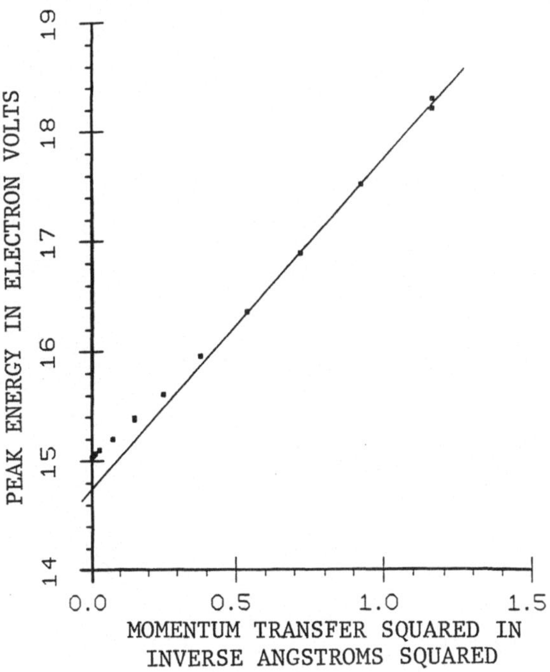

Fig. 9. Al plasmon dispersion.

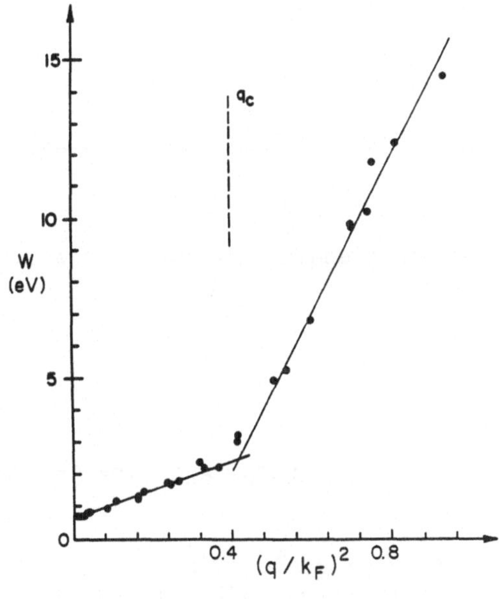

Figure 10. Width of the Al plasmon peak versus q^2.

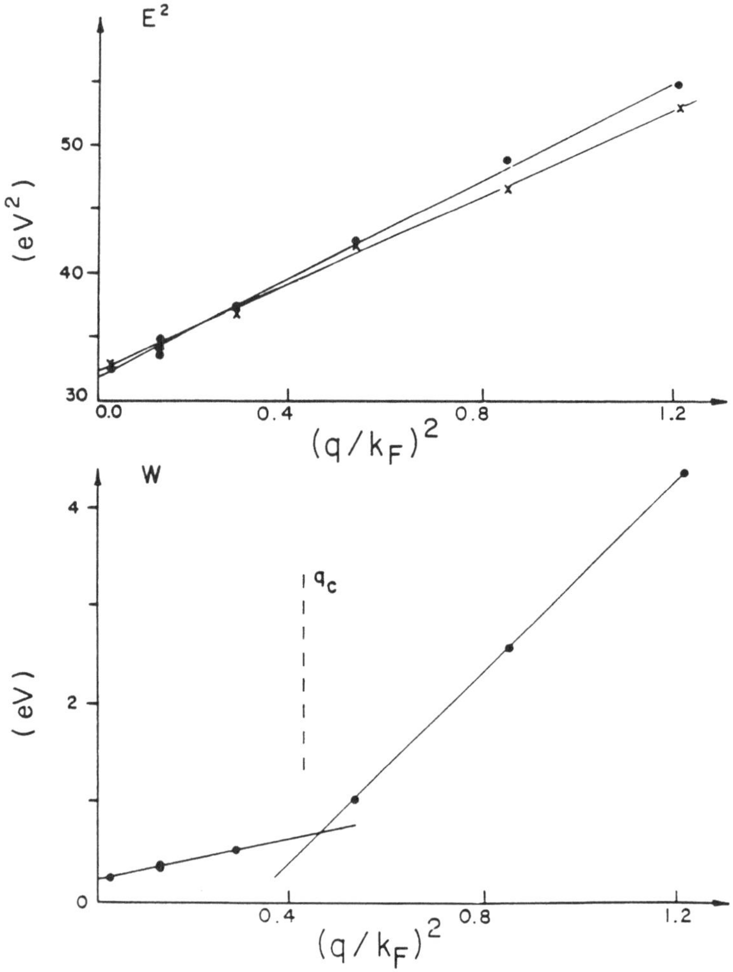

Figure 11. Na plasmon energy and width versus q^2.

the two q values are equal, each being one half of the value set by the spectrometer.

Figure 13 shows an example of multiple scattering for the case of Al. Two sample thicknesses, 500 Å and 1100 Å should show differing amounts of multiple scattering. The thermal diffuse peak at 15 eV is apparent in both measurements, and both show a broad asymmetric structure peaked around 23 eV, which is the single

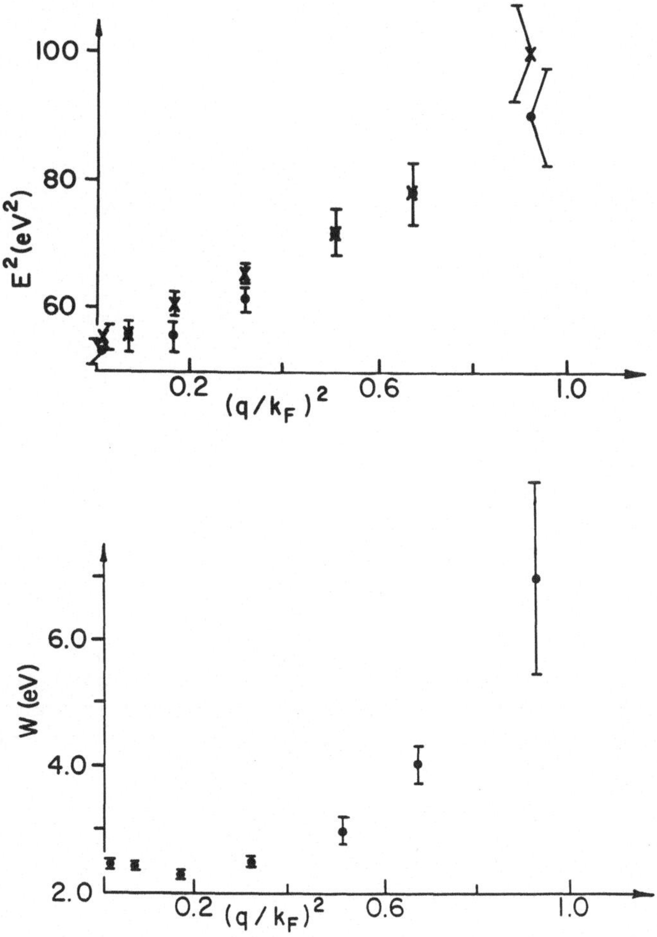

Figure 12. Li plasmon energy and width versus q^2.

scattering structure we desire. At higher energies the thicker sample shows a larger contribution which is due to double inelastic scattering. Comparison of different sample thicknesses serves as a check on the algorithm used in removing double scattering from the measured data.

Since the overall cross section varies as $1/v^2$ where v is the fast electron velocity it is clearly advantageous to use as high a beam energy as possible. In addition rather thin sample

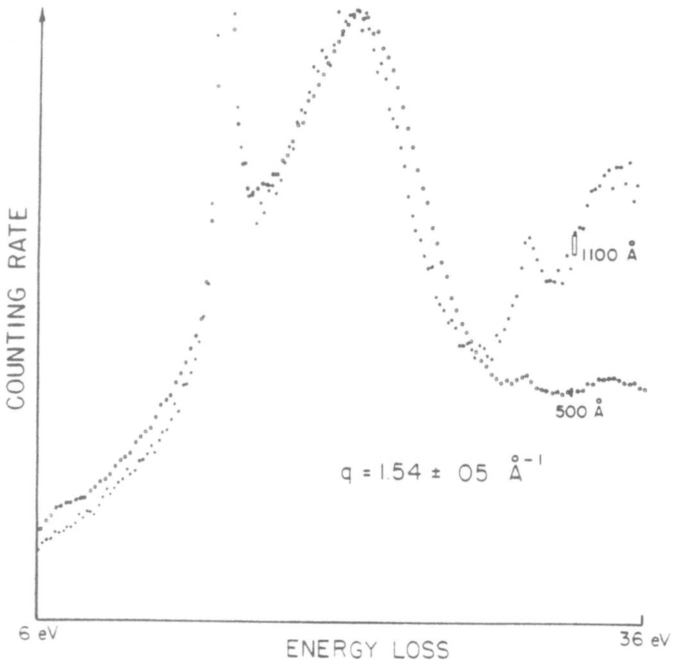

Figure 13. Al plasmon peak for q = 1.54 Å^{-1} using samples 500 Å and 1100 Å thick. The effect of double inelastic scattering can be seen near the high energy end of the region shown.

films are required to further reduce multiple scattering. What counts is the product $1/v^2\ell$ where ℓ is the sample thickness.

BEYOND THE RANDOM PHASE APPROXIMATION

Various phenomenological modifications of the Lindhard function have been proposed to account for some of the major differences between the RPA result and measurements such as those shown above. The most obvious difference is the finite damping manifest in the data for $q < q_c$. Mermin has shown how to include damping in the RPA in a manner that does not violate charge conservation [10].

Another important effect which we have virtually ignored to this point is the indistinguishability of the electrons. Electrons with parallel spins cannot have the same orbital quantum numbers. Generally speaking this has the effect of keeping parallel spin electrons from getting closer to each other than roughly $1/k_F$. This is similar to the effect of screening described above. In reality then the parallel spin electrons don't need to be screened from each other as much as antiparallel spin electrons since the Pauli principle tends to keep them apart anyway. When averaging over all electrons the net result is that the effective screening does not need to be as strong as the Lindhard function indicates because the Pauli principle is helping out.

One means of approximately including the effects of the electron-electron interaction beyond RPA is by modifying the Lindhard dielectric function ε_L using the expression :

$$\varepsilon(q,\omega) - 1 = \frac{\varepsilon_L(q, \;) - 1}{1 - G(q,\omega)(\varepsilon_L(q,\omega) - 1)}$$

where $G(q,\omega)$ is an arbitrary function. This has the form of a Clausius-Mossotti type local field correction. This form is intuitively reasonable since presumably the RPA gives the correct uniform response to the applied perturbation while it is the local field in the vicinity of a given electron that is desired. In addition this form is suggested by perturbation treatments using short range interactions as well as density functional approaches [11].

With $G(q,\omega)$ an arbitrary function this form can clearly be adjusted to fit any measured data. What is desired is to find a simple form for G which is capable of describing measurements and then seek a microscopic origin for this form. An approach which has been used is to allow G to be complex but to depend only on q [9]. Then at a given q value there are three free parameters : Im E, the imaginary part of the energy, ReG and ImG, the real and imaginary parts of G. Using a non-linear least squares fitting routine, measurements on Al and Na were fit with the above modified Lindhard function [9].

An example of the quality of the fit is shown in figure 14, for Al at $q = 1.47$ Å^{-1}. The thermal diffuse peak has been ignored and double inelastic scattering has been estimated and subtracted as shown. The dots are the measurements and the solid, dashed an dot-dashed lines are fits using the modified Lindhard function. Curiously, the measured data are sharper near the peak than any of the fit functions.

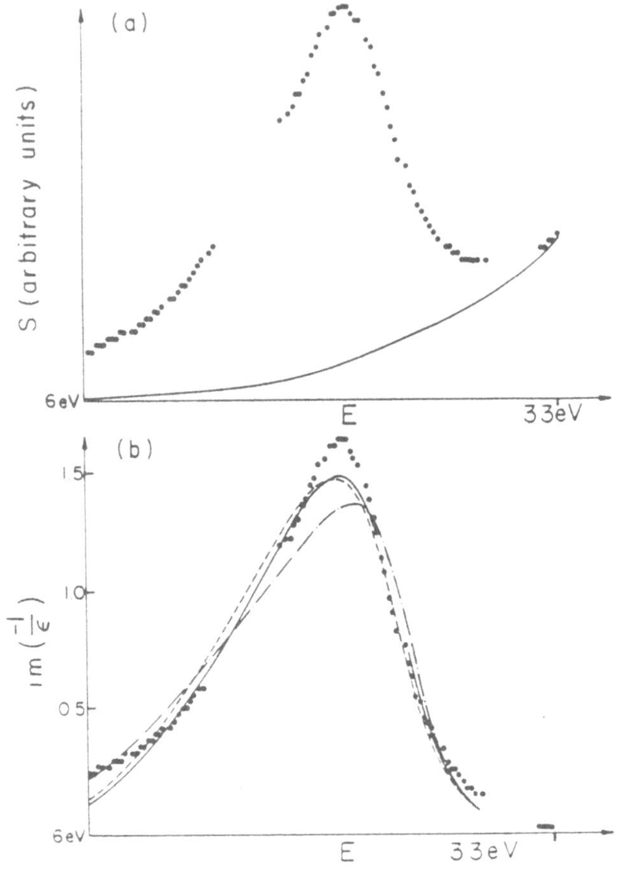

Figure 14. Al plasmon peak for q = 1.47 $Å^{-1}$. Upper curve shows the raw data with the thermal diffuse peak removed, and an estimate of double inelastic scattering drawn as a solid line. The lower curves are fits using the modified Lindhard function described in the text after removal of double scattering.

Figures 15 and 16 show values of ImE, ReG, and ImG as a function of q for Al and Na respectively. ReG is linear in q^2 with a slope about twice that obtained by Vashishta and Singwi [11], and ImE and ImG are piecewise linear in q^2. For $q > q_c$ both

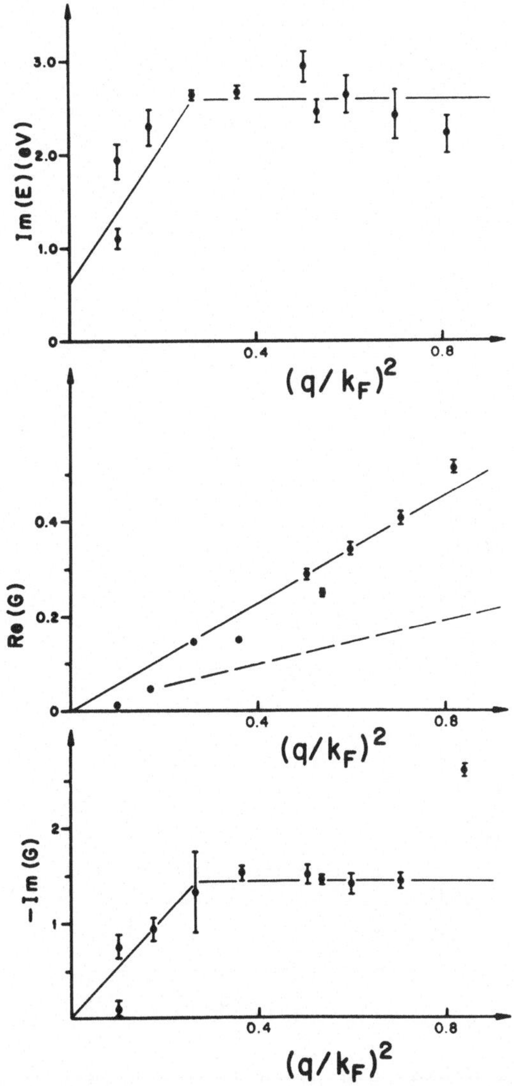

Figure 15. Modified Lindhard parameters versus $(q/k_F)^2$ for Al.

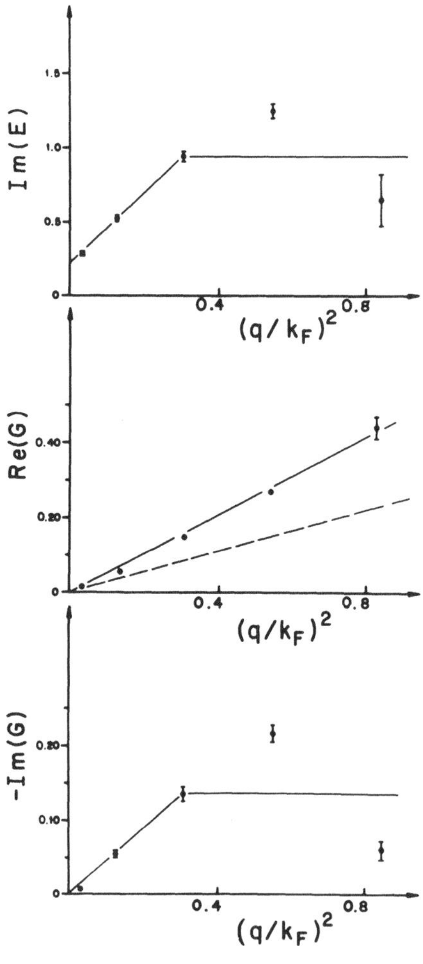

Figure 16. Modified Lindhard parameters versus $(q/k_F)^2$ for Na.

imaginary functions remain constant while ReG continues to increase. In other words beyond q_c no increase in damping is needed beyond that generated at lower q values. The fact that ReG continues to increase apparently means that the magnitude of the screening continuous to need reduction even at large q.

PHYSICAL MEANING OF THE FUNCTION PARAMETERS

In the past few years Sturm [12] and Gibbons [13] have both studied the effects of the periodic field due to the ion cores on plasmon damping. They have found that interband transitions influence both the q = 0 plasmon width and its q-dependence for $q < q_c$. Figure 17 shows the plasmon width of the alkali metals plotted against the square of the pseudopotential which most influences the fermi surface shape. Clearly the strength of the pseudopotential contributes to the plasmon width. In addition they have found that as q is increased, substantial variations in width can occur due to band structure effects. Generally the width increases due to these effects but it can also decrease. The range in q values over which these changes occur depends on the details of the band structure parameters. This may well be the explanation of the observed decrease in plasmon width for Li at small q.

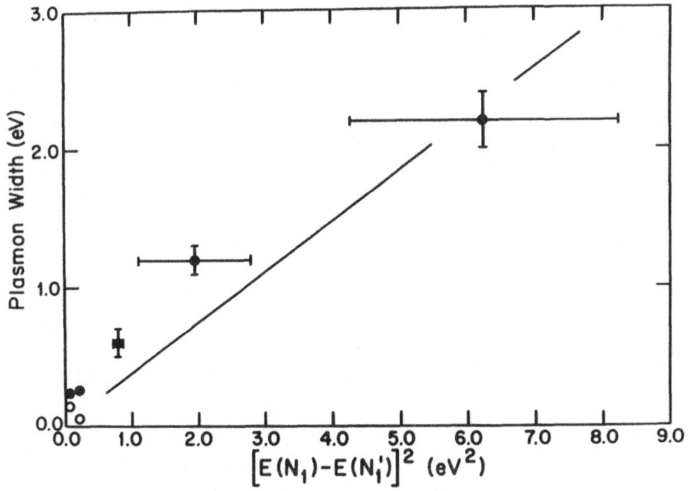

Figure 17. Plasmon width versus the square of the nearest zone boundary gap for the alkali metals.

Of particular interest is Gibbons' suggestion [13] that for intermediate q values in some of the alkali metals the increase in width with q is due directly to electron-electron scattering. If this is true it could provide an important test of our ability to evaluate electron-electron effects for finite values of q and ω.

The effect of band structure on plasmon dispersion has not been studied, so it is not yet known whether the observed change in slope for Al at small q values is due to such effects.

In addition the large value of ReG over a wide range of q values presently has no accepted physical explanation.

Recently S. Ichimaru and colleagues [14] at the university of Tokyo have generated a more sophisticated modification of the Lindhard dielectric function than that described here. They include independent short and long time relaxation rates which are adjusted to satisfy a third moment sum rule. All parameters can be evaluated from a knowledge of the pair distribution function g(r) which they have evaluated and tabulated. They then use the resulting $\varepsilon(q,\omega)$ to evaluate numerous properties of electronic systems such as the plasmon dispersion, ground state energy, and the density at which the Wigner transition takes place.

HIGH q MEASUREMENTS

Above we have concentrated on q values in the range of 0 to somewhat above q_c. For $q \gg q_c$ the scattering process is in the extreme single particle or Compton limit, in which an incident electron or photon scatters from a single electron transferring a large amount of momentum and energy to it. Such measurements can provide a determination of the momentum distribution of electrons in the solid. We understand the q = 0 limit reasonably well, and the $q \gg q_c$ limit quite well. The question remains, does anything interesting happen in the transition between the two regimes?

A number of measurements using both photon and electron scattering have been carried out in this intermediate regime. Perhaps the most provocative of these measurements is that of Platzman and Eisenberger who used inelastic x-ray scattering [15]. They found that for $q \simeq 2k_F$ an apparent splitting of the plasmon peak occurred. (See figure 18). Part of the observed scattering strength seemed to stop dispersing near this point while the rest continued on through the single particle continuum. A physical explanation of this feature was proposed by the authors. They suggested that near $q \simeq 1/\langle r\rangle$ where $\langle r\rangle$ is the mean interparticle spacing the instantaneous structure factor $S(q) = \int_0^\infty S(q,\omega)d\omega$ has

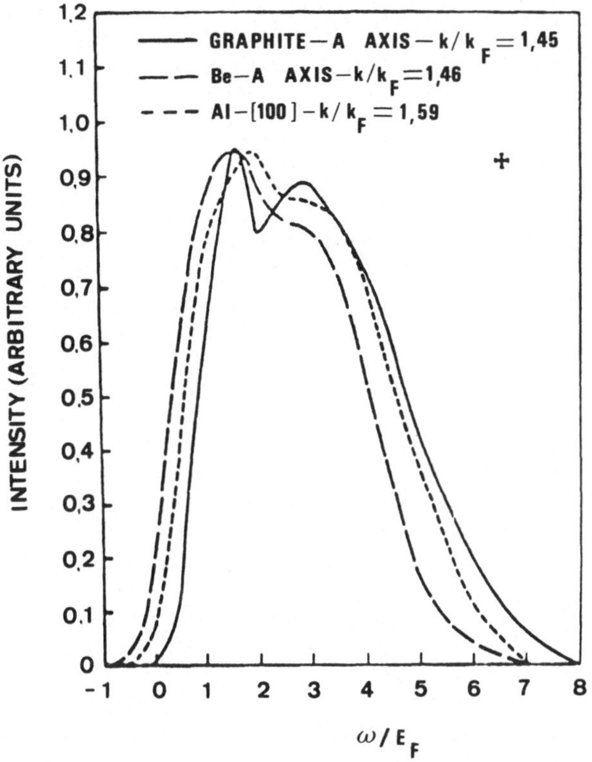

Figure 18. X-ray inelastic scattering from Al, Be and graphite for $q/k_F \simeq 1.5$.

a local maximum, just as it does in a liquid. Such a maximum will tend to flatten the plasmon dispersion, or even cause a dip, as in the case of the roton dispersion curve in liquid He^4. Later measurements of S(q) using a synchrotron light source and an edge filter were not inconsistent with this proposed explanation [16]. Another explanation involving the energy dependence of the one-electron lifetime has also been proposed to explain this result [17]. So far electron scattering measurements have not been able to reproduce this behavior. Larger multiple scattering corrections are needed for electron scattering than for x-ray scattering measurements in this range of momentum transfers so the overall uncertainties are greater at present.

If this peaking in S(q) is real, and if it increases as the electron density decreases, then it may represent a precursor to the Wigner crystallization or other static plasma instability which is predicted to occur at sufficiently low electron density. The shape of S(q,ω) is roughly the same in this momentum transfer range for graphite, Be and Al which have widely differing periodic field strengths. Nevertheless, further measurements of this effect using other metals with a wider range of electron densities and periodic field strengths must be made to clearly distinguish between possible electron-electron and electron-lattice origins of the phenomenon.

STRONGER PERIODIC FIELDS

The periodic field due to the ion cores may play a role in all the above phenomena. It is useful therefore to survey what is known about the dielectric response of electrons in solids with stronger periodic fields than the simple metals discussed above.

As the strength of the periodic field is increased, eventually a gap in the energy level distribution appears and the enormous ground state degeneracy which dominates so many metallic properties is lost. A finite minimum energy is now needed to create an electron-hole pair. The kinds of electronic excitations that can be created now include excitons just below the gap, interband transitions above the gap, and plasmons at higher energies.

When both excitons and plasmons are present the two are coupled via the dipole-dipole interaction. The result is to push the plasmon higher in energy. It is possible to obtain a broad-brush view of excitons and plasmons in semi-conductors and insulators using the Phillips-Van Vechten universal semiconductor model [8]. This plus a few assumptions leads to the graph shown in figure 18 which depicts the plasmon (ω_+) and exciton (ω_-) energies as a function of the ionicity f. Figure 19 shows the ratio of the strength of the longitudinal exciton and interband transitions to the plasmon strength as a function of f. It is clear that the higher energy plasmon dominates the spectrum for semiconductors and insulators just as it does for metals.

In the presence of a periodic field the dielectric response function is no longer a scalar function of q, ω, but a tensor. An applied perturbation with wave vector q can produce a response at q+G where G is any reciprocal lattice vector. Assuming a weak periodic field this problem can be described in the same way as that of non-interacting electrons in a periodic field : Bragg scattering occurs when a plasmon approaches the Brillouin Zone boundary, and gaps open up near the zone boundary [18] (See figure 21).

Figure 22 shows an example of measurements in a single crystal sample of LiF. The exciton at around 13.5 eV disperses upward with q in a manner which depends on the orientation of q relative to the crystal axis. (See figure 23). To higher energies peaks due to interband transitions move around with changing q.

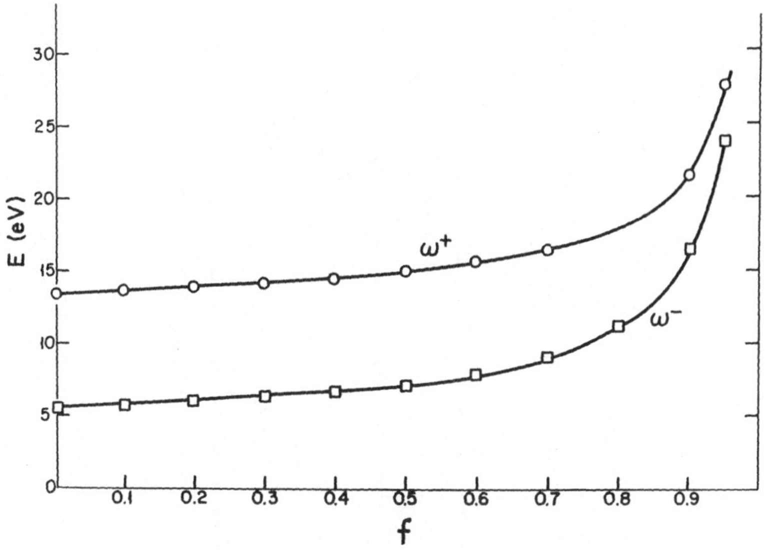

Figure 19. Plasmon (ω^+) and exciton (ω^-) energies as a function of Phillips ionicity f.

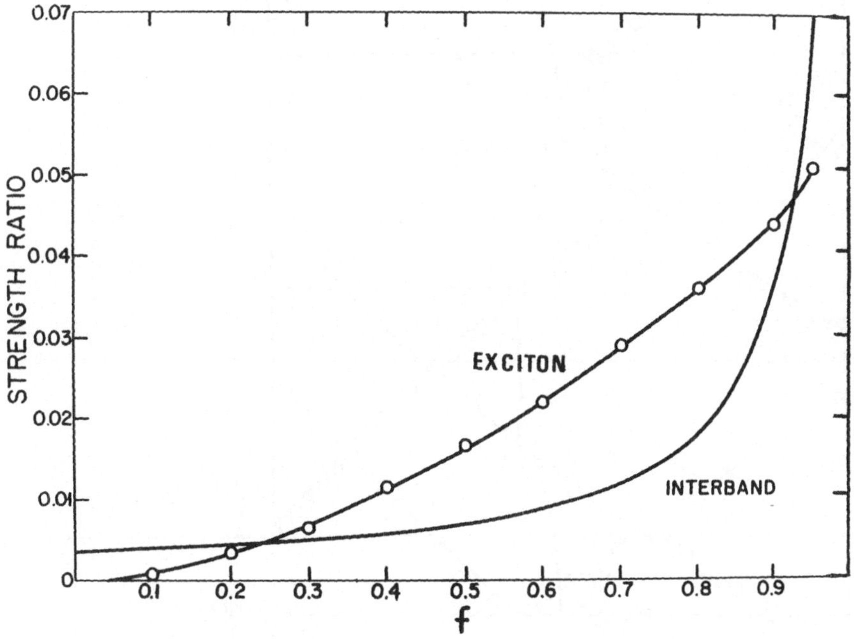

Figure 20. Ratio of exciton and interband strength to that of the plasmon as a function of ionicity f.

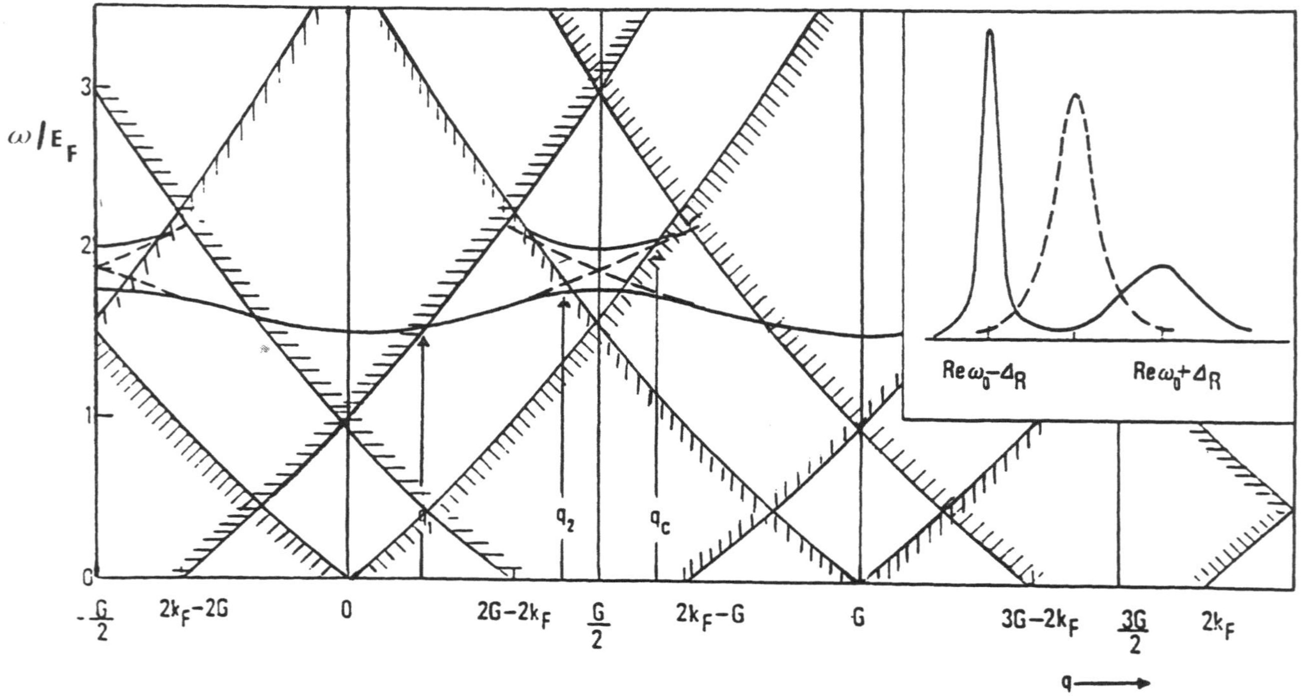

Figure 21. Modification of plasmon dispersion due to interaction with a weak periodic field.

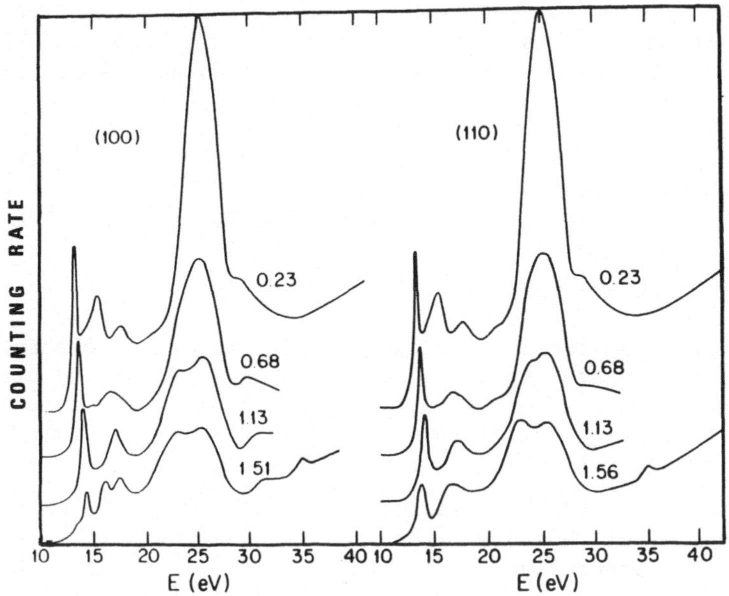

Figure 22. IES measurements on single crystal LiF. The highest of values indicated are near the zone boundary in both the (100) and (110) directions.

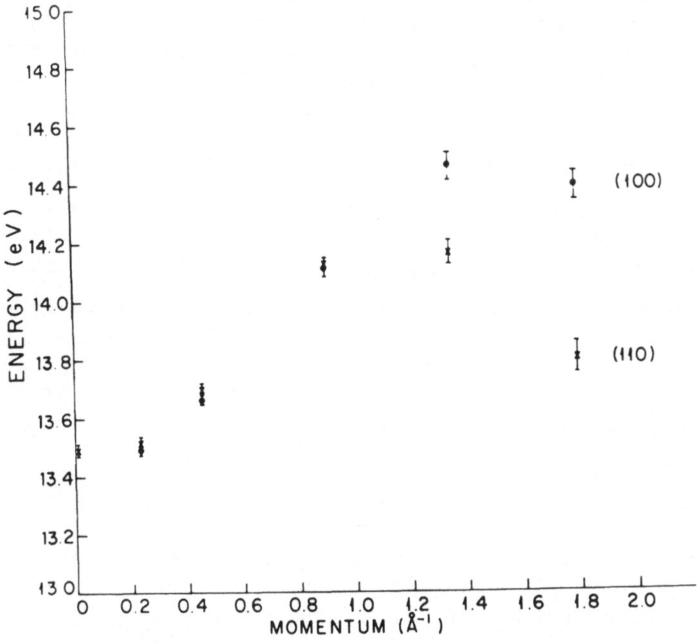

Figure 23. Exciton dispersion in LiF determined with IES measurements.

Centered near 25 eV is the plasmon, easily the dominant feature in the spectrum. As q is increased it hardly disperses at all, but broadens and near $q \simeq G/2$ in both the (110) and (100) directions, appears to split. This appears to be a good candidate for the predicted plasmon band gap in crystals. The shape of the plasmon peak qualitatively resembles that observed by Platzman and Eisenberger described above. Whether the physical origin of these splittings is the same will have to be determined by further study.

COLLECTIVE EFFECTS IN ATOMS : A TEST CASE

The importance of collective motions of the electrons in individual atoms has long been a question of interest and often of considerable dispute. Here we present an example which is the continuum excitations of electrons in the 4d shell of atoms near Xe in the periodic table.

Consider for example Te. The threshold for 4d excitation in Te is 41 eV. At this energy only a very weak structure is seen in the optical absorption coefficient α (or in ε_2 or Im $(\frac{-1}{\varepsilon})$). At considerably higher energy there is a large peak centered near 80 eV which is undoubtedly caused by the 4d electrons. See figure 24 and 25. One of the striking features of this peak is that the oscillator strength under it corresponds to nearly 10 electrons - all that the 4d shell contains. The total oscillator strength of the 4d shell is certainly somewhat greater than 10 due to the Pauli exclusion of downward transitions but nevertheless it is remarkable that so much of the strength of the 4d shell appears under just one feature in the spectrum. By contrast, about 20 % of the oscillator strength of the hydrogen 1s state goes to the 2p level, 40 % to the remaining discrete levels, and 40 % to the continuum.

In nuclear physics such a concentration of oscillator strength into a single feature is well known. In virtually all nuclei a peak in the photo-absorption spectrum occurs with energy given by $\bar{E} \simeq 70\ A^{-1/3}$ MeV. It is referred to as a "giant dipole resonance" because integration of the peak shows that the dipole oscillator strength of all the protons in the nucleus are involved. Figure 26 shows an example. The large width of these peaks indicate a very rapid decay rate.

These features are understood as being due to a collective motion of the nucleus in which the proton and neutron clouds move out of phase with each other. Since all the protons move together the intensity of the dipole moment is proportional to Z^2 rather than to Z resulting in the fast decay rate. This example of the coherence of the collective motion giving rise to a strong dipolar coupling is similar to both the superradiant state of an atomic gas, and the strong optical absorption features due to coupled surface plasmons in thin metal films.

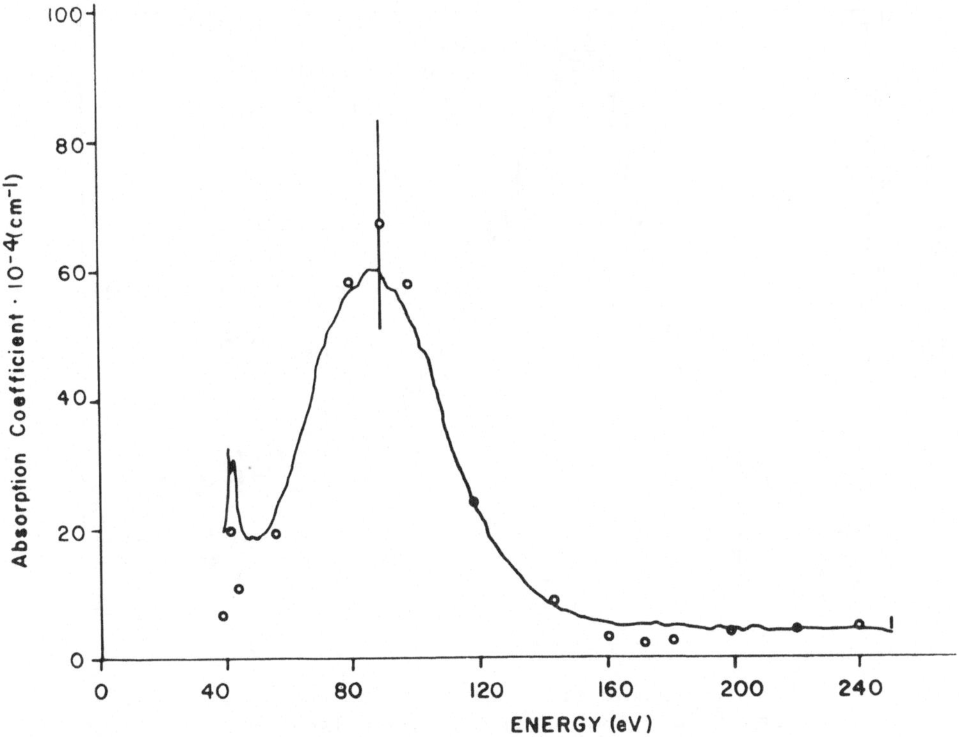

Figure 24. Absorption coefficient for Te as a function of energy in the region of the 4d maximum.

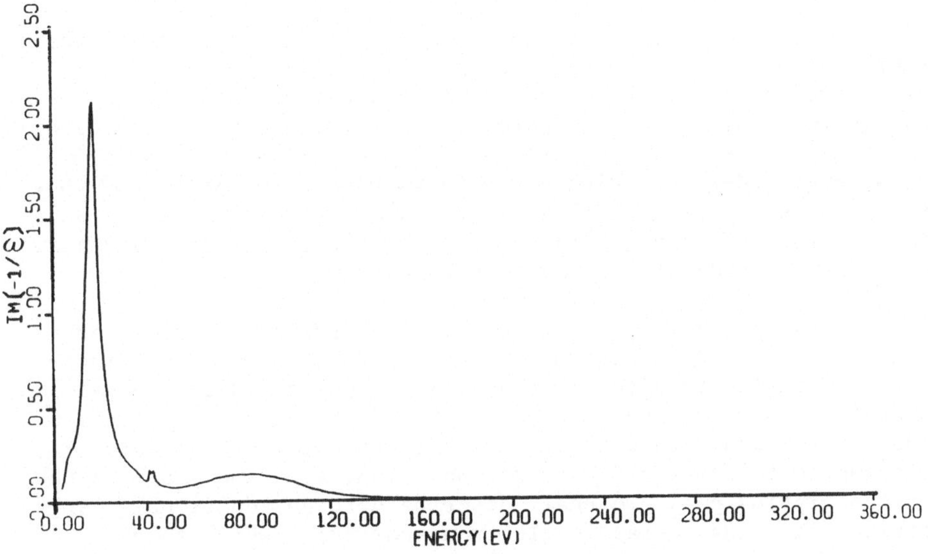

Figure 25. Im $(\frac{-1}{\varepsilon})$ for Te.

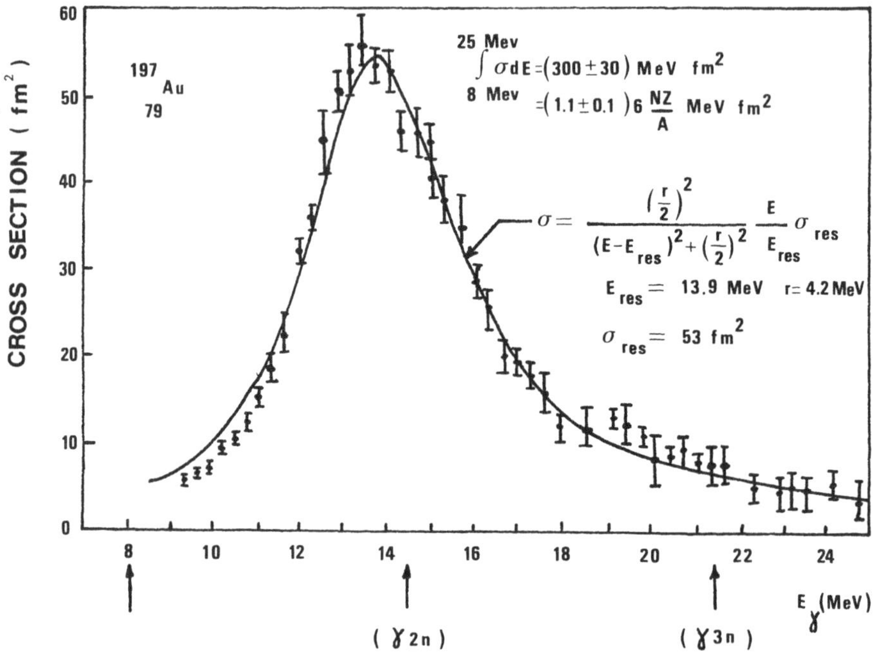

Figure 26. Photoabsorption cross section for the Au nucleus. Taken from A. Bohr and B.R. Mottelson, Nuclear Structure Vol.II (Benjamin, Reading, Mass., 1975).

The strength of the photoabsorption feature due to the atomic 4d shell suggests that perhaps it too is due to a collective motion of the 10 4d electrons. Another possibility is that it is simply an atomic potential effect. If the correct atomic potential is used then perhaps most of the oscillator strength will be gathered under this single continuum hump, similar to the case of F centers in alkali halides where more than 90 % of the ground state oscillator strength is taken up by the first excited state. Both types of theoretical models have been proposed.

In general, collective models of photoabsorption work well in the continuum region between core thresholds, and single particle models are definitely needed to understand the thresholds themselves. Here we deal with an intermediate case. Both types of models have been used and both are able to obtain reasonable agreement with experiment for the position, size and shape of the photoabsorption peak. We shall describe here an additional test carried out using inelastic electron scattering which shows promise for providing a distinction between single particle and collective models.

The most successfull single-particle-like calculation is that of Kelly [19]. He solves the non-local Hartree-Fock equations using a potential for the ejected electron calculated for all states of the system having 1P character. He then uses many-body perturbation theory in low order. This results in a different absorption strength for the length and velocity forms of the dipole operator which gives some indication of the overall errors involved. The geometric mean between the length and velocity curves is in good agreement with measurements.

Wendin uses a quite different approach. He begins with the same non-local Hartree-Fock equations and calculates a basis set for the ejected electron by averaging over the angular quantum numbers of the core hole [20]. With this basis set he then uses the Random Phase Approximation. The only difference between this use of the RPA and that described above for simple metals is that a different basis set is involved. Wendin finds a zero in the atomic dielectric response function indicating a collective resonance, an "atomic plasmon". The imaginary part of the dielectric function is quite large near the zero of the real part so the plasmon is strongly damped. To obtain reasonable agreement with measurement, Wendin found it necessary to modify the RPA result by including some degree of core relaxation. Figure 27 shows a comparison between the above two calculations and experiment for Ba.

Another collective approach was used by Amus'ya [21]. He used the same basis set as Kelly, and then carried out the RPA calculation, finding good agreement with experiment.

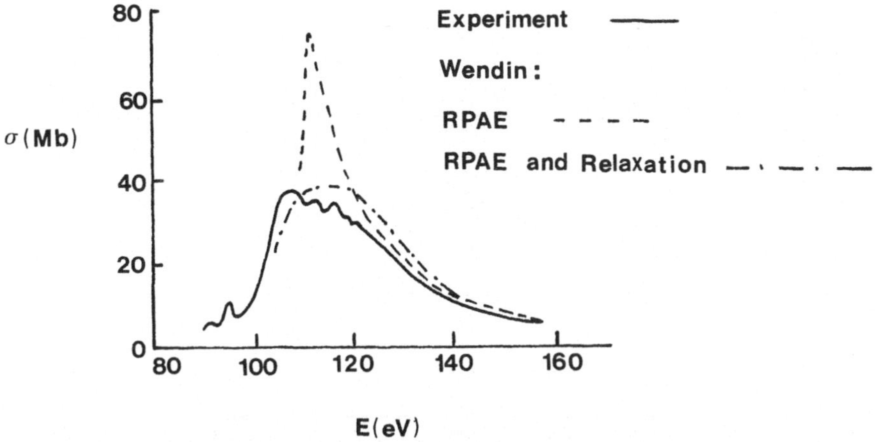

Figure 27. Comparison between calculations and measurements for atomic Ba using models described in the text.

So we appear to have rather different theoretical approaches giving rise to comparable agreement with experiment. This makes it difficult to say what the dominant physical processes are which give rise to these strong broad resonances in the 4d shell photoionization curves.

Recently an experiment was carried out in an attempt to distinguish between the different theoretical approaches to this problem [22] . The quantity measured was the dispersion of the resonance, that is the mean resonance energy $\bar{E}$ as a function of q, the momentum transferred to the atom during 4d ionization by a fast incident electron.

Thinking in terms of a collective model, we expect dispersion because transferring momentum to the atom creates shorter wavelength collective states which have higher energy. In terms of a single particle model, increasing q brings in terms in the transition operator which are beyond the dipole approximation. New final states with different angular momentum are accessed, so again the mean energy can change.

Figure 28 shows results of measurements for Te. The mean energy of the resonance increases linearly with q^2. The dispersion coefficient defined by $\bar{E} = E_0 + \beta q^2$ is $\beta = 4.5 \pm 1$ for Te. Somewhat lower values were obtained for Sb and Ba.

Now the question is, what do the various models discussed above predict for this dispersion coefficient? Unfortunately none

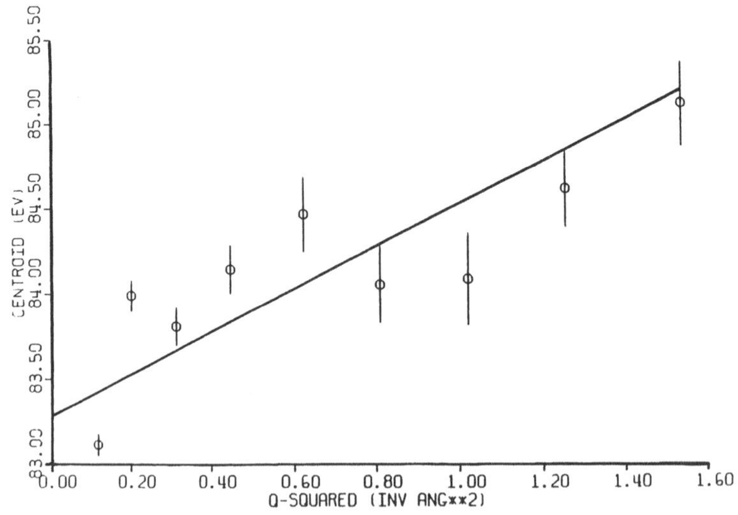

Figure 28. Dispersion of the 4d maximum for Te.

of the above model calculations has been repeated to date using the transition operator $e^{iq.r}$ instead of the dipole operator, so we do not yet have a direct answer to this question. In an attempt to provide some insight into this question, however, let us use some very simple models to inquire whether the dispersion coefficient can distinguish between the collective and single particle origins of these peaks.

The simplest possible single particle model is that of a stationary unbound target electron. Energy and momentum conservation then require that the energy taken up by the target be equal to $\hbar^2 q^2/2m$ where q is the momentum transfer. The dispersion coefficient is then just $\hbar^2/2m = 3.8$ eV Å^2. This is just the result for Compton scattering. For very large momentum transfers we expect the distribution of scattered electrons to be centered about this value.

A somewhat more realistic model we have used is a hydrogenic 4d shell treated in the Hartree approximation. The result for this model is $\beta = 2.3$ eV Å^2. Other simple one-electron models with various potentials have all yielded values of β between zero and the free electron value of 3.8 eV Å^2. It is as though the effect of a localizing potential is to reduce the tendency of the absorption strength to move up in energy.

The simplest possible collective model would be a metallic sphere of uniform charge density with radius equal to the mean 4d shell radius. (0.49 Å for Te). It might be questioned whether a plasma model of this type is at all appropriate for bound electrons. The usual criteria for determining how plasma-like a system of charged particles is involves comparing the mean kinetic energy with the mean potential energy for the particles. The mean kinetic energy for the 4d electrons in Te is 450 eV. The mean coulomb potential between one of the electrons and the other nine is 208 eV. So the ratio of kinetic to potential energies is a little larger than 2. This is a very favorable ratio for plasma-like behavior.

So from the point of view of electron density there is no difficulty thinking of the 4d electrons in Te as being a degenerate plasma. The principal difficulty in doing so arises from the fact that the electrons are highly localized so that a long wavelength collective oscillation of the electrons is impossible. The small size of the atom forces the longest wavelength oscillation to be comparable to the mean 4d shell radius. This means that Landau damping is present for all modes. We do not expect sharp resonances but broad ones similar to plasmons with $q > q_c$.

Since there is an energy gap in the spectrum of excitations of the 4d electrons the longitudinal resonance energy is given by

$$E_\ell^2 = (h\omega_p)^2 + E_g^2$$

where ω_p is the plasma frequency corresponding to the electron density in the sphere and E_g is the core binding energy.

Both surface and volume plasma modes must be included. Calculations indicate that for low q values, the $\ell = 1$ surface wave dominates, followed by the bulk mode. For higher q (e.g. $q \simeq 1\ \text{Å}^{-1}$) the $\ell = 1$ and $\ell = 2$ surface modes dominate. A weighted average of the various contributions at each q value allows the dispersion coefficient to be evaluated. We find $\beta = 11.2\ \text{eV Å}^2$.

Figure 29 shows the measured dispersion coefficient for Sb, Te and Ba along with simple model predictions. As the figure shows, the measured values fall closer to the single particle model results than to the collective model value. All the models described here are over-simplified. A final resolution of the question of the ability of β values to distinguish between single-particle and collective modes of atomic structure features will have to await recalculations using the more realistic models mentioned earlier.

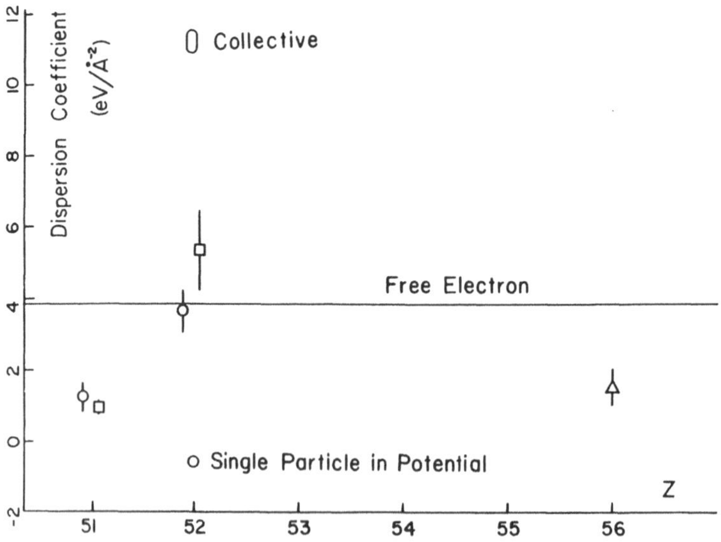

Figure 29. Measured and calculated dispersion coefficients plotted vs. atomic number.

REFERENCES

[1] P. Debye and E. Hückel, Phys. Z. 24, 185 (1923).

[2] D. Brydges, Commun. Math. Phys. 58, 313 (1978).

[3] See for example, "Elementary Excitations in Solids", D. Pines, W.A. Benjamin, Inc., New York, 1964.

[4] See for example, "Waves and Interactions in Solid State Plasmas", P.M. Platzman and P.A. Wolff, Vol.135 of Solid State Physics, H. Ehrenreich, F. Seitz, D. Turnbull, Eds. Academic Press, New York, 1973.

[5] J. Lindhard, K. Dan. Vidensk. Selsk, Mat.-Fys. Medd. 28 ¶8 (1954).

[6] J.D. Cohen and C.P. Slichter, Phys. Rev. Lett. 40, 129 (1978).

[7] E. Shiles, T. Sasaki, M. Inokuti and D. Smith, Phys. Rev. B22, 1612 (1980).

[8] "Inelastic Electron Scattering", S.E. Schnatterly, Vol.34 of Solid State Physics, H. Ehrenreich, F. Seitz, D. Turnbull, Eds. Academic Press, New York, 1979.

[9] P.C. Gibbons, S.E. Schnatterly, J.J. Ritsko and J.R. Fields, Phys. Rev. B13, 2451 (1976).

[10] N.D. Mermin, Phys. Rev. B1, 2362 (1970).

[11] P. Vashishta and K.S. Singwi, Phys. Rev. B6, 875 (1972).

[12] K. Sturm, Z. Phys. B27, ¶1 (1977).

[13] P.C. Gibbons, Phys. Rev.

[14] K. Utsumi and S. Ichimaru, Phys. Rev. B22, 1522 (1980) and 4 subsequent papers in preprint form.

[15] P.M. Platzman and P. Eisenberger, Phys. Rev. Lett. 33, 152 (1974).

[16] P. Eisenberger, W.C. Marra and G.S. Brown, Phys.Rev.Lett. 45, 1439 (1980).

[17] G. Mukhopadhyay, R.K. Kalia, and K.S. Singwi, Phys. Rev. Lett. 34, 950 (1975).

[18] R. Girlanda, M. Parrinello and E. Tosatti, Phys. Rev. Lett. 36, 1386 (1976).

[19] H.P. Kelly, "Photoionization Cross Sections and Auger Rates Calculated by Many-Body Perturbation Theory" in Photoionization and Other Probes of Manh-Electron Interactions, F.J. Wuilleumier, ed. (Plenum, N.Y. 1976) p.83.

[20] G. Wendin, "The Random Phase Approximation with Exchange" : ibid. p.61.

[21] M. Ya Amus'ya, N.A. Cherepkov and L.V. Chernysheva, Sov. Phys. JETP 33, 90 (1971).

[22] C.P. Franck, Thesis, Princeton University, 1978, C.P. Franck and S.E. Schnatterly, submitted to Phys. Rev.

CHARGE DENSITY WAVE PHENOMENA

IN POTASSIUM

A.W. Overhauser

Department of Physics
Purdue University
West Lafayette, IN 47907

I. THE MYSTERIES OF THE SIMPLE METALS

Introduction

The importance of many-electron effects in metals was realized in 1929 by F. Bloch.[1] He showed that exchange interactions cause a significant enhancement of the spin susceptibility χ_s over the value χ_p derived by Pauli for free electrons. The enhancement factor (in the Hartree-Fock approximation) is given by

$$\chi_s/\chi_p = (1-r_s/6.03)^{-1} . \quad (1)$$

r_s is the radius of a sphere containing one electron and is related to the density n by

$$n = [\frac{4\pi}{3} (r_s a_B)^3]^{-1} , \quad (2)$$

where a_B is the Bohr radius. For metals this factor would range from 1.5 to 6.

In 1938 E. Wigner[2] pointed out that electron-electron correlations caused by the dynamical scattering of opposite-spin electrons, would reduce the enhancement given by (1). Since then more than fifty theorists have attempted to calculate the effect of correlations on χ_s. The results have been displayed dramatically by Kushida, et al.[3] The collection of theoretical curves for χ_s vs r_s given in their Fig. 6 looks like a plate of spaghetti.

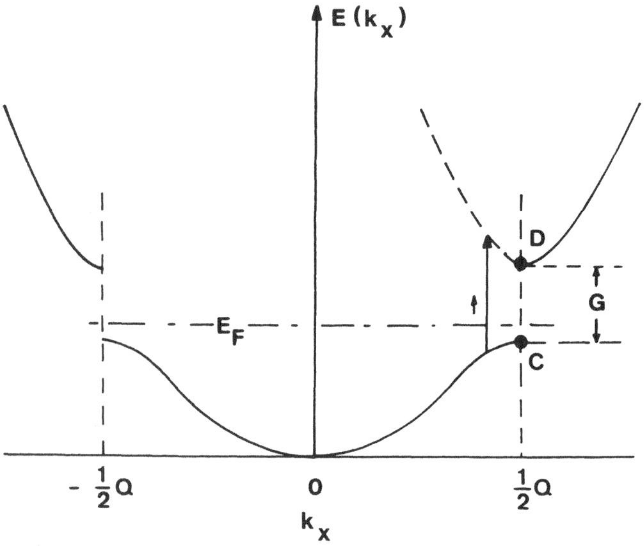

Fig. 1. Electronic energy $E(k_x)$ against k_x for a CDW structure. The CDW energy gap is G, and the Fermi energy E_F lies between the gap edges C and D. The optical transition t is allowed for a CDW structure.

The purpose of this presentation is to summarize experimental evidence that exchange and correlation effects lead to a broken-symmetry state in a simple metal. Theoretical arguments, which we feel are convincing, will be presented later. This strategy seems best in view of the widely different approaches current in the study of many-electron effects (with different results as cited above).

The most widely studied, simple metal is potassium. Lithium and sodium experience phase transitions on cooling to 4°K, and this destroys single crystals. Rubidium and cesium are very soft and consequently difficult to handle. De Haas-van Alphen experiments on potassium, e.g. Shoenberg and Stiles[4], indicate an isotropic Fermi surface - to about one part in 10^3. Nevertheless, as we will review, other experiments show its Fermi surface is not even simply connected.

Charge-Density-Wave Structure

A CDW is a broken-symmetry, ground state of a metal in which the conduction-electron density has a small sinusoidal modulation,[5]

$$\rho = \rho_o[1 + p\cos(\vec{Q}\cdot\vec{r} + \phi)], \qquad (3)$$

(over and above any modulation caused by the positive-ion lattice). In general the wavevector $\vec{Q}$ is incommensurate with the lattice, so the phase ϕ is an arbitrary parameter which labels the infinite number of degenerate ground states. The one-electron wavefunctions

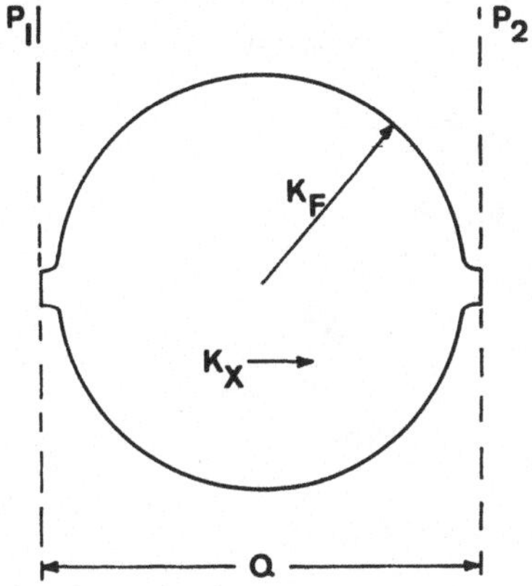

Fig. 2. Fermi surface of an ideal metal with a CDW ground state. The surface is distorted from that of a sphere of radius k_F near the energy-gap planes p_1 and p_2.

are amplitude modulated by an exchange and correlation potential,

$$V = -G \cos(\vec{Q}\cdot\vec{r} + \phi), \qquad (4)$$

in Schrodinger's equation. The theoretical problem addresses the question whether ρ and G are selfconsistent and, if so, whether the CDW state has lower energy than the normal one. Since a modulation arises only when occupied, free-electron states $\vec{k}$ are mixed with empty ones $\vec{k}\pm\vec{Q}$, the optimum $|\vec{Q}|$ is $\sim 2k_F$, the diameter of the Fermi surface.

An obvious consequence of the potential V is the creation of two new energy gaps in k-space along planes passing through $\pm 1/2\ \vec{Q}$. See Fig. 1. This leads to a small distortion of the Fermi surface - possibly to the formation of necks, as shown in Fig. 2. As mentioned above, such effects have not been reported in de Haas-van Alphen studies, a failure which will be discussed below.

Mayer-El Naby Optical Anomaly

The first indication that potassium had a CDW structure was the discovery by Mayer and El Naby[6] of an intense interband, optical transition with a threshold at 0.6 eV, followed by an asymmetric peak at 0.8 eV, as shown in Fig. 3. The ordinary interband absorption - caused by the Brillouin-zone energy gaps - has a threshold of 1.3 eV. A quantitative account of the anomalous absorption succeeds merely by letting the CDW gap G = 0.6 eV and by calculating the optical transition rate across these new energy gaps.[7] The solid curve in Fig. 3 shows exceptional agreement.

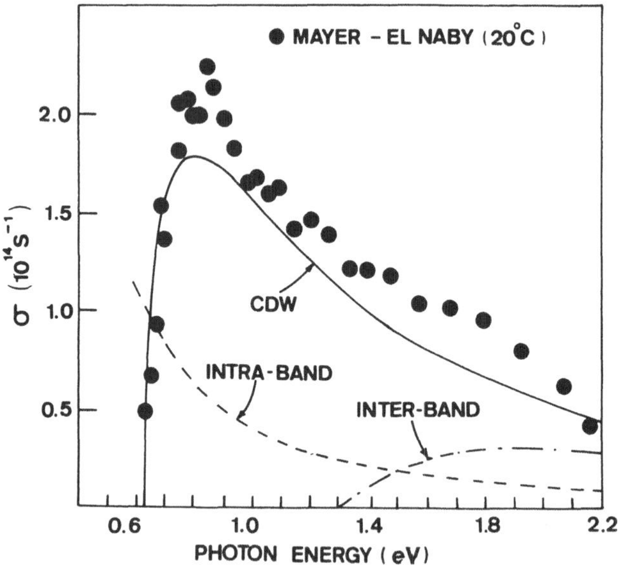

Fig. 3. Anomalous optical absorption spectrum of potassium. The intra-band conductivity (dashed curve) has been subtracted from the experimental data before being plotted. The solid curve shows the theoretical absorption introduced by a CDW structure. A 'normal' metal would exhibit only the inter-band absorption with a threshold, as shown, at 1.3eV.

Ordinarily potassium, which is body-centered cubic, would have isotropic optical properties. However, the new absorption occurs only from the component of the photon polarization vector $\vec{\varepsilon}$ parallel to $\vec{Q}$. The CDW optical absorption is uniaxial! This is crucial to an explanation of why the Mayer-El Naby anomaly is not obtained in measurements on evaporated films. Such films, deposited on smooth, amorphous substrates, always have a [110] cyrstal direction normal to the surface. It turns out that the optimum direction of $\vec{Q}$ is then normal to the surface too. Since an infrared photon incident on the film will have its polarization vector $\vec{\varepsilon}$ parallel to the surface, <u>inside</u> the metal (even at oblique incidence), anomalous optical absorption <u>cannot</u> occur. At bulk-metal vacuum surfaces, the Mayer-El Naby anomaly has been reproduced by three workers. Such anisotropy requires a CDW.

Low-Temperature Magnetoresistance

It is easy to understand that a metal with a spherical Fermi surface will have no magnetoresistance. Suppose that a magnetic field is in the z direction, and that the electric field E is in the x-y plane. Let $\vec{P}$ be the total momentum of the electron Fermi sea, and suppose that the units employed allow the following equations for Newton's laws:

$$dP_x/dt = E_x + \omega_c P_y - P_x/\tau \ , \tag{5a}$$

$$dP_y/dt = E_y + \omega_c P_x - P_y/\tau \ , \tag{5b}$$

The second term on the right hand side is the Lorentz force, with $\omega_c \equiv eH/mc$. The relaxation term was added to take account of scattering by impurities, which allows the electrons to reach equilibrium, $\vec{P} = 0$. In the non-equilibrium steady state $d\vec{P}/dt = 0$. Furthermore if the wire confines the current to the $\hat{x}$ direction, $P_y = 0$. Eq.(5a) can be solved immediately for the current.

$$P_x = \tau E_x, \tag{6}$$

which is Ohm's law. The resistance is independent of the magnetic field! Lifshitz et al.[8] proved that this is true for any simply-connected Fermi surface whenever $\omega_c\tau >> 1$.

Experimentally all workers have found that $\rho(H)$ for the alkali metals increases with H, without evidence of saturation, even at fields where $\omega_c\tau > 300$. The only intrinsic mechanism which allows this behavior is the presence of open orbits. These occur only if the Fermi surface is multiply connected, which would result from a CDW structure. Extrinsic mechanisms for a linear magnetoresistance, such as voids, have been ruled out by measurements which show there are no voids.[9]

The size of the magnetoresistance in potassium varies from sample to sample, and from run to run on the same sample. This behavior, found perplexing by the experimentalists, is easily explained by CDW's. The $\vec{Q}$ direction has some optimum axis (α,β,γ) energetically. There are 24 cubically equivalent (α,β,γ) axes. Consequently a macroscopic sample will be divided into $\vec{Q}$ domains, for which the direction of $\vec{Q}$ varies from domain to domain. Each time a sample is cooled to 4°K from room temperature it will have a different $\vec{Q}$-domain distribution. This will cause a different open-orbit distribution and, consequently, a different magnetoresistance.

Induced-torque Measurements

Inductive techniques were applied to resistance measurements on alkali metals to avoid problems that could arise from placement of the potential probes. The induced-torque method was first used by Schaefer and Marcus[10] to study the orientation dependence of the magnetoresistance of sodium and potassium. Not only did they confirm the non-saturating magnetoresistance but they found that it was highly anisotropic - as much as 15 to 1. There were always four peaks in a 360° rotation, even if a threefold [111] axis were the rotation axis. These results were repeated about 200 times

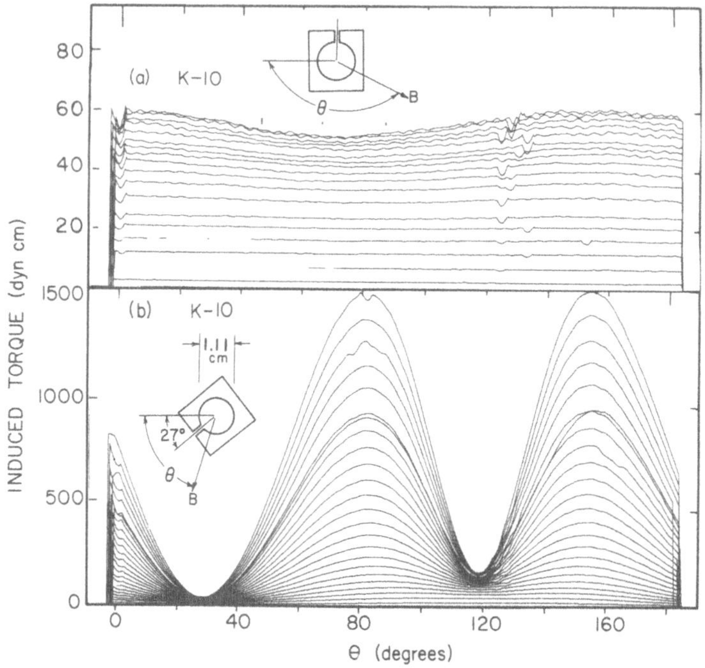

Fig. 4. Induced-torque against magnetic field direction θ for a potassium sphere 1.11 cm in diameter. In (a) the field B is rotated about the growth axis. The curves shown are for 0.5, 1,2,3,...,kG. In (b) the plane of rotation contains the growth axis, and B = 1,2,3,...kG. The data are from Holroyd and Datars (1975).

during a two-year period prior to publication. Their results were reproduced by Holroyd and Datars,[11] who found one specimen with an anisotropy (at 23 kG) of 45 to 1. See Fig. 4. These results prove that potassium (and sodium) have neither cubic symmetry nor simply-connected Fermi surfaces.

The Oil Drop Effect

Holroyd and Datars were able to obtain isotropic torque patterns by cleaning off all the oil (in which the spherical samples were grown). The samples were mounted in the apparatus by resting them on cotton. However if these same samples were later fastened to their supports with a drop of oil (which freezes on cooling to 4°K), they then displayed the familiar four-peaked anisotropy. $\vec{Q}$-domain structure, present if a CDW is present, provides a reason-

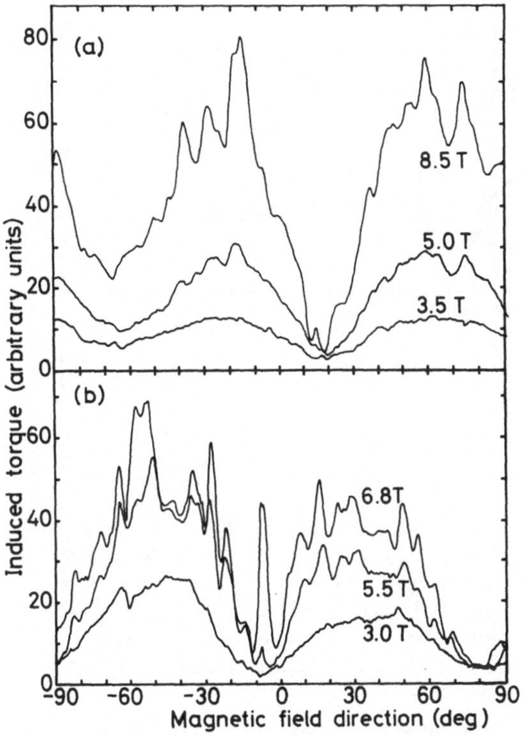

Fig. 5. Magnetoresistance of K vs $\vec{H}$ (Coulter and Datars). (a) Single crystal grown in oil, $\vec{H}$ in a (211) plane. (b) Single crystal grown in a mold, $\vec{H}$ in a (321) plane.

able explanation. Randomly oriented Q-domains will exhibit macroscopic isotropy (as long as H < 30 kG). The thermal stress which arises from frozen oil (on cooling) will break the 24-fold degeneracy of the $\vec{Q}$-domains and produce a preferred orientation. This leads to an anisotropic resistivity.

Other Anomalous Phenomena

Many other properties exhibit extraordinary behavior and require that potassium have a CDW ground state. Splitting of the conduction-electron spin resonance,[12] anisotropy of the residual resistance,[13] direct observation of the phase excitations in point-contact spectroscopy,[14] discrepancies in the Hall coefficient from its required value,[15] etc. Some of these are discussed in a review by Overhauser.[16] Even de Haas-van Alphen measurements under

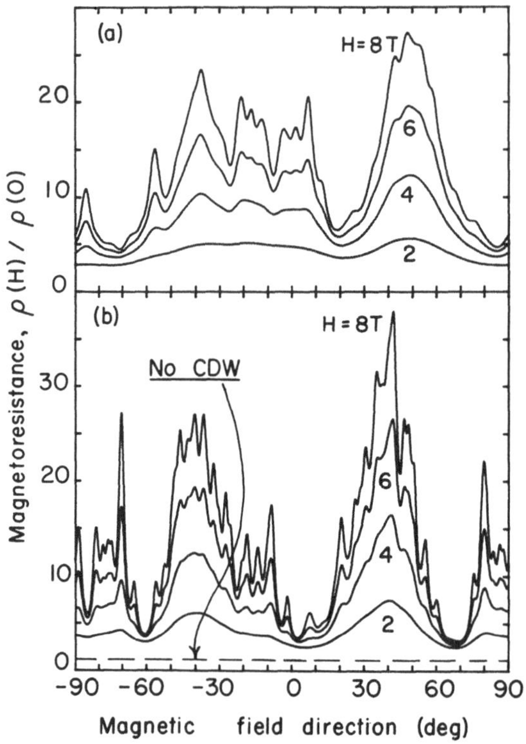

Fig. 6. Magnetoresistance of potassium vs. $\vec{H}$ computed by Huberman and Overhauser using the CDW model.

pressure require anomalous structure (to escape non-conservation of electron number), as reported by Altounian and Datars.[17] The recently observed open-orbit torque peaks by Coulter and Datars[18] at 80 kG are conclusive. They are shown in Fig. 5. Theoretical magnetoresistance spectra, based on the CDW model, were computed by Huberman and Overhauser.[19] They are shown in Fig. 6.

I believe the failure to observe anomalous behavior in ordinary de Haas-van Alphen experiments is a consequence of sample preparation leading to $\vec{Q}$-domain sizes smaller than the diameter of a cyclotron orbit. Specimens are typically 1 mm in size, and etching is avoided. Possibly larger specimens (5-10 mm), heavily etched so that [110] surface facets become visible to the naked eye, will lead to larger Q-domains. Some authors have admitted discarding data which did not behave "properly". Perhaps these cases were in fact the important ones.

II. PHASONS: WHAT THEY ARE AND WHAT THEY DO

Introduction

Charge-Density-Wave (CDW) structure in an isotropic metal is a broken-symmetry groundstate caused exclusively by exchange and correlation effects. Since translation symmetry is broken, there will be a continuous spectrum of collective excitations (reaching zero frequency if the CDW is incommensurate).[20] The occurrence of new excitations when new order parameters arise has been well-known in solid state physics for many decades (e.g. spin waves). In elementary particle theory this has become known as Goldstone's theorem.

A CDW is described by a sinusoidal modulation of the conduction electron density:

$$\rho = \rho_o[1 + p\cos(\vec{Q}_o\cdot\vec{r} + \phi)]. \qquad (7)$$

The new order parameters are p, $\vec{Q}_o$ and ϕ. The latter can take on any value between 0 and 2π, and provides a label for the infinite number of degenerate ground states. The ground state energy will lie at the bottom of a parabolic, four-dimensional valley, corresponding to the components of $\vec{Q}_o$ and p. The change in energy caused by small deviations from the equilibrium values, $\delta\vec{Q}$ and δp, can be written,

$$E = E_o + \alpha(\delta Q_{||})^2 + \beta(\delta Q_{\perp})^2 + \gamma(\delta p)^2. \qquad (8)$$

For simplicity we have assumed axial symmetry along the $\vec{Q}$ axis. I will use this expression as an energy density when $\delta\vec{Q}$ is allowed to be a slowly varying function of $\vec{r}$.

Phase Modulation

One may discover that the local direction of $\vec{Q}$ varies sinusoidally in space and time if one allows the phase ϕ in Eq. (7) to be,

$$\phi = \phi_q \sin(\vec{q}\cdot\vec{r} - \omega t), \qquad (9)$$

instead of a constant. This may be seen readily by taking the amplitude, ϕ_q, of the phase modulation to be small, and by expanding the sine function for small $\vec{q}\cdot\vec{r}$ and t = 0. Then Eq.(7) becomes

$$\rho = \rho_o \{1 + p[\cos(\vec{Q}_o + \phi_q\vec{q})\cdot r]\}. \qquad (10)$$

If $\vec{q}$ is parallel to $\vec{Q}_o$, the <u>local</u> magnitude of $\vec{Q}$ is changed. If $\vec{q}$ is perpendicular to $\vec{Q}_o$, the <u>local</u> direction of $\vec{Q}$ is rotated. In general, the phase modulation given by Eq.(9) corresponds to a sinusoidal modulation of both the local direction and magnitude of $\vec{Q}$. Actually we will assume that

$$|\vec{q}| << |\vec{Q}_o| \quad , \tag{11}$$

and will discuss below the theoretical limits of $\vec{q}$.

The collective modes will be harmonic oscillators, so we call them "phasons". The potential energy of a phase excitation can be calculated from Eq.(8),

$$U = \int \Delta E \; [\vec{Q}_{local}(\vec{r})] \; d^3r \tag{12}$$

The kinetic energy of a phase excitation arises from the motion of the positive ions. It must be noted that a CDW cannot occur unless the metal remains electrically neutral. This means that the positive ions undergo a sinusoidal lattice displacement from their periodic crystal sites $\vec{L}$. Accordingly,

$$\vec{U}(\vec{L}) = \vec{A} \sin \, [\vec{Q}_o \cdot \vec{L} + \phi_q \; \sin(\vec{q} \cdot \vec{L} - \omega t)]. \tag{13}$$

The displacement amplitude $\vec{A}$ is proportional to p, and its direction depends on the direction of $\vec{Q}$ and the elastic moduli.[21] The kinetic energy is easily computed from,

$$T = \frac{1}{2} \Sigma_{\vec{L}} \; M[\dot{\vec{U}}(\vec{L})]^2 \quad , \tag{14}$$

where M is the ionic mass. For an harmonic oscillator the time average of U and T are equal. This requirement leads to the phason dispersion relation:[20]

$$\omega = [c_{||}^2 q_{||}^2 + c_{\perp}^2 q_{\perp}^2]^{\frac{1}{2}} \quad . \tag{15}$$

The principal-axis values, $c_{||}$ and $c_{\perp}$, of the phason velocity are proportional to $\alpha^{1/2}$ and $\beta^{1/2}$ respectively.

Relation Between Phasons and Phonons

When advantage is taken of the infinitesimal value of ϕ_q, Eq.(13) can be written,

$$\vec{U}(\vec{L}) \cong \vec{A} \sin \vec{Q}_o \cdot \vec{L} + \frac{1}{2} \vec{A} \; \phi_q \; \sin[(\vec{q} + \vec{Q}) \cdot \vec{L} - \omega t]$$
$$+ \frac{1}{2} \vec{A} \; \phi_q \; \sin[(\vec{q} - \vec{Q}) \cdot \vec{L} - \omega t]. \tag{16}$$

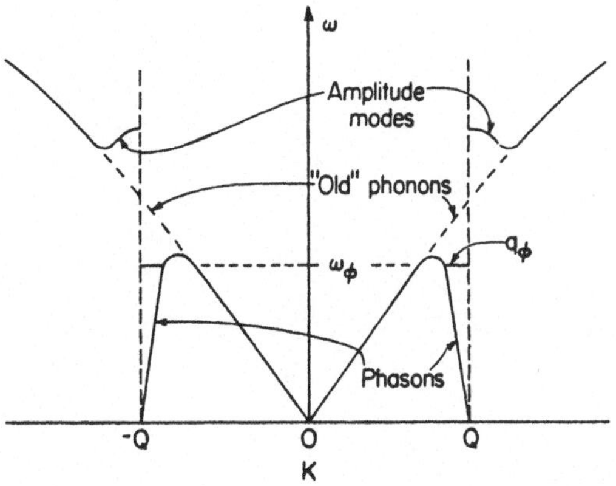

Fig. 7. Schematic illustration of the vibrational modes in a metal having a CDW structure. The frequency of the phason branch goes to zero at $\pm \vec{Q}$. A phason is a linear superposition of two "old" phonons, and the amplitude modes are the orthogonal linear combination. Phason and amplitude modes quickly merge into the old phonon spectrum, as indicated.

Clearly, a phason is a linear combination of the two "old" phonon modes: $\vec{q}+\vec{Q}$ and $\vec{q}-\vec{Q}$. When these "old" modes are coupled together, one linear combination is pushed down (the phasons) and the orthogonal linear combination is pushed up. One can show that the latter mode corresponds to an amplitude modulation of the CDW.[22] See Fig. 7.

Phasons have three important physical consequences because of their linear dispersion relation. One may visualize the phason "cone" as the result of "pressing down" on the lowest phonon dispersion surface near $\vec{Q}_o$ (or $\vec{Q}_o-\vec{G}$, to get into the Brillouin zone) with an anisotropic ice cream cone until the point touches $\omega=0$. The phason cutoff frequency ω_ϕ is essentially the frequency of the (old) phonon mode at $\vec{Q}_o$. See Fig. 8.

The three important phason parameters that should be measured are $c_{||}$, $c_\perp$ and ω_ϕ. They enter all calculations of observable properties.

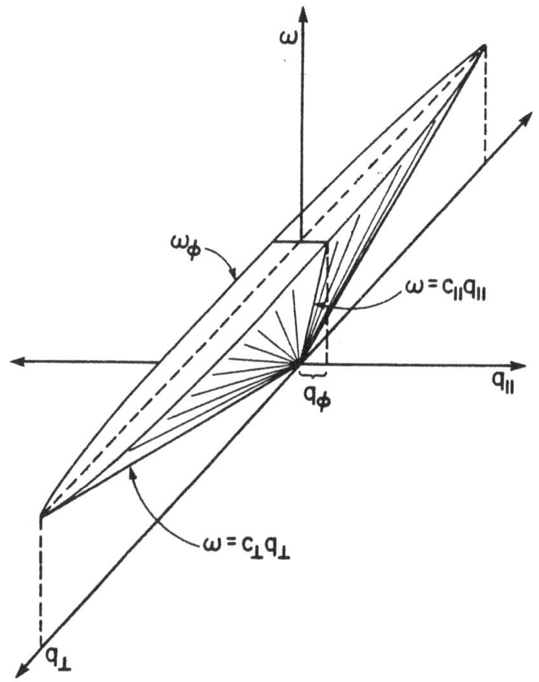

Fig. 8. Anisotropic cone of the phason spectrum, showing the longitudinal and transverse phason velocities $c_{||}$ and $c_{\perp}$. ω_ϕ is the frequency cutoff of the phason spectrum and q_ϕ is the wave-vector cutoff along Q.

The Phason Heat Capacity

A low temperature anomaly will appear in the heat capacity of a metal having a CDW. It can be calculated only if one knows how the phasons and amplitude modes blend into the phonon branch as $|\vec{q}|$ increases. A simple model employs a 2x2 dynamical matrix:[23]

$$\begin{vmatrix} \omega_\phi^2-\omega^2 & \omega_\phi^2\,F(q) \\ \omega_\phi^2\,F(q) & \omega_\phi^2-\omega^2 \end{vmatrix} = 0 \,. \tag{17}$$

It is easy to verify that,

$$F(q) \cong 1-\left(\frac{c_q q}{\omega_q}\right)^2 + \ldots \,, \tag{18}$$

if ω is to approach 0 linearly at q = 0. c_q is the phason velocity for the $\vec{q}$ direction. The function $F = \exp[-(c_q q/\omega_q)^2]$

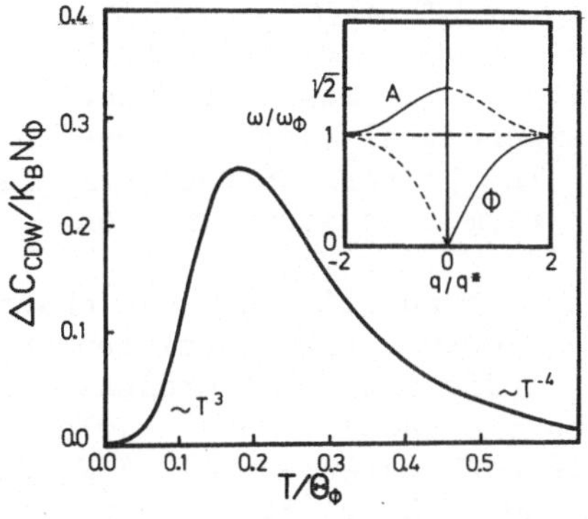

Fig. 9. Heat-capacity anomaly caused by an incommensurate CDW. The inset shows the vibrational spectrum near $\vec{Q}'$. ϕ is the phason branch and A the amplitude-mode branch. The horizontal line, at $\omega = \omega_\phi$, is the phason cutoff frequency--the phonon frequency at $\vec{Q}'$ with no CDW.

satisfies this requirement, and is a convenient guess. It leads to the phason (and amplitude mode) spectrum shown in the inset of Fig. 9. The old phonon frequencies have been set equal to the constant ω_ϕ for simplicity. If $R(\vec{q})$, $S(\vec{q})$ and P are the Einstein heat capacities of a phason, amplitude mode, and a phonon mode (of frequency ω_ϕ), then the change in heat capacity caused by a CDW is:

$$\Delta C_{CDW} = \sum_{\vec{q}} [R(\vec{q}) + S(\vec{q}) - 2P]. \quad (19)$$

The shape of ΔC_{CDW} is universal and has its peak near $0.18\theta_\phi$, ($\hbar\omega_\phi \equiv k_B\theta_\phi$). See Fig. 9.

An anomaly of this type was first seen by Sawada and Satoh[24] in $LaGe_2$ and shown to be associated with a CDW structure. Recently, Amarasekara and Keesom have observed a similar anomaly in potassium.[25]

Low Temperature Resistivity

The low temperature resistivity of a metal such as potassium is expected to have four contributions:

$$\rho = \rho_o + AT^m E^{-(T_o/T)} + BT^2 + CT^5. \qquad (20)$$

The residual resistivity ρ_o is caused by impurities (and other imperfections). The exponential term is caused by the freeze out of umklapp processes. The T^2 term arises from electron-electron scattering, and the T^5 term from acoustic phonon scattering.

A term approximately quadratic in T was observed in potassium by van Kempen et al.[26] However the coefficient B varied widely from sample to sample, indicating that ordinary electron-electron scattering could not be the cause. Later Rowlands et al.[27] found a similar result. This work extended to 0.4°K and had an experimental scatter of only 10^{-15} Ωcm. It fit a power law of $T^{1.5}$. Power law fits proportional to T or T^2 were well outside the experimental scatter. A theory of electron-phason scattering[28] fits extremely well, as can be seen in Fig. 10. (The arrows at the two highest temperatures show where the points would fall if the umklapp tail were subtracted. Phason scattering is important at low temperatures, unlike acoustic phonons, since (as can be seen from Eq.(16)) they cause large momentum transfers, $\sim \pm \vec{Q}$, even though q is very small.[28]

Variability of the data from sample to sample is also explained (as an intrinsic property) since the large momentum transfers occur along $\pm \vec{Q}$. Thus the average resistivity will depend on the orientation distribution of $\vec{Q}$ domains.

Point Contact Spectroscopy

Two metal points touched together in a small contact (size << mean free path) create a junction similar to a superconducting tunnel junction. Spectra of d^2V/d^2I vs. voltage V reveal phonon spectra (weighted by a square matrix element and a geometric factor). The geometric factor arises in this case only because an electron must be scattered back through the contact it has just crossed in order for the subsequent phonon event to contribute to the contact resistance. Accordingly low energy acoustic phonons, which scatter electrons at the Fermi surface only through small angles, contribute a vanishing contribution at voltages comparable to their energy.

Nevertheless, data on potassium by Jansen et al.[29] show a striking anomaly below 1 meV, where there should be no contribution at all. Recently Ashraf and Swihart[30] have found that electron-phason scattering explains the anomaly - its shape and size - without any adjustable parameters. Data on the phason spectrum were taken from the fit to the resistivity data mentioned above.[28]

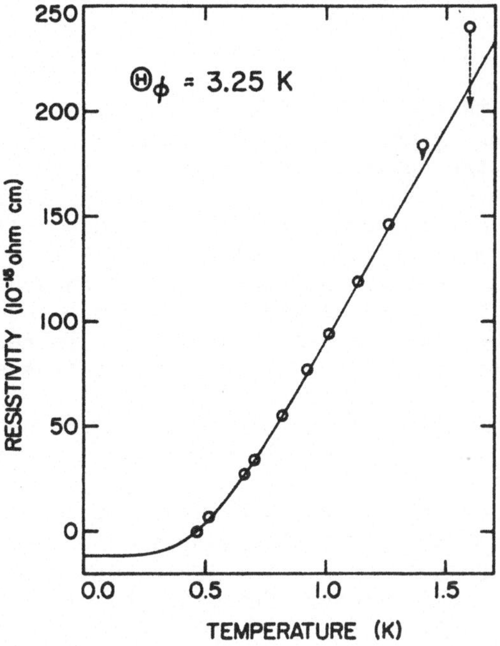

Fig. 10. Plot of resistivity versus temperature for the data of sample K2c of Rowlands, Duvvury, and Woods indicated by circles. The curve and arrows are described in the text.

Phason Thermal Diffuse Scattering

A CDW structure leads to new diffraction peaks. Each reciprocal lattice vector $\vec{G}$ acquires two satellites, for which the Laue equation is

$$\vec{k}' = \vec{k} + \vec{G} \pm \vec{Q}. \tag{21}$$

The phasons cause a thermal diffuse scattering around each satellite:

$$\vec{k}' = \vec{k} + \vec{G} \pm \vec{Q} \pm \vec{q}_{phason} . \tag{22}$$

This diffuse scattering will produce a pancake-shaped cloud because of the phason anisotropy. In electron diffraction it shows up as streaks, as seen in TaS_2.[31]

III. THEORY OF CHARGE DENSITY WAVES

Introduction

The purpose of this section is to show that a CDW is the anticipated electronic ground state of a simple, isotropic metal. We will assume that the positive-ion background has been replaced by a uniformly charged jelly, having no mechanical rigidity. The last assumption - a key property of the deformable jellium model - guarantees that all microscopic electric fields will be zero. (There will be no Hartree term in the one-electron Schrodinger equation.) A CDW arises exclusively from exchange and correlation, as we will show. The alkali metals, in the aspects relevant here, correspond well to this model. Their elastic stiffness is two orders of magnitude less than that of copper, and the pressure dependence of their moduli show no evidence of a Born-Mayer ion-ion repulsion.

The fact that potassium has a CDW structure not only confirms the theoretical arguments, but at the same time proves that some treatments of exchange and correlation are fundamentally incomplete, and sometimes dramatically wrong. This will be discussed at length in the concluding section.

SDW-CDW Instability Theorem

The exact ground-state energy of an electron system is,

$$E \equiv E_{HF} + E_{corr} , \tag{23}$$

where E_{HF} is the Hartree-Fock ground state. This equation is a definition of the correlation energy. The spin-density wave instability theorem,[32] which I proved in 1962, showed that the normal (paramagnetic) state always has an instability of the SDW type. The theorem is rigorous and applies for all electron densities in the Hartree-Fock approximation. If one allows the jellium to be deformable, then the instability can be any admixture of SDW and CDW.[5]

$$\rho_{\pm} = \frac{1}{2} \rho_o [1 + p \cos(\vec{Q} \cdot \vec{r} \pm \phi)] , \tag{24}$$

where $\rho_{\pm}$ are the electron densities for up and down spin. See Fig. 11. The HF energy is invariant to the choice of ϕ as long as the jellium background cancels all the charge modulation from the electrons. The kinetic energy and the exchange interactions (up with up and down with down) do not care about the relative phase of the up-spin and down-spin modulations.

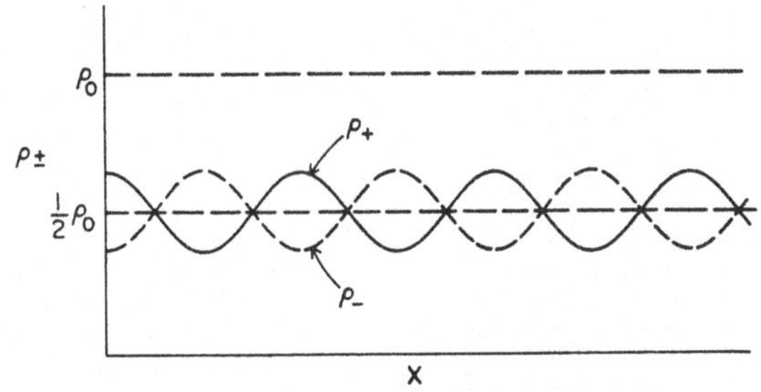

Fig. 11. Electronic charge density against x of up- and down-spin electrons for an ideal metal with a SDW ground state. The total charge density is a constant, ρ_0. In a CDW, ρ_+ and ρ_- are in phase.

Proof of this theorem is intricate. However the strategy is simple. One chooses $Q = 2k_F$ and lets the energy gap G be a variable. This gap is caused by the exchange potential, and is created by the amplitude modulation of the one-electron wavefunctions. One considers only those states near the energy gap at point C in Fig. 1 - the ones displaced $\sim 1/2$ G in energy from the free electron parabola and $\sim$100% amplitude modulated. These are in a spherical cap of height $\sim G$ and radius $\sim G^{1/2}$; see Fig. 12. Their number N is accordingly, $\sim G^2$. The kinetic energy "investment" to deform these states from plane waves to,

$$\psi \cong e^{i\vec{k}.\vec{r}} + \alpha^+ e^{i(\vec{k}+\vec{Q})\ r} + \alpha^- e^{i(\vec{k}-\vec{Q}).\vec{r}}, \qquad (25)$$

is their number N times the extra kinetic energy of ψ (compared to a pure plane wave). This is $\sim G$. Therefore,

$$\Delta T \sim G^3 \qquad (26)$$

New contributions to the exchange energy arise from the interactions of the states just described with virtually excited states associated with the spherical cap on the opposite side of the Fermi sphere. These states lie above k_F in Fig. 12. They contribute an energy "dividend",

$$\Delta E_{ex} \sim - \sum_{\vec{k},\vec{k}'} \frac{4\pi e^2}{|\vec{k}'-\vec{k}|^2} \sim - G^3 \ln\left(\frac{E_F}{G}\right) . \qquad (27)$$

This can always be made larger in magnitude than Eq.(26) by choosing G small enough, which proves the theorem. But one must show too (as it has been)[32] that Eq.(27) is larger than the loss of exchange energy resulting from the admixtures in (25). Consequently, a SDW or CDW state always lies lower than the normal (plane wave) state

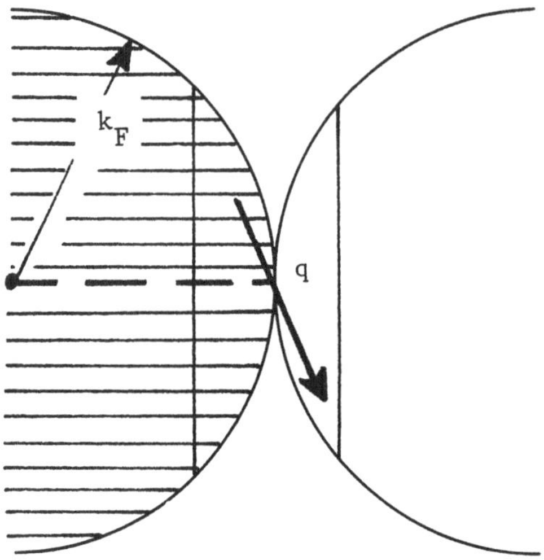

Fig. 12. The new exchange interactions that cause SDW or CDW instabilities are those between momentum states at the head and tail of the arrow $\vec{q}$. (States $\vec{k}$ in the spherical cap enclosing the head of the arrow, which have energy > E_F, are "virtually" occupied by their interaction with states $\vec{k}-\vec{Q}$ having energy < E_F.)

in the Hartree-Fock approximation, as shown on the left side of Fig. 13.

The Correlation Energy Correction

The algebraic sign of the correlation energy correction for SDW and CDW states is now the crucial issue. We will show that they have opposite sign - in fact a SDW state is destabilized and a CDW is made more stable.

The correlation energy can be calculated semi-quantitatively from

$$E_{corr} \cong - \sum_i \frac{\langle i|V_{eff}|o\rangle^2}{\Delta E_i} \qquad (28)$$

where ΔE_i is the two-particle two-hole excitation energy relative to the supposed ground state, $|o\rangle$. What is the meaning of V_{eff}? The dominant contributions to Eq.(28) are the virtual scattering of opposite-spin electrons from below the Fermi energy to empty states above. This proceeds by a screened interaction V_{scr}.

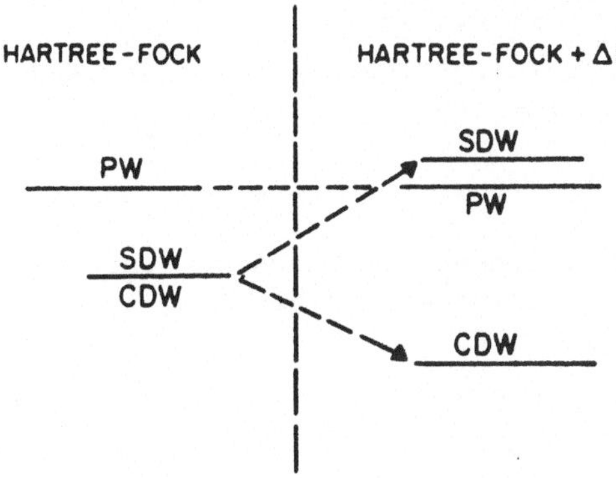

Fig. 13. Energy-level diagram for the plane-wave (PW), SDW and CDW states of an ideal metal. In the Hartree-Fock approximation the SDW and CDW states are degenerate and have lower energies than the plane-wave state. The degeneracy is split, as shown on the right, by correlation-energy corrections.

$$\psi' = C \{\psi_o - \sum_i \frac{<i|V_{scr}|o>}{\Delta E_i} \psi_i\}, \tag{29}$$

where C is a normalizing factor. The new energy is found by taking the expectation value of the exact Hamiltonian:

$$E' = <\psi'| \sum_\lambda \frac{p_\lambda^2}{2m} + \sum_{\lambda<\mu} \frac{e^2}{r_{\lambda\mu}} | \psi'> \quad . \tag{30}$$

If one works this out, one finds[5] Eq.(28), provided

$$<i|V_{eff}|o>^2 \equiv 2<i|V|o> <i|V_{scr}|o> - <i|V_{scr}|o>^2, \tag{31}$$

where $V \equiv e^2/r$. This procedure was first used in correlation energy calculations by W. Macke.[33]

The changes in correlation energy associated with the amplitude-modulated states was worked out in detail by Overhauser[5]. We present here a simplified version which gives accurately the same results. For the SDW case,

$$\phi_\pm \simeq e^{i\vec{k}\cdot\vec{r}} (1 \pm p \cos \vec{Q}\cdot\vec{r})^{1/2}, \tag{32}$$

where $p << 1$. The matrix element for virtual scattering of two opposite-sign electrons is,

$$m = \langle (\vec{k}+\vec{q})_+ \; (\vec{k}' - \vec{q})_- | V_{eff} | \vec{k}_+ \vec{k}'_- \rangle \; . \tag{33}$$

Two of the four factors in (32) have plus signs, and the other two have minus. It follows that:

$$m \approx m_o \left(1 - \frac{1}{2} p^2\right), \tag{34}$$

where m_o is the matrix element for the same event with $p = o$. As a consequence the correlation energy becomes, to order p^2,

$$E^{SDW}_{corr} \cong - \sum_i \frac{m_o^2}{\Delta E_i} (1 - p^2). \tag{35}$$

Every term is reduced in magnitude by the factor $(1-p^2)$; so the SDW state is raised relative to the (correlated) plane wave state.

For the CDW case all four factors in (32) have a positive sign, say. Accordingly,

$$E^{SDW}_{corr} \cong - \sum_i \frac{m_o^2}{\Delta E_i} (1+p^2) \; . \tag{36}$$

Each term in the correlation energy is enhanced in magnitude and this adds stability to the CDW state. (There are high energy virtual excitations - to states above the CDW energy gaps - for which the above remarks are not true; but they are relatively less important.)

It is easy to understand why electron-electron scattering splits the SDW and CDW degeneracy. For the CDW case up-spin and down-spin electrons are stratified in alternate laminar layers, and have less probability of scattering from one another. A CDW coaxes both spin types into the same layers, and increases their virtual scattering. These effects of correlation are shown on the right side of Fig. 13.

Analogy with Uniform Deformations

E. Wigner was the first to point out[2] that electron correlations reduce the enhancement of the spin susceptibility χ_s (compared to the Pauli value χ_p) caused by exchange. The straight line in Fig. 14 intersecting the horizontal axis near $r_s = 6$ is the Hartree-Fock value. As one approaches ferromagnetism the number of antiparallel-spin pairs approaches zero. So the correlation energy becomes smaller with larger spin polarization, and this cancels part of the exchange enhancement. Therefore the exact value of χ_s/χ_p lies above the Hartree-Fock line (as shown in Fig. 14).

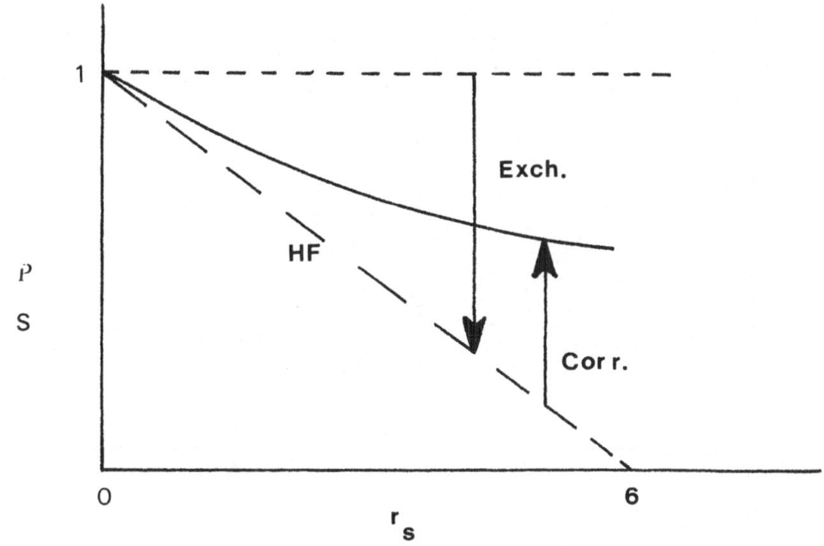

Fig. 14. Spin susceptibility of an electron gas compared to the Pauli value χ_p versus r_s. r_s is defined by Eq.(2).

The compressibility K of an electron gas is also enhanced by exchange - in fact, by the same factor as the spin susceptibility. However, the correlation energy enhances K <u>even</u> further. All calculations of E_{corr} vs. r_s show that the second derivative of E_{corr} with respect to density contributes to this. The general conclusion is that exchange and correlation reinforce one another for charge modulations (as shown in Fig. 15). The results we obtained earlier for the $Q = 2k_F$ case are similar to the uniform, $Q = 0$, examples just discussed.

CONCLUSIONS

A simple metal, with a deformable positive-ion background, should exhibit a CDW instability. Alkali metals are prime candidates for illustrating this phenomenon. Because they are monovalent, their (nearly spherical) Fermi surfaces do not extend to the Brillouin zone boundaries. Anomalous magnetotransport phenomena, clearly ascribable to a multiply connected Fermi surface can then be attributed to a broken-symmetry CDW state. The many striking anomalies of potassium,[16] and especially the direct ob-

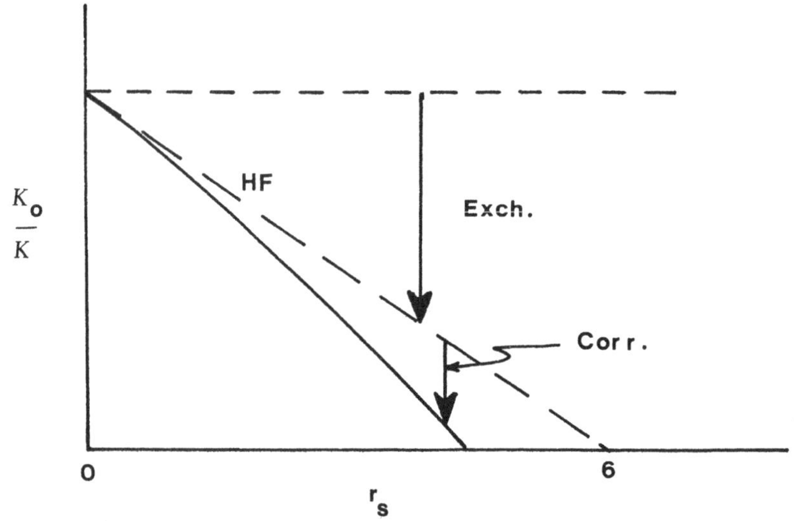

Fig. 15. Compressibility K of an electron gas compared to K_o (for a non-interacting Fermi gas) versus r_s.

servation of open orbits by Coulter and Datars,[18] confirm this behavior.

Several frequently used approximations in many-electron theory are shown to be generally invalid by the CDW structure of potassium. A common view is that electrons in metals can be treated as quasi-particles interacting via short-range, screened interactions. This is permitted for scattering processes, but should be forbidden for use in Hamiltonians. For example, if screened exchange were used (and it has been) to calculate the electron gas compressibility, one would find a result given by the behavior of the spin susceptibility (shown in Fig. 14) instead of that given by the second derivative of the total energy with respect to volume (as shown in Fig. 15). In other words, a screened-exchange treatment leads to a correction (relative to Hartree-Fock) having the wrong sign. Similarly, screened-exchange interactions do not permit a CDW structure in potassium.

Local density approximations are commonly used for exchange and correlation potentials. This may be a satisfactory method when they are used with an adjustable parameter to fit experimental data. But they are not predictive. For example, the CDW structure in potassium requires an exchange potential,

$$V_x = -G \cos \vec{Q}\cdot\vec{r} \quad . \tag{37}$$

where G = 0.6 eV. The resulting charge modulation is

$$\rho = \rho_o \; (1+p \cos \vec{Q}\cdot\vec{r}) \; . \tag{38}$$

G and p must be self consistent. If one calculates p from the observed G, assuming V_x is local, p = 0.17. If one now uses a Kohn-Sham local-exchange approximation to calculate G from p, one finds G = 0.2 eV. The inconsistency is half an order of magnitude. Unfortunately, the Kohn-Sham scheme is supposed to be best when applied to the ground-state problem of a free-electron metal.

The failure of local-exchange approximations to predict the correct ground state of potassium is caused by the fact that CDW states are supported by the non-local, dynamical effects of exchange. This can be easily illustrated by considering a uniform electron gas with an infinitesimal fractional modulation ε. The Slater local-exchange operator is then

$$A_s = -\frac{e^2 k_F}{\pi} \; (1 + \varepsilon \cos q \; x)^{1/3} \; . \tag{39}$$

Since ε is small, we have:

$$A_s \cong -\frac{e^2 k_F}{3\pi} \; \varepsilon \cos q \; x \; . \tag{40}$$

The dynamical properties of an operator are displayed by calculating its off-diagonal matrix elements. The only such elements for (40) are,

$$\langle \vec{k} \pm \vec{q} | A_s | \vec{k} \rangle = -\frac{e^2 k_F}{6\pi} \quad . \tag{41}$$

Observe that this result is independent of q and independent of $\vec{k}$. (It is <u>local</u> in both senses of the word).

Consider now the exact exchange operator A. It is defined by the equation,

$$A \; \psi(\vec{r}) = - \sum_i \left[\int \phi_i^+(\vec{s}) \psi(\vec{s}) \frac{e^2}{|\vec{r}-\vec{s}|} d^3 s \right] \phi_i(\vec{r}), \tag{42}$$

where $\{\phi_i\}$ are the occupied levels. For this illustration these can be taken to be the perturbed electron states associated with a potential such as (37), and leading to a fractional polarization p=ε. The <u>exact</u> off-diagonal matrix elements of A have been calculated.[34] The ratios of these to the local ones from A_s are shown in Fig. 16 as a function of q, and for four values of $\vec{k}$ The non-

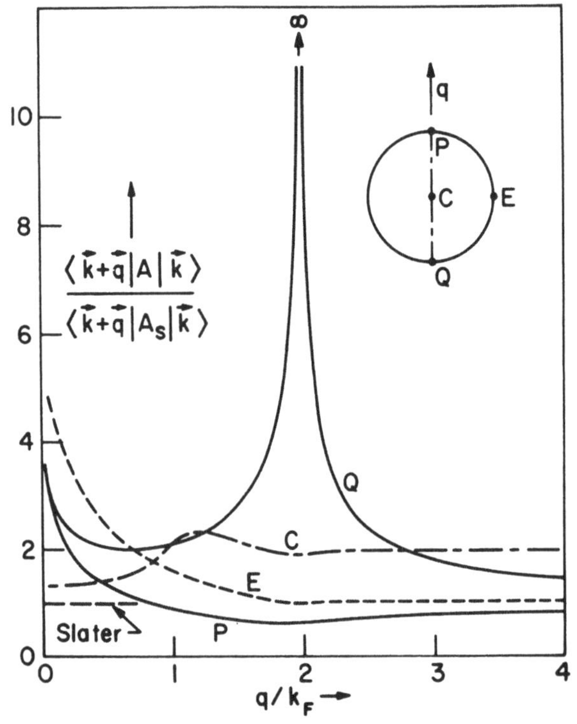

Fig. 16. Ratio of the off-diagonal matrix elements of the exact exchange operator to the Slater, local-density approximation for a small charge modulation of wave vector $\vec{q}$. Results for four initial states $\vec{k}$ are shown.

local effects are spectacular. Clearly a non-local approximation (constant, horizontal line at unity or 2/3) discards important dynamical properties. The singularity at $q = 2k_F$ is the cause of CDW phenomena.

It seems that progress in many-electron physics may have been impeded during the last fifteen years by the wide use of local, density-functional approximations to exchange and correlation.

REFERENCES

1. F. Bloch, Z. Physik 57, 545 (1929).
2. E. Wigner, Trans. Faraday Soc. 34, 678 (1938).
3. T. Kushida, J.C. Murphy, and M. Hanabusa, Phys. Rev. B13, 5136 (1976).
4. D. Shoenberg and P.J. Stiles, Proc. Roy. Soc. A 281, 62 (1964).

5. A.W. Overhauser, Phys. Rev. 167, 691 (1968).
6. H. Mayer and M.H. El Naby, Z. Physik 174, 269 (1963).
7. A.W. Overhauser, Phys. Rev. Lett. 13, 190 (1964).
8. I.M. Lifshitz, M. Ya Azbel, and M. Kagonov, Zh. eksp. teor. Fiz. 31, 63 (1956); English translation: Soviet Phys. JETP 4 41 (1957).
9. G. Stetter, W. Adlhart, G. Fritsch, E. Steichele, and E. Lüscher, J. Phys. F8, 2075 (1978).
10. J.A. Schaefer and J.A. Marcus, Phys. Rev. Lett. 27, 935 (1971).
11. F.W. Holroyd and W.R. Datars, Can. J. Phys. 53, 2517 (1975).
12. A.W. Overhauser and A.M. de Graaf, Phys. Rev. 168, 763 (1968).
13. M.L. Bishop and A.W. Overhauser, Phys. Rev. B18, 2447 (1978).
14. M. Ashrof and J.C. Swihart, unpublished.
15. D.E. Chimenti and B.W. Maxfield, Phys. Rev. B7, 350 (1973).
16. A.W. Overhauser, Adv. Phys. 27, 343 (1978).
17. Z. Altoumian and W.R. Datars, Can. J. Phys. 58, 370 (1980).
18. P.G. Coulter and W.R. Datars, Phys. Rev. Lett. 45, 1021 (1980).
19. M. Huberman and A.W. Overhauser, Phys. Rev. Lett. 47, 682 (1981).
20. A.W. Overhauser, Phys. Rev. B3, 3173 (1971).
21. G.F. Giuliani and A.W. Overhauser, Phys. Rev. B20, 1328 (1979).
22. G.F. Giuliani and A.W. Overhauser, Phys. Rev. B23, 3737 (1981).
23. G.F. Giuliani and A.W. Overhauser, Phys. Rev. Lett. 45, 1335 (1980).
24. A. Sawada and T. Satoh, J. Low Temp. Phys. 30, 455 (1978).
25. C.D. Amarasekara and P.H. Keesomn Phys. Rev. Lett. 47, Nov. 2 (1981).
26. H. van Kempen, J.S. Lass, J.H.J.M. Ribot, and P. Wyder, Phys. Rev. Lett. 37, 1574 (1976).
27. J.A. Rowlands, C. Durvury, and S.B. Woods, Phys. Rev. Lett. 40, 1201 (1978).
28. M.F. Bishop and A.W. Overhauser, Phys. Rev. B23, 3638 (1981).
29. A.G.M. Jansen, J.H. vanden Bosch, H. van Kempen, J.H.J.M. Ribot, P.H.H. Smeets, and P. Wyder, J. Phys. F10, 265 (1980).
30. M. Ashraf and J.C. Swihart, to be published.
31. J.A. Wilson, F.J. DiSalvo, and S. Mahajan, Adv. Phys. 24, 117 (1975).
32. A.W. Overhauser, Phys. Rev. 128, 1437 (1962).
33. W. Macke, Z. Naturf. 5a , 192 (1950).
34. A.W. Overhauser, Phys. Rev. B2, 874 (1970).

ELECTRON-HOLE LIQUID: RÔLE OF CORRELATIONS*

K. S. Singwi

Department of Physics and Astronomy
Northwestern University
Evanston, IL 60201

INTRODUCTION

The principal objective of this Advanced Study Institute is to discuss the role of electron correlations in solids, molecules and atoms. Keeping this aim in mind, I have chosen for my lectures the topic of correlations in electron-hole liquid (EHL) -a liquid which can be produced in the laboratory in a semiconductor like Ge and which consists of degenerate electrons and holes. The other reasons for my choice are: (a) The EHL is the simplest of all Fermi liquids whose constituents are elementary particles interacting via the coulomb law of force; (b) It constitutes an ideal jellium model -a model on which are based many, if not all, of our many-body calculations; (c) It provides experimental data which can then be compared directly with theory thus enabling one to judge the relative merits of many different approximations which one adopts in the many-body calculations of correlation energy. We shall see that EHL is indeed an interesting system both from experimental and theoretical points of view. The discussion which follows will be limited to T = 0 K. The more fundamental aspects of electron correlations in solids will be discussed by other lecturers at this Institute.

---

*A series of lectures delivered at the NATO International Advanced Study Institute, University of Antwerp (UIA), Belgium. July 20-31 (1981).

II. WHAT IS AN ELECTRON-HOLE LIQUID?

To understand what an EHL is let us consider the model of an ideal semiconductor which consists of a single isotropic conduction and valence bands that are separated by an energy gap E_g. Normally the valence band is full and the conduction band is empty. Light quanta of energy greater than E_g can excite electrons from the valence to the conduction band, leaving behind positively charged holes. The attractive coulomb interaction between an electron-hole pair leads to a bound state, called an exciton, which lies in the gap. The binding energy E_{ex} of an exciton is much less than that of a hydrogen atom:

$$E_{ex} = -\frac{\mu e^4}{2\hbar^2 K^2} \;; \qquad a_{ex} = \frac{K h^2}{\mu e^2} \tag{1}$$

$$\simeq 10^{-3}\ \text{ev} \;; \qquad \simeq 100\ \text{Å} \;,$$

whereas for a hydrogen atom,

$$E_H = 13.6\ \text{ev} \;, \qquad a_H = 0.5\ \text{Å} \;,$$

where K is the static dielectric constant of the medium and the reduced mass $\mu = m_e m_h/(m_e+m_h)$, m_e and m_h being the effective electron and hole masses. Note the different scales of energy and length in the present context.

In an indirect band semiconductor the life time of an exciton is $\sim 10^{-6}$sec, after which the electron and the hole recombine emitting a characteristic radiation with energy $E_g-|E_x|$. Under the influence of intense illumination e.g., the laser light, one generates a high density of excitons. When the density of excitons is high enough such that the mean separation between them is less than the excitonic Bohr radius (~100 Å), the excitons lose their identity and the system undergoes a phase transition to a new state of matter, called the electron-hole liquid (EHL). This is a first-order phase transition very much like the gas-liquid transition with a latent heat which can be measured experimentally. It is important to realize that EHL is a degenerate plasma of electrons and holes. The electrons and holes recombine in this liquid giving a characteristic broad radiation the width of which is a measure of their Fermi energies. A study of the line shape of this luminescence provides information about the properties of the plasma (See Fig. 1a and 1b). The plasma occurs in the form of tiny droplets (EHD) something like drops of water in a fog. Let me just say that such a phase transition has been seen experimentally[1] in semiconductors such as Ge and Si and phase diagrams have been mapped out.

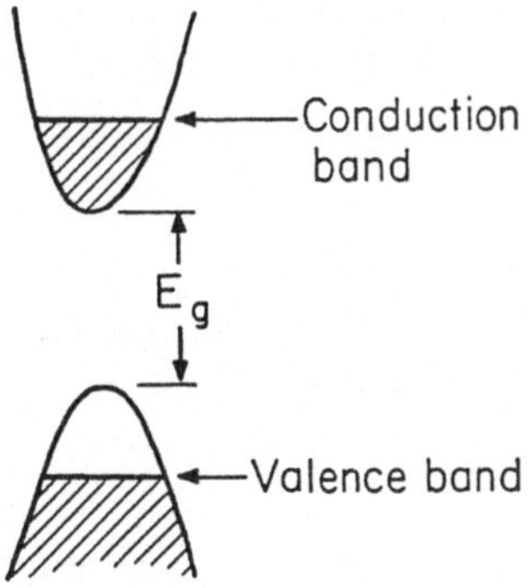

Fig. 1a. Schematic representation of a model system consisting of isotropic conduction and valence bands that are separated by an energy gap, E_g. The number of electrons in the conduction band are equal to the number of holes in the valence band.

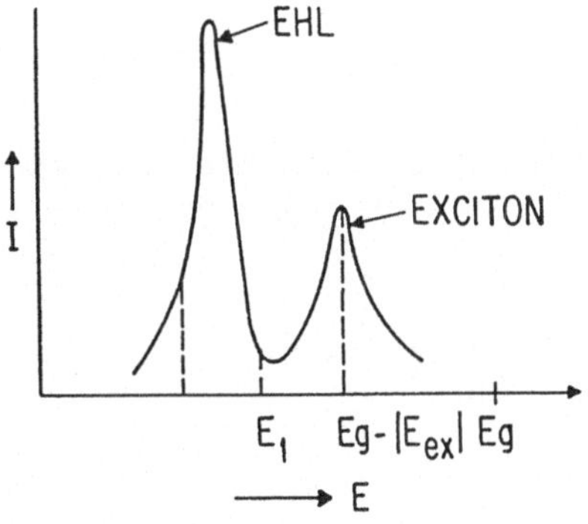

Fig. 1b. Typical recombination radiation spectrum in a semiconductor like Ge and Si.

III. IS THE PLASMA PHASE MORE STABLE THEN THE EXCITONIC PHASE?

It is interesting to inquire whether at T = 0 K, the plasma phase is more stable than the excitonic phase in the model system. At T = 0, the condition for thermodynamic equilibrium is that the pressure be 0 i.e.,

$$\left(\frac{\partial E}{\partial \Omega}\right)_N = 0 \;, \tag{2}$$

where E is the total energy of N e-h pairs in a certain volume Ω. Since $E = N[\varepsilon(n) + E_g]$, Eq. 2 becomes

$$\frac{\partial \varepsilon}{\partial n} = 0, \tag{3}$$

where n is the number density and ε is the energy per e-h pair. The problem thus reduces to that of calculating ε as a function of density, the minimum in the curve gives the ground-state energy and the equilibrium density n_0.

It is convenient to express various physical quantities in reduced units. We shall take the unit of length to be the excitonic Bohr radius a_{ex} and the unit of energy as excitonic Rydberg, 1 Ry = E_{ex}. For the ideal system the Fermi momentum k_F is defined by

$$k_F^3 = 3\pi^2 n. \tag{4}$$

We shall also define a dimensionless parameter r_s, which is a measure of interparticle separation, through the relation

$$\frac{1}{n} = \frac{4\pi}{3} r_s^3 a_{ex}^3 \tag{5}$$

The contributions to $\varepsilon(n)$ arise from kinetic, exchange and correlation. The kinetic and exchange energies per pair are, respectively, given by

$$t(r_s) = \frac{3}{10} \hbar^2 \left(\frac{1}{m_e} + \frac{1}{m_h}\right) (3\pi^2 n)^{2/3}$$

$$= \frac{2.21}{r_s^2} \text{ Ry} \tag{6}$$

$$\varepsilon_x = \frac{-3e^2}{2\pi K} (3\pi^2 n)^{1/3} = -\frac{1.83}{r_s} \text{ Ry} \tag{7}$$

Calculation of the correlation energy is a much more involved task. For the moment, we shall use Wigner's expression

$$\varepsilon_c = -\frac{A}{B+r_s} \quad Ry \quad , \tag{8}$$

where

$$A = 0.88, \ B = 7.8.$$

Different contributions to ε are shown in Fig. 2. The ground-state energy of the system is -0.55 Ry which is considerably higher than the excitonic binding energy -1 Ry. Hence the model system is not bound with respect to the excitonic state. Obviously, the stability of the e-h liquid phase depends critically on the competition between kinetic energy on one hand and the exchange-correlation energy on the other.

Experimentally[1] it is known that the liquid phase is bound both in Ge and Si. The location and structure of conduction and valence bands play an important role in the stability of the condensed phase. The anisotropy and the multiplicity of the bands tend to lower the kinetic energy and consequently increase the binding energy of EHL. Furthermore, in indirect band semiconductors, the recombination is accompanied by emission of phonons to conserve momentum and hence the life time of the condensed phase is longer. It is for these reasons that semiconductors like Ge and Si are ideal for the formation of electron-hole liquid.

Before we proceed to calculate the ground state energy of EHL in Ge, let us examine what important information can be extracted from the observed luminescence spectra (Fig. 1b). The energy corresponding to the upper edge of the plasma line ε_1 is the energy required to remove an e-h pair from the plasma.

$$\begin{aligned}\varepsilon_1 &= E(N) - E(N-1) = \left(\frac{\partial E}{\partial N}\right)_\Omega \\ &= \varepsilon(n) + n\frac{\partial \varepsilon(n)}{\partial n} + E_g\end{aligned}$$

At equilibrium since $\frac{\partial \varepsilon(n)}{\partial n} = 0$, therefore

$$\varepsilon_1 = \varepsilon(n) + E_g \tag{9}$$

The difference between the exciton line $E_g - |E_x|$ and ε_1 gives the condensation energy ϕ of an e-h pair. Knowing ϕ and adding it to E_x one obtains the ground-state energy of the plasma, which can then be compared directly with calculation. The lower edge of the EHL line gives the renormalized band gap. The renormalization occurs because the conduction and valence bands come closer together as a result of many-body effects.

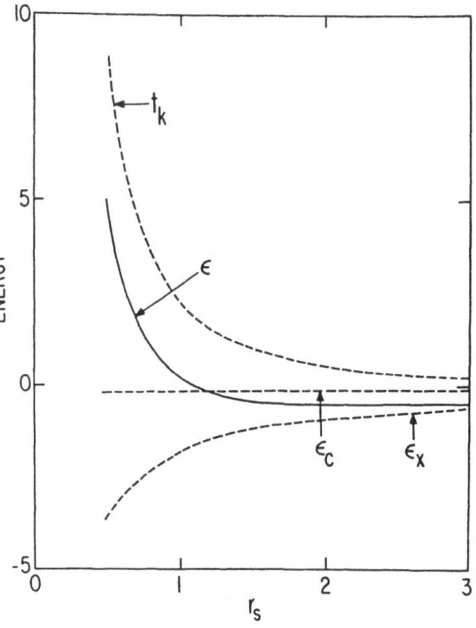

Fig. 2. r_s dependence of kinetics (t_k), exchange (ε_x), correlation (ε_c) and total energies (ε), each measured in unit of an excitonic rydberg. The correlation energy per e-h pair is $\varepsilon_c = \varepsilon^{ee} + \varepsilon^{hh}$, where ε^{ee} and ε^{hh} are taken from the results of an electron liquid.

IV. GROUND-STATE ENERGY OF EHL IN Ge

(i) Band Structure of Ge.

The band structure of Ge is shown in Fig. 3. It has a valence band maximum at the center of the Brillouin zone. Due to spin-orbit coupling, the valence level splits into a four-fold Γ_8^+ level and a two-fold Γ_7^+ level. Because of the large splitting between Γ_8^+ and Γ_7^+ (0.29 eV), the latter is usually ignored in calculating the properties of EHL. Away from the center of the Brillouin zone, the Γ_8^+ level splits into a heavy hole (hh) and light hole (lh) bands. The energy surface of these bands can be described by[2]

$$\varepsilon_{lh,hh}(\vec{k}) = Ak^2 \pm [B^2k^4 + C^2(k_x^2k_y^2 + k_y^2k_z^2 + k_z^2k_x^2)]^{1/2}, \qquad (10)$$

where (+) and (−) refer respectively to light and heavy hole bands. The parameters A, B and C are obtained from cyclotron resonance measurements. The density of states mass for light and heavy bands are defined through

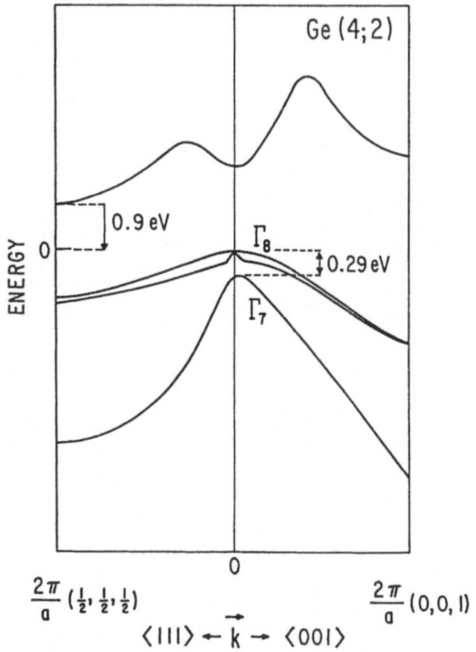

Fig. 3. A sketch of the band structure of Ge.

$$m_{lh,hh}^{3/2} = \pi^2\hbar^3(\frac{2}{\varepsilon})^{1/2} \sum_k \delta(\varepsilon-\varepsilon_{lh,hh}(\vec{k})). \tag{11}$$

In most theoretical calculations the complex band structure is simplified to isotropic bands by substituting $B^1 = (B + \frac{C^2}{6})^{1/2}$ for B and setting C = 0 in Eq. (10). This considerably simplifies the calculation of exchange - correlation energy.

The conduction bands are ellipsoids. In Ge the four-fold conduction minima lie at the zone boundary $\frac{2\pi}{a}$ (1,1,1). The transverse and longitudinal masses of electrons are known accurately from cyclotron resonance experiments. Their values along with the density of states and optical masses are listed in Table 1.

Table 1. Table of constants for Ge and Si

	m_{ei}	m_{et}	m_{oe}	m_{de}	A	B	C	$m_{\ell h}$	m_{hh}	E_{ex} (meV)	
Ge	1.58	0.082	0.120	0.22	13.38	8.48	13.15	0.042	0.347	15.36	2.65
Si	0.9163	0.1905	0.2588	0.32	4.28	0.75	4.85	0.154	0.523	11.4	12.85

The masses are in units of the bare electron mass m; A, B and C are in units of $\hbar^2/2m$.
The values of A, B and C are taken from the latest cyclotron resonance work of Hensel and Suzuki[2].

Values of band parameters for Ge(1;1) and Si(2;1). The masses are given in units of the bare electron mass. m_{de}, m_{dh} are density of states masses for electrons and holes, respectively, m_{oe} and m_{oh} are their respective optical masses. μ is the reduced optical mass of electron and hole, and E_{ex} is the excitonic rydberg

	m_{el}	m_{et}	m_{de}	m_{oe}	m_{hl}	m_{ht}	m_{dh}	m_{oh}	μ	E_{ex} (meV)
Ge(1;1)	1.580	0.082	0.2198	0.120	0.040	0.130	0.088	0.075	0.046	2.65
Si(2;1)	0.9163	0.1905	0.3216	0.2588	0.154	0.523	0.2354	0.2336	0.1228	12.85

$$m_{el}^{-1} = A + (B^2 + C^2/3)^{1/2},\ m_{et}^{-1} = A - \frac{1}{2}(B^2 + C^2/3)^{1/2} \ : \ Ge(1;1)$$

$$m_{el}^{-1} = A + B,\ m_{et}^{-1} = A - B/2 \ : \ Si(2;1)$$

(ii) Kinetic Energy

The kinetic energy per electron is given by

$$tc(n) = \frac{3}{5}\ E_F^e = \frac{3}{5}\ \frac{\hbar^2 k_e^2}{2m_{de}} = \frac{3}{10}\ \frac{\hbar^2}{m_{de}}\ (\frac{3\pi^2 n}{\nu_e})^{2/3}$$

$$= \frac{2.21}{r_s^2}\ \frac{\mu}{m_{de}\nu_e^{2/3}}\ Ry\ , \qquad (12)$$

$$\nu_e = 4 \text{ for Ge, } m_{de} = (m_{et}^2\ m_{el})^{1/3}\ .$$

The quantity k_e is the equivalent spherical momentum related to the electron density n by $k_e = (3\pi^2 n/\nu_e)$. It is evident from Eq. (12) that the degeneracy and anisotropy of bands reduce the kinetic energy (compare with Eq. 6).

The light and heavy holes have the same Fermi energy. Therefore

$$t_{lh} = t_{hh} = \frac{3}{5}\ E_F^h\ ,$$

where

$$E_F^h = \frac{\hbar^2}{2m_{lh}}\ (3\pi^2 n_{lh})^{2/3} = \frac{\hbar^2}{2m_{hh}}\ (3\pi^2 n_{hh})^{2/3}$$

$$t_h = t_{lh} = t_{hh} = \frac{2.21}{r_s^2}\ \frac{\mu}{m_{hh}}\ \frac{1}{(1+\eta_h{}^{3/2})^{2/3}} \qquad (13)$$

where

$\eta_h = m_{lh}/m_{hh}$. The reduces mass μ is defined by

$$\frac{1}{\mu} = \frac{1}{m_{0e}} + \frac{1}{m_{0h}}\ , \qquad (14a)$$

where the optical masses of electrons ahd holes m_{0e} and m_{0h} are given by

$$\frac{1}{m_{0e}} = \frac{1}{3}\ (\frac{2}{m_{et}} + \frac{1}{m_{el}})$$

$$\frac{1}{m_{0h}} = \frac{1}{2}\ (\frac{1}{m_{lh}} + \frac{1}{m_{hh}}) \qquad (14b)$$

(iii) Exchange-Correlation Energy.

Inspite of the complications arising from the band characteristics of Ge such as the anisotropy of the conduction bands, the coupling between the valence bands at k = 0 and their warping, the exchange energy can be evaluated in the same way as the exchange energy of electrons in an electron liquid. This however, is not the case with the correlation energy. As a result of sophisticated calculations[3] of the exchange-correlation energy ε_{xc} for many model systems, one important conclusion that has emerged is that ε_{xc} is a function of the density (i.e., r_s) of the plasma only and is independent of the band structure of the semiconductor. Although exchange and correlation individually depend on the band structure the sum does not. This important observation, although as yet empirical, has been responsible for considerable simplification in calculating the energetics and thermodynamics of EHL.

Since the correlation energy ε_c of the condensed phase is a sizeable part of the ground-state energy, we shall consider it here in somewhat greater detail. Approximate schemes that have been used to calculate ε_c are the same as those used earlier in the case of an electron liquid except that these schemes have been generalized to deal with a multicomponent system like EHL. These approximations go under the names: (i) RPA[4] (ii) Hubbard[5] (iii) Nozieres-Pines interpolation scheme[6] and (iv) Self-consistent scheme of Singwi, Tosi, Land and Sjolander (STLS).[7] For reasons which will be clear later we shall concentrate on the last scheme.

(a) Self-consistent Scheme for One-component System

We shall here summarize only the basic results of the scheme and refer the reader for details to the original paper.[7] Consider a homogeneous system of electrons of density n_0 on a rigid, uniform, positive background in the presence of a weak external potential $V_{ext}(r,t)$. The induced density $\delta n(q,\omega)$ in the Fourier space is related to the external potential $V_{ext}(q,\omega)$ by

$$\langle\delta n(\vec{q},\omega)\rangle = \chi(\vec{q},\omega)\, V_{ext}(\vec{q},\omega) , \tag{15}$$

where χ is the density-density response function. In the so called generalized RPA one usually writes

$$\langle\delta n(\vec{q},\omega)\rangle = \chi_0(\vec{q},\omega)\, [V_{ext}(\vec{q},\omega) + \psi(\vec{q},\omega)\, \langle\delta n(\vec{q},\omega)\rangle] \tag{16}$$

where χ is the polarizability of the noninteracting gas and $\psi\delta n$ is the polarization potential. As it stands Eq. (16) is quite general as long as ψ remains undefined. A physically more reasonable form is the one in which χ_0 is replaced by χ_{sc}, the latter accounting for mass renormalization and self-energy effects. In the STLS scheme, $\psi(q,\omega)$ is static and is given by

$$\psi(\vec{q},\omega) = v(\vec{q})\, (1 - G(\vec{q})), \tag{17}$$

where $v(q) = 4\pi e^2/q^2$ and

$$G(\vec{q}) = -\frac{-1}{n_0}\int \frac{d\vec{q}'}{(2\pi)^3}\cdot\frac{\vec{q}\cdot\vec{q}'}{q'^2}\,[S(\vec{q}-\vec{q}')-1] \tag{18}$$

S(q) is the structure factor. Using (15) - (17), we have

$$\chi(\vec{q},\omega) = \frac{\chi_0(q,\omega)}{1 - v(\vec{q})\,[1 - G(\vec{q})]\,\chi_0(\vec{q},\omega)} \tag{19}$$

The relation between the dielectric function and $\chi(q,\omega)$ is

$$\frac{1}{\varepsilon(\vec{q},\omega)} = 1 - v\,(\vec{q})\quad \chi(\vec{q},\omega). \tag{20}$$

The T = 0 fluctuation-dissipation theorem is

$$S(q) = \frac{\hbar}{\pi n_0}\int_0^\infty d\omega\ \text{Im.}\ \chi(\vec{q},\omega) \tag{21}$$

Equations (18), (19) and (21) constitute a self-consistent set. Starting with some form for S(q), one determines G(q) from Eq. (18), and then substituting the latter in (19) one evaluates $\chi(\vec{q},\omega)$ The Im. $\chi(\vec{q},\omega)$ is substituted in (21) and S(q) is calculated for the next iteration. This procedure is repeated until a self-consistent solution is obtained.

The usual pair-correlation function is obtained from $s(\vec{q})$:

$$g(\vec{r}) - 1 = \frac{1}{n_0}\int \frac{d\vec{q}}{(2\pi)^3}\ e^{i\vec{q}\cdot\vec{r}}\ [S(\vec{q}) - 1] \tag{22}$$

The RPA, which is a special case of the self-consistent scheme, corresponds to taking G(q) = 0, The Hubbard approximation[5] (HA) is recovered by substituting for S(q) in (18) its Hartree-Fock value, namely

$$S_{HF}(\vec{q}) = \frac{-2}{(2\pi)^3 n_o}\int d\vec{k}\int d\vec{k}'\ \delta(\vec{k} - \vec{k}' + \vec{q}) \tag{23}$$
$$|\vec{k}| \leqslant k_F;\ |\vec{k}'| \leqslant k_F,$$

which then leads to

$$G(q) = 1/2\ \frac{q^2}{q^2 + k_F^2}\quad , \tag{24}$$

where k_F is the Fermi momentum; $k_F^3 = 3\pi^2 n_o$

The superiority of the self-consistent scheme lies in the fact that g(r) remains positive definite for small interparticle separation for $r_s \leqslant 4$, whereas it is not so in the RPA or HA. This negative behavior of g(r) for small r is due to the neglect of short range correlations.

The above dielectric formulation enables us to calculate the correlation energy of the electron gas. The interaction energy can be expressed[6] in terms of the dielectric function:

$$E_{int} = \sum_{\vec{q}} \int_0^\infty \frac{d\omega}{(2\pi)} \text{ Im.} \frac{1}{\varepsilon(\vec{q},\omega)} + \frac{2\pi Ne^2}{q^2}, \tag{25}$$

which on using Eq. (21) can be written as

$$E_{int} = \sum_{\vec{q}} \frac{2\pi e^2}{q^2} [S(\vec{q}) -1] \tag{26}$$

or equivalently

$$E_{int} = -\left(\frac{4}{\pi r_s}\right)\left(\frac{9\pi}{4}\right)^{1/3} \gamma(r_s) \text{ Ry}, \tag{27}$$

where

$$1 \text{ Ry} = e^2/2a_0 \; ; \; n^{-1} = \frac{4\pi}{3} r_s^3 a_0^3$$

$$\gamma(r_s) = -\frac{1}{2k_F} \int_0^\infty dq[S(q) -1] \tag{28}$$

Knowing the interaction energy, it is easy to calculate [6] the ground state energy E using a theorem due to Pauli according to which

$$E(e^2) = E(0) + \int_0^{e^2} \frac{d\lambda}{\lambda} E_{int}(\lambda), \tag{29}$$

λ being the interaction parameter and E(0) is the energy in the absence of any interaction and therefore corresponds to the noninteracting kinetic energy. The correlation energy, E_c is the difference between the ground-state energy and the H-F energy (noninteracting kinetic + exchange). Therefore

$$E_c = \int_0^{e^2} \frac{d\lambda}{\lambda} E_{int}(\lambda) - E_x, \tag{30}$$

where E_x is the exchange energy per electron. Changing the integration variable λ to r_s through the transformation $h^2 k_F u = \left(\frac{9\pi}{4}\right)^{1/3} m\lambda$ and substituting Eq. (27) in (30) we have

$$E_c = \frac{1}{r_s^2} \int_0^{r_s} du \, [- \frac{4}{\pi} (\frac{9\pi}{4})^{1/3} \gamma(u) + 0.9163] \text{ Ry} \tag{31}$$

Singwi et al.[7] using their self-consistent values of S(q) have calculated E_c and their results are given in Table 2 together with the results obtained in other approximations. In the last row we have given the most recent numerical results of Ceperley and Alder[8] obtained by Monte Carlo variational calculations (See also Fig. 4). The remarkable agreement between these values and those calculated in the self-consistent scheme give us confidence in the latter.

(b) Generalization to EHL

Generalization[9] of the self-consistent scheme to the case of a multicomponent plasma of electrons and holes is straightforward and we shall not go into it here. However, the calculational procedure is considerably more complex because it involves solutions of a set of coupled nonlinear equations for the partial structure factors $S_{ee}(q)$, $S_{eh}(q)$ and $S_{hh}(q)$. Once the structure factors are known, the correlation energy can be calculated straightforwardly in the same way as in RPA.

Table 2. Correlation Energy (Ryd/electron)

r_s	1	2	3	4	5	6	10	20
Wigner	-0.100	-0.090	-0.082	-0.075	-0.069	-0.064	-0.049	-
RPA	-0.157	-0.124	-0.105	-0.094	-0.085	-0.078	-	-
Hubbard	-0.131	-0.102	-0.086	-0.076	-0.069	-0.064	-	-
Nozierés and Pines	-0.115	-0.094	-0.081	-0.072	-0.065	-0.060	-	-
STLS	-0.124	-0.092	-0.075	-0.064	-0.056	-0.050	-0.036	-0.022
Ceperley and Alder	-0.119	-0.090	-0.074	-0.064	-0.057	-0.051	-0.038	-0.023

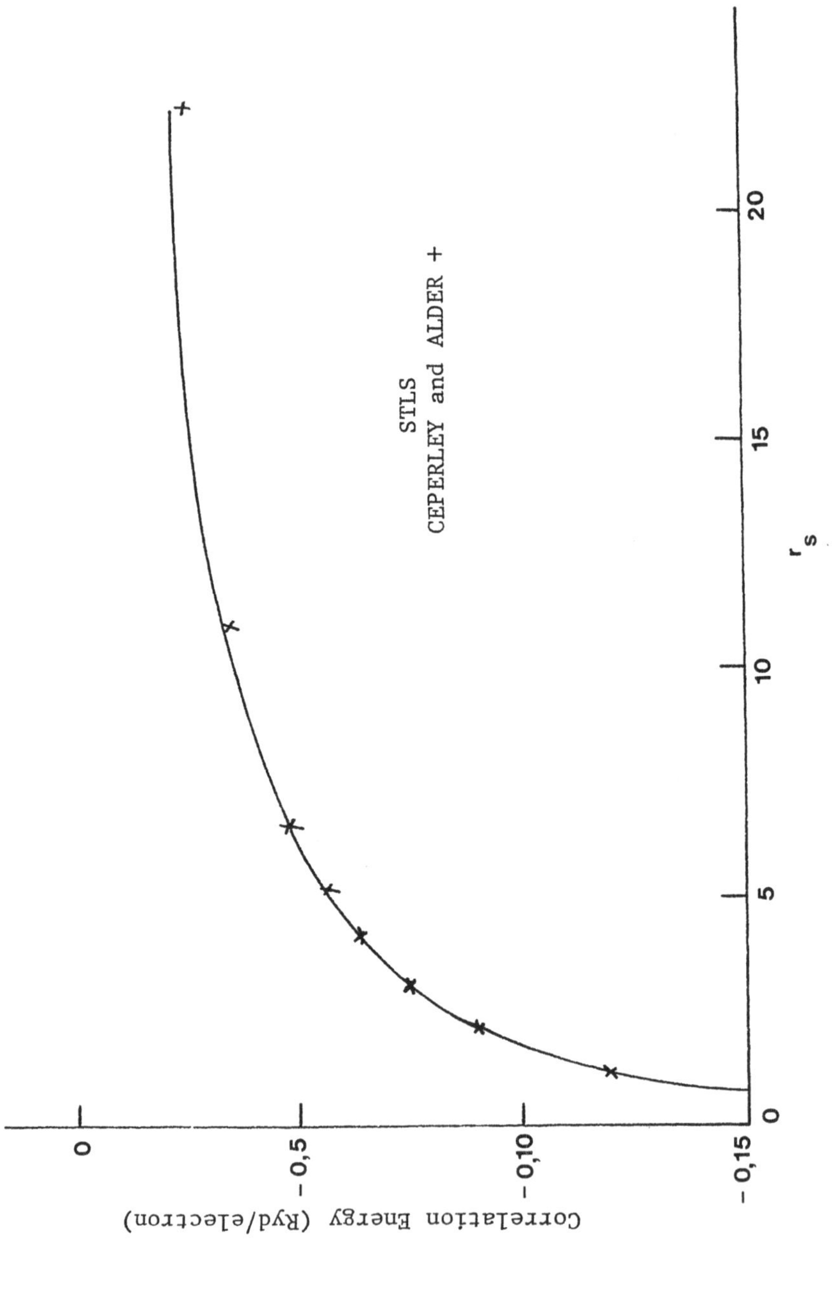

Figure 4. Correlation Energy (Ryd/electron)

We had remarked earlier that the sum of the exchange and the correlation energy is independent of the band characteristics and depends only on the density i.e., r_s. In this connection it should be clearly borne in mind that the exchange and correlation effects should be treated on an equal footing; if a certain band structure effect is ignored in the calculation of the correlation energy it should also be excluded in exchange energy. Over a wide range of r_s values, one finds that ε_{xc} can be fitted to the following simple form[3]

$$w_{xc}(r_s) = \frac{A_0 r_s + A_1}{r_s^2 + A_2 r_s + A_3} , \qquad (32)$$

The coefficients A_0-A_3 are obtained from a fit to the actual numerical values of ε_{xc} for a number of systems, and their values are given in Table 3. A plot of ε_{xc} versus r_s is shown in Fig. 5. For purposes of comparison are also shown the numerical values[9] of ε_{xc} in FSC approximation for Si (6;2), Ge(4;2) and a model system of isotropic conduction band and an identical valence band. One advantage of a simple form such Eq. (32) is that it greatly facilitates the evaluation of the energetics of EHL.

(c) Ground-State Energy of EHL in Ge

The energy per particle ε can be written simply as

$$\varepsilon = \frac{\alpha}{r_s^2} + \frac{A_0 r_s + A_1}{r_s^2 + A_2 r_s + A_3} \ \text{Ry} , \qquad (33)$$

(FSC).

Table 3. The coefficients A_0, A_1, A_2, and A_3 of the fit to the FSC values of exchange-correlation energies ε_{xc} per e-h pair. The form of the fit to ε_{xc} is given in Eq. 33).

A_0	-5.0879
A_1	-4.8316
A_2	3.0426
A_3	0.0152

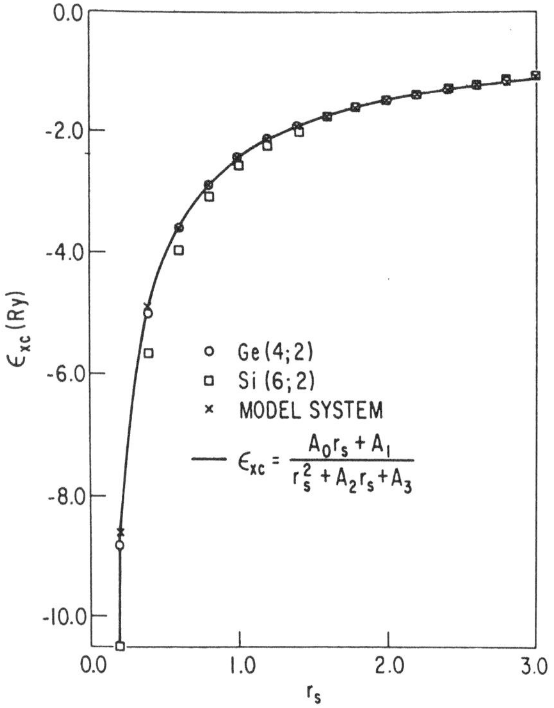

Fig. 5. The r_s dependence of exchange-correlation energy, ε_{xc}, for various systems evaluated in the fully self-consistent approximation (FSC). The calculated values are fitted to the form $(A_0 r_s + A_1)/(r_s^2 + A_2 r_s + A_3)$ which is expected to be applicable in all the semiconductors.

where $\alpha = 0.4697$ for Ge. Figure 6 shows the r_s dependence of ε in Ge. In the figure we make a distinction between HA and the results of Brinkman et al.[2,10], even though the latter have evaluated the correlation energy in Hubbard approximation. In Ref. 10 the complex band structure effects are included only in exchange and not in correlation energy. The results marked HA are obtained in the collapsed band model[10] for both exchange and correlation by ignoring the effect of valence band coupling. While the differences between HA and BRAC are substantial, the FSC and HA results are in good agreement especially for $r_s \leqslant 0.6$. At higher values of r_s the larger differences between FSC and HA are due to effects of (e,e), (e,h) and (h,h) multiple scatterings and short-range coulomb correlations, both of which are included in an approximate way only in FSC approximation, but not in HA. Detailed calculations of the ground-state, energy of EHL in Ge and Si were first made by Combescot and Nozières[11] (CN).

The minimum of the curve in Fig. 6 gives the ground-state energy ε and the equilibrium density n_o. These results for Ge and Si are given in Table 4. The experimental study of the luminescence line shape directly yields the binding energy ϕ of the condensed phase. The measured value[2] of ϕ is 1.8 meV. The theoretical value of ϕ is given by $\varepsilon - |E_x|$, which gives 1.73 meV remembering that $\varepsilon = -2.22$ Ry (1 Ry = 2.65 meV) and taking the most reliable value of the excitonic binding energy to be 4.15 meV. The equilibrium density is $2.22 \times 10^{17}/cm^3$ which is also in excellent agreement with experiment. The same is true in the case of EHL in Si.

Because of the high equilibrium density ($r_s \simeq 0.6$ for Ge), the RPA and HA also yield values of ϕ and n_o which are in good agreement with experiment. Thus the calculation of the ground-state energy for such low values of r_s is not a very sensitive test of the different many-body approximations. This was expected. However, there are differences between FSC approximation on one hand, and HA and RPA on the other in the behaviour of pair-correlation functions. To illustrate this point, we show in Fig. 7 the behaviour of e-h pair correlation function around the equilibrium r_s in the EHL in Ge. Here since the pair correlation function $g_{eh}(r)$ has been evaluated in the collapsed band model for holes, $g_{eh}(r)$ is the same for the light and heavy holes. The important point to note in Fig. 7 is that the differences between $g^{RPA}(r)$ and $g^{HA}(r)$ are negligible, which however is not the case with $g^{FSC}(r)$. The large difference for small interparticle separation between $g^{FSC}(r)$ and $g^{HA}(r)$ is due to the fact that in the self-consistent theory short range correlations are treated better. The effect of correlations becomes more important as r_s increases. To illustrate this point, we show in Fig. 8 the enhancement factor $\rho = g_{eh}(0)$ as a function of r_s in different approximations. Such a behaviour of $g^{FSC}_{eh}(0)$ has indeed been seen experimentally[12] in <111> -stressed Ge.

Table 4. Theoretical results for the equilibrium density, n_o, ground state energy, ε_o, compressibility, χ, and enhancement factor ρ in three different approximations.

		n_o ($10^{17}cm^{-3}$)	ε_o (meV)	χ ($meV^{-1}cm^{-3}$)	ρ
Ge(4;2)	HA	1.8	5.3	5.5×10^{-18}	1.8
	CN	2.0	6.1(5.65)		
	FSC	2.2	5.9	5.0×10^{-18}	2.3
Si(6;2)	HA	34	20.4		2.2
	CN	31.(36)	21.(19.4)		
	FSC	32	22	9.6×10^{-20}	3.5
Ge(1;1)	HA	0.12	2.8		2.6
	CN	0.13	2.6		
	FSC	0.11	3.1	2.2×10^{-16}	6.8
Si(2;1)	HA	5.3	13.3		2.8
	CN				
	FSC	4.5	14.7	9.6×10^{-19}	7.4

V. EHL in Stressed Ge

Application of uniaxial stress can considerably alter the band structure. For example, in Ge compressional stress along <111> direction removes the four-fold degeneracy of conduction bands by moving three of them relative to the fourth one, and also decouples the light and heavy hole bands at k = 0. The separation between the split conduction bands and the decoupled valence bands increases with stress, whereas the energy gap decreases. We shall here not discuss the energy-wave number relationship of the perturbed hole bands.

Certain unique situations occur at specific values of stress (see Fig. 9). At 2.8 kg/mm^2 the Fermi energy of electrons is

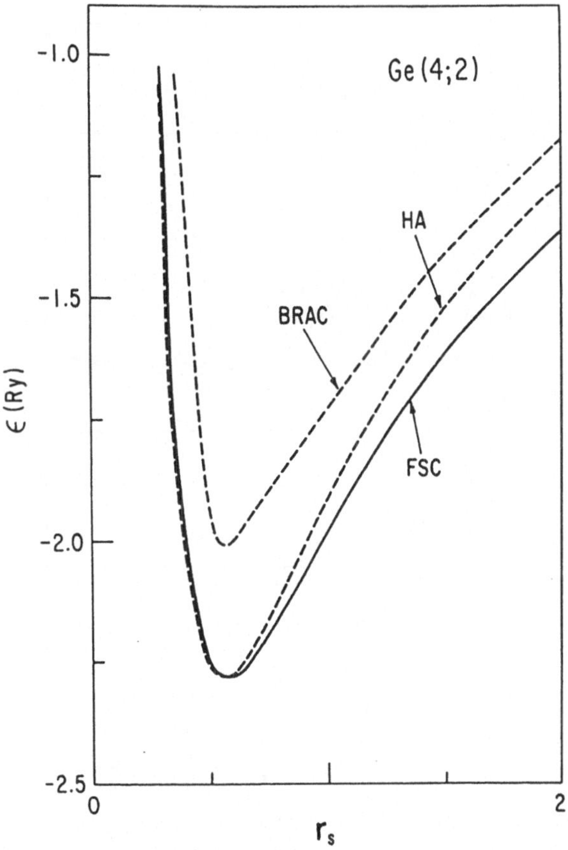

Figure 6. Ground state energy ε as a function of r_s in Ge(4;2) in three different approximations; 1) BRAC represents the results of Brinkman et al. (ref.2); 2) the curve marked HA corresponds to Hubbard approximation but it differs from the results of BRAC in that both the exchange and correlation energies are evaluated in the collapsed band model; 3) the curve FSC is obtained by using the fit (Eq.33), for the exchange-correlation energy.

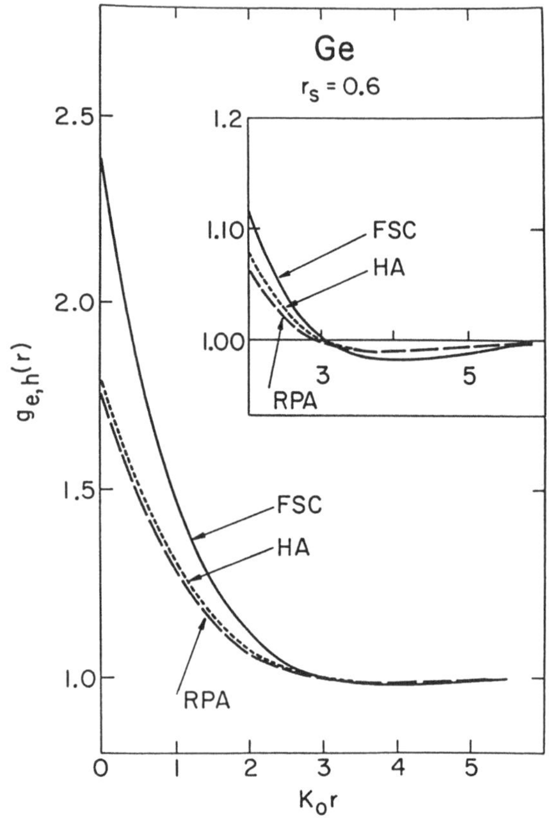

Figure 7. The spatial dependence of e-h pair correlation function $g_{eh}(r)$ in Ge(4;2) around the equilibrium r_s. Note the fully self-consistent (FSC) results differ substantially from those in Hubbard (HA) and random phase (RPA) approximations. The value of $g_{eh}(r)$ at $r = 0$ gives the enhancement factor ρ. The inset shows Friedel oscillations in pari correlation function for large values of r.

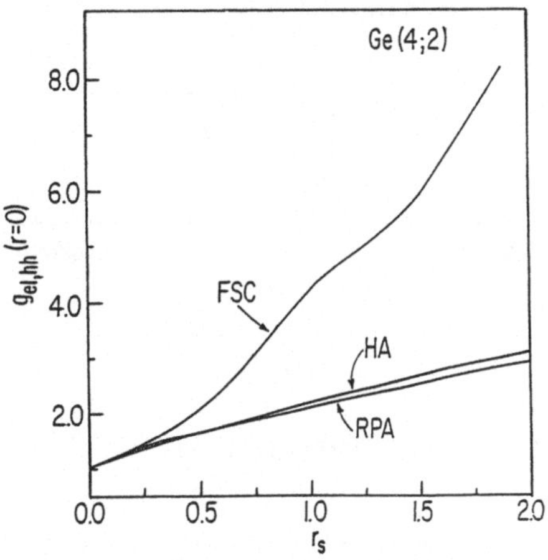

Fig. 8. The enhancement factor ρ as a function of r_s in Ge(4;2). ρ in the fully self-consistent approximation differs substantially from the values in Hubbard (HA) and random phase (RPA) approximations.

just equal to the splitting between the conduction bands, and the electrons therefore occupy only the lower conduction band, The valence band splitting is such that both the light and heavy hole bands are still occupied. The system is denoted by Ge(1;2). At very large values of stress (> 10 kg/mm^2), the splitting between the two hole bands is so large that only the upper hole band is occupied, the system is then denoted by Ge(1;1). The occupied band is an ellipsoid with transverse and longitudinal masses given in Table 1. The conduction band masses remain unaltered.

Calculation of the ground-state energy in Ge under very large <111> stress (i.e., Ge(1;1)) presents an interesting problem for two reasons: (a) the calculation is relatively much simpler since one is dealing with only one conduction and one valence band whose structure is simple and (b) since the electrons (and holes) are all in one band their kinetic energy contribution is large and therefore the condensation into EHL will critically depend on the value of the exchange - correlation energy.

In Ge (1;1) kinetic energy per e-h pair is

$$t(r_s) = \frac{1.616}{r_s^2} \tag{34}$$

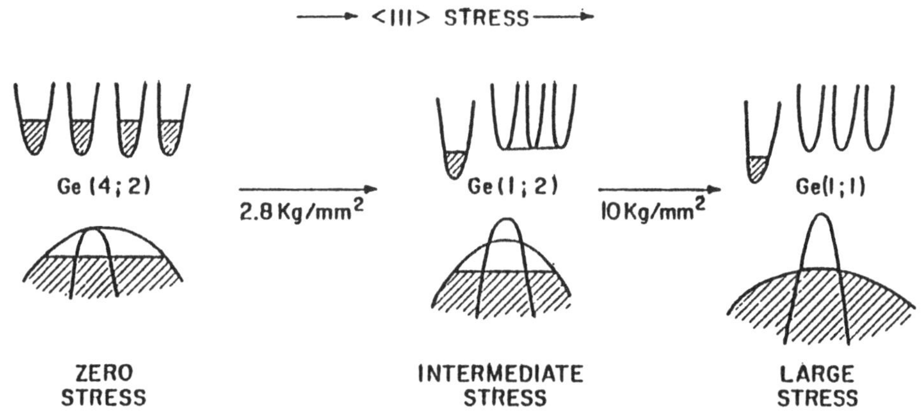

Fig. 9. A schematic diagram of the conduction and valence bands in Ge under <111> uniaxial stress. The conduction bands split into a single and a three fold set. At a stress of 2.8 kg/mm^2 the splitting between the two sets of conduction bands is equal to the Fermi energy of electrons and the splitting between the light and heavy hole bands is smaller than the Fermi energy of the holes.

The total energy ε is given by the sum of (34) and (33). Minimization of ε with respect to r_s gives the ground-state energy ε_0 and the equilibrium density n_0. These results are given in Table 4. One sees that the difference between HA and FSC approximations is quite appreciable in this case compared to that in normal Ge(4;2). This again is the result of correlations playing a more important role for larger r_s and being treated better in FSC approximation. Note that the enhancement factor ρ is almost a factor of three larger than that given by HA. Recent experimental estimates[12,13] of ρ in Ge (1;1) and Si(2;1) are in better agreement with the FSC values than HA.

In Ge (1;1) the calculation based on HA predicts that the binding energy of an e-h pair in the condensed phase is only 0.07 E_{ex}, which is just marginal for condensation, whereas the value of ϕ in FSC is 0.13 E_{ex}. The condensed phase in Ge(1;1) is thus stable and its existence has now been fully confirmed. The theory also predicts condensation in Si(2;1), and it has been confirmed by experiment.[15]

VI. PHASE SEPARATION OF "HOT" AND "COLD" LIQUIDS

Under the influence of a <111>-uniaxial stress three of the four conduction valleys move up while the remaining valley moves down in energy. We shall refer to the electrons in the upper valleys as "hot" electrons and those in the lower valley as "cold" electrons. Experimental observations indicate that the intervalley scattering time for the electrons is of the order of one microsecond. This time is long enough for the hot electrons to thermalize within the upper valleys, although their Fermi level need not be the same as that of the "cold" electrons in the lower valley. Also this time is comparable with the life time of EHL. If one ignores the slow transfer of electrons between valleys, one can treat the hot and cold electrons as two separate species in quasi-thermal equilibrium with each other and with the holes.

By varying the experimental conditions, it is possible to vary the relative numbers of hot and cold electrons in the electron-hole drop (EHD). Thus the EHD in <111>-stressed Ge provides a unique opportunity to study the thermodynamics of a degenerate two-component coulomb Fermi liquid.

Recently, Kirczenow and Singwi[16] have suggested that the EHD system in Ge <111> containing hot and cold electrons can exist in two forms. There exists a critical value X_{crit} of the ratio $X = N_h/N_c$ of the number of hot and cold electrons such that for $X > X_{crit}$ the EHD is homogeneous (i.e., contains both the hot and the cold electrons) while for $X < X_{crit}$ a separation into two coexisting electron-hole liquid phases is predicted. One phase consists entirely of cold electrons (and holes) while the other contains both hot and cold electrons (and holes) . If verified experimentally, this would constitute a unique case of a phase separation of two degenerate Fermi liquids. From a theoretical stand point it is an interesting problem since, unlike the situation of phase separation into two classical liquids, here it would be possible to calculate the coexistence curve from first principles. The most familiar example in Physics of the phase separation into two quantum liquids is the mixture of He^3 and He^4 II, one of them being a Fermi liquid and the other a Bose liquid. There are no first principle calculations for the latter case.

Let N_h, N_e and N_H be the numbers of hot and cold electrons and of holes respectively. Let E be the total energy of the system. Let n_h, n_e and n_H be the corresponding number densities and ε be the energy density. At T = 0, the condition for thermodynamic equilibrium is

$$\left(\frac{\partial E}{\partial V}\right)_{N_h,N_c,N_H} = 0 \tag{35}$$

For a homogeneous system (35) becomes

$$\varepsilon - n_h \left(\frac{\partial \varepsilon}{\partial n_h}\right)_{n_c} - n_c \left(\frac{\partial \varepsilon}{\partial n_c}\right)_{n_h} = 0, \tag{36}$$

where derivatives are taken keeping $n_H = n_h + n_c$. The chemical potentials for hot and cold electrons are defined by

$$\mu_h = \left(\frac{\partial E}{\partial N_h}\right)_{V,N_c,N_c+N_h=N_H} = \left(\frac{\partial \varepsilon}{\partial n_h}\right)_{n_c} \tag{37}$$

$$\mu_c = \left(\frac{\partial E}{\partial N_c}\right)_{V,Nh,Nc+Nh=NH} = \left(\frac{\partial \varepsilon}{\partial n_c}\right)_{n_h}$$

A complete differential of Eq. (36) gives

$$n_h\, d\mu_h + n_c\, d\mu_c = 0 \tag{38}$$

or $$X\, d\mu_h = -\, d\mu_c\ , \tag{39}$$

where

$$X = n_h/n_c$$

Integrating (39) we have

$$\mu_c(II) - \mu_c(I) = -\int_I^{II} X\, d\mu_h\ , \tag{40}$$

where I and II refer to the two phases, i.e., cold and hot liquids respectively. One sees from (40) that a Maxwell construction in μ_h-X space can be used to take care of the equality of the chemical potential μ_c in phase equilibrium. If one plots μ_h versus X, several cases occur as shown in Fig. 10. In case (c) a phase separation occurs if $X_I < X < X_{II}$. In case (b) a phase separation occurs if $X < X_c$, where X_c is given by the equal area construction. Here one phase corresponds to X = 0 and the other to $X = X_c$. Since the X = 0 phase contains no hot electrons, the chemical potential conditions for phase equilibrium is $\mu_h(X = 0) > \mu_h(X_c)$ and $\mu_c(X = 0) = \mu_c(X = X_c)$. In case (a) and for $X > X_c$ in case (b) there is no phase separation.

The main problem consists in the evaluation of ε which is a function of X and n. The most difficult part is the calculation of the kinetic energy of holes for different stress values because of the complex band structure of holes. The exchange-correlation contribution of both electrons and holes is straight forward to calculate using Eq. (32), if one assumes tht it depends only on the total density n. A detailed numerical calculation of the kinetic energy contribution of holes taking all the complexity of their band structure has been done by Kirczenow and Singwi.[16] Their results for the chemical potentials μ_h and μ_c as a function of concentration X for various values of stress are shown in Fig. 11. The dependence of μ_h on X is similar to curve b of Fig. 10. It is

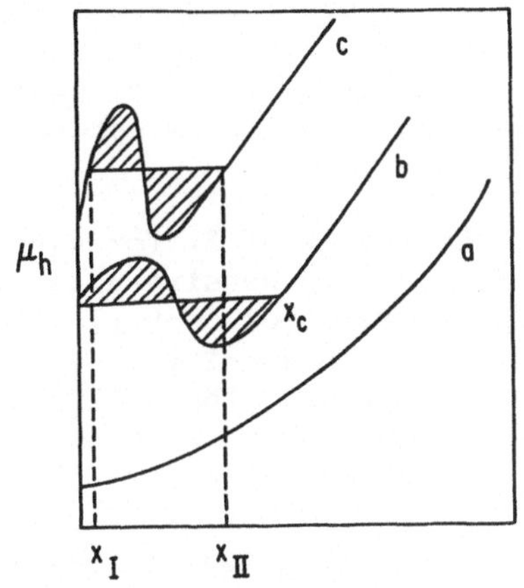

Fig. 10. A schematic representation of variations of the chemical potential μ_h with $x = N_h/N_c$, where h and c refer to "hot" and "cold" electrons, respectively. a) no phase separation; b) the phase separation will occur for $0 < x < x_c$; and c) the phase separation will occur for $x_1 < x < x_{11}$.

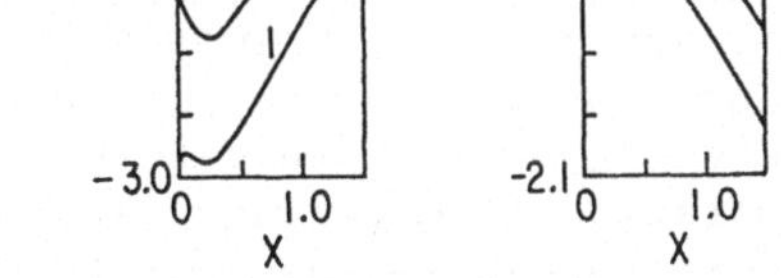

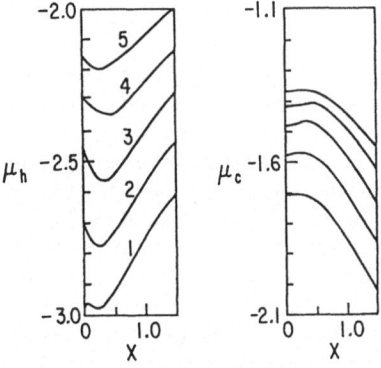

Fig. 11. Results of Kirczenow and Singwi (ref. 16) for the variations of μ_h and μ_c with x. The different curves correspond to 1, 2, 3, 4, and 5 meV stress induced splitting between the valence bands in Ge under <111> uniaxial stress.

seen that at intermediate values of stress the minima in μ_h versus X although shallow, allow a Maxwell construction of equal areas, and so the phase separation occurs for $X = X_c$. Note that when phase separation is seen for $X = X_c$ in μ_h versus X, it is also found that $\mu_c\,(X = 0) = \mu_c(X = X_c)$. At very small and large values of stress, the equal area construction is not possible and so there is no phase separation.

In Table 5 we list the values of X_c for various values of stress and also the total electron density n in phases I and II for a given model of exchange-correlation energy. Note that the density n_I of phase I is just the density at T = 0 of the EHD which does not contain any hot electrons. Phase II which contains both hot and cold electrons is almost three times as dense as phase I. The situation is shown schematically in Fig. 12.

A similar calculation has been made by Kirczenow[17] in <110>-stressed Ge and he finds tht in this system there is no phase separatin for any value of the stress. The main difference between <110>-Ge and <111>-Ge is that in the former one has two hot and two cold valleys whereas in the latter there are three hot valleys and one cold valley. In both cases the perturbed hole band structure is nearly the same. The above result tells us that the occurrence of the phase separation of the electron-hole liquid is particularly favored if the conduction band in the stressed crystal has markedly unequal numbers of hot and cold valleys. The phase separation is likely to occur for small ratios of hot to cold electrons. This rule is easy to undersand intuitively. The more hot electron valleys there are, the smaller the increase in the kinetic energy when the hot electrons condense into a fraction of the volume which they would otherwise occupy if the electron-hole liquid was homogeneous. But such a condensation can only be favored if the number of hot electrons in the EHL is low, since otherwise the magnitude of the exchange-correlation energy would not increase sufficiently to offset the increase in kinetic energy which would occur in the condensation process. In Ge-<111> the nonparabolicity of the stress-split valence band is an important factor for the phase sepration to occur. At very large and very small stresses the extra hole kinetic energy which would be gained is sufficient to prevent a phase separation from taking place. However, at intermediate values of stress where the hole Fermi level is close to the valence band splitting, the hole density of states near to the Fermi level is strongly enhanced by the stress-induced valence-band nonparabolicity. This effect (which is absent at very large and very small stress) reduces the hole kinetic energy cost of condensation making the phase separation possible.

Experimental evidence for the possible existence of such a phase separation has come from two independent and different types

Table 5. Total electron density (sum of hot and cold electron densities) n_I of phase I and n_{II} of phase II, and concentration $x_c = n_h/n_c$ for phase II of the EHD at T = 0. The concentration x of phase I is zero. S_H is the stress-induced splitting of the valence band at k = 0. It is assumed that the EHD does not occupy the entire crystal so that p = 0.

		$E^o_{\chi_c}$			$E^o_{\chi_c}$	
S_H (meV)	χ_c	η_I (10^{16} cm^{-3})	η_{II}	χ_c	η_I (10^{16} cm^{-3})	η_{II}
1.5	0.31	8.7	16.6	0.28	6.4	11.8
2.0	0.38	8.0	17.6	0.37	5.6	12.5
2.5	0.46	7.1	18.5	0.47	4.7	13.1
3.0	0.54	6.2	19.0	0.54	4.2	12.8
3.5	0.60	5.7	18.6	0.54	3.8	11.3
4.0	0.60	5.3	16.9	0.48	3.4	9.4
4.5	0.46	4.9	12.3	0.50	3.1	8.8
5.0	0.45	4.5	11.4	0.53	2.8	8.3

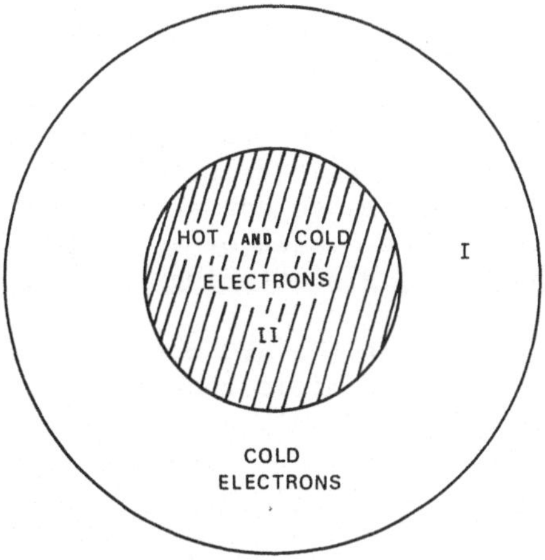

Fig. 12. Schematic representation of phase separation of EHL in <111> stressed Ge. Phase I contains cold electrons and holes. Phase II contains cold and hot electrons and holes. Phase II is three times as dense as phase I. Experimental indications[19] are that phase II is surrounded by phase I.

of experiments. One is from the experiments of Bajaj et al.[18] These authors study the line shapes of the time-resolved luminescence spectra of EHL in Ge under <111> uniaxial stress. They find that after the lapse of 0.4 μsec of the initial laser pulse i.e., corresponding to $X \leq X_c$, it is not possible to fit the line shape with the assumption of a homogeneous liquid phase consisting of hot and cold electrons. They find the critical value of X below which the phase separation occurs to be 0.63 for a stress of 6.8 kg/mm^2 - a value in reasonable agreement with the theoretical prediction of Kirczenow and Singwi.[16] The other evidence comes from the experiments of Timusk and Zarate.[19] These authors study the far infra red absorption of EHD in <111> Ge (Stress: 5 to 12 kg/mm^2) at 1.2°K. They observe two plasmon absorption lines corresponding to the densities of the two liquid phases.

VII. REMARKS ON CORRELATIONS IN A MODEL e-h SYSTEM

Let us consider a model system consisting of a single parabolic conduction and valence bands. Theoretically this is the simplest system to treat in fully self-consistent approximation, and was first discussed by Vashishta et al.[9] (VBS). The idea underlying this discussion is to examine: (i) the behaviour of the ground-state energy as a function of r_s for the mass ratio m_h/m_e = 1 and (ii) the density i.e., r_s at which the minimum of the energy occurs for a varying mass ratio m_h/m_e.

Case (i) The results of the calculation of the ground state energy by VBS in different approximations are shown in Fig. 13. There are several points to note: (a) For $r_s \ll 1$, the difference between various approximations are, as expected, negligible; however, for $r_s > 1$ these differences become appreciable. (b) Compared to HA, RPA over estimates correlations and (c) FSC approximation gives much lower value of ε compared to HA and RPA, which again is due to multiple scattering between ee, eh and hh.

For $r_s > 4$, VBS find that the SFC method does not converge thus showing that the STLS scheme is unable to predict the formation of excitons for low densities of the e-h plasma. On the basis of the simple Mott criterion for the metal-nonmetal transition, we should expect the formation of a bound excitonic state at r_s around 10. STLS scheme, therefore, needs to be improved to take care of this lacuna. In a one-component plasma, as we have seen before, this scheme gives extremely reliable values of the correlation energy for values of r_s as large as 20. It is therefore somewhat of a puzzle as to why the scheme fails in a two-component e-h plasma. The answer to this riddle perhaps lies

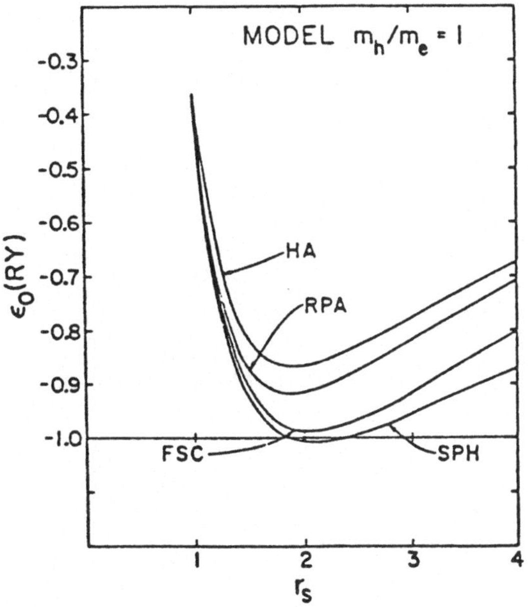

Fig. 13. Ground-state energy per e-h pair ε_0, in excitonic rydbergs, vs r_s in four approximations for the model system.

in the fact that in a two-component plasma, three particle correlations need to be taken into account to prevent large piling up of electrons on a hole which the STLS theory gives for lower densities. This is also the reason for the failure of the theory to predict correct values for the life time of positrons in metals for $r_s \geqslant 4$.

Case (ii) In Fig. 14 we give the VBS results for the ground state energy as a functin of r_s for different values of the mass ratio m_h/m_c. Here again there are some interesting points to note: (a) In the FSC approximation one does not obtain a metallic binding at least up to $m_h/m_e = 6$ and the value of the energy minimum almost remains constant with increasing value of this ratio. (b) In the Hubbard approximation one obtains a metallic binding for $m_h/m_e > 10$. (c) The r_s value at which the minimum in ε occurs shifts very slightly to lower values for increasing mass ratio m_h/m_e. In both the approximations the minimum occurs at nearly the same place. VBS point out that if one assume that this trend continues for larger mass ratios, we might conjecture that metallic hydrogen (m_H/m_e ~2000) in the liquid phase should exist for r_s around 1.8.

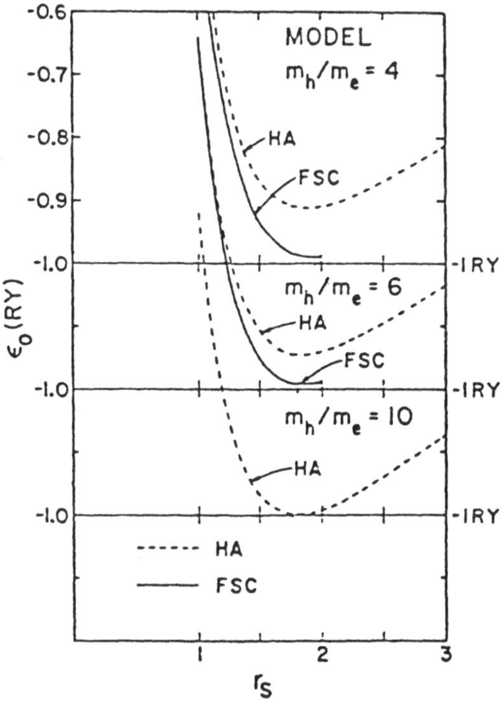

Fig. 14. Ground-state energy per e-h pair ε_0, in excitonic rydbergs, vs rs in Hubbard and fully-self-consistent approximations for m_h/m_e = 4,6 and 10. Note the shift in the vertical scale for the three cases.

The above conjecture of Vashishta et al.[9] seems to be borne out by a very recent calculation of Chakravarty and Ashcroft[20] who have obtained a variational upper bound to the ground-state energies of metallic hydrogen and concluded that the possibility of a T = 0 liquid-metallic phase can not be ruled out for r_s = 1.64.

An interesting possibility which offers itself in an e-h liquid and which needs to be investigated is whether it would become superconducting. If a suitable indirect band semiconductor with a mass ratio $m_h/m_e > 10$ can be found, it is quite possiible that the e-h liquid in it might become superconducting at very low temperatures. This question is now under investigation. The problem is analogous to that of the superconductivity of liquid metallic hydrogen but much easier to realize in practice.

ACKNOWLEDGEMENTS

I am thankful to Rajiv Kalia, and P. Vashishta for their help in preparing these lectures. This work was supported by the NSF Grant No. DMR 77-09937.

REFERENCES

No attempt has been made to give all the references. For an exhaustive list of references on the subject see the following review articles:[1-3]

1. See, Y.E. Pokrovsky; Phys. Status Solidi (A), 384-(1972).
2. See, T.M. Rice in Solid State Physics, edited by H. Ehrenreich, F. Seitz and D. Turnbull (Academic Press, New York, 1977), Vol. 32. J.C. Hensel, T.G. Phillips and G.A. Thomas, Ibid.
3. See, P. Vashishta, P.K. Kalia and K.S. Singwi in Electron-Hole Liquid edited by L.V. Keldish and D.C Jeffries (North Holland, Amsterdam, to be published.)
4. J. Lindhard, K. Dan. Vidensk Selsk, Mat-fys. Medd. 28, 8 (1954). D. Pines and D. Bohm, Phys. Rev. 85, 338 (1952); 92, 609 (1953).
5. J. Hubbard, Proc. Roy. Soc. (London) A243, 336 (1957).
6. See D. Pines and P. Nozieres, The Theory of Quantum Liquids (Benjamin, Amsterdam, 1969).
7. K.S. Singwi, M.P. Tosi, R.H. Land and A. Sjölander, Phys. Rev. 176, 589 (1968).
8. D.M. Ceperley and B.J. Alder, Phys. Rev. Lett. 45, 566(1980).
9. P. Vashishta, P. Bhattacharya and K.S. Singwi, Phys. Rev. B10, 5108 (1974); P. Bhattacharya, V. Massida, K.S. Singwi and P. Vashishta, Phys. Rev. B10, 5127 (1974).
10. W.F. Brinkman and T.M. Rice, Phys. Rev. B7, 1508 (1973).
11. M. Combescot and P. Nozières, J. Phys. C5, 2369 (1972).
12. H-h. Chou and G.K. Wong, Phys. Rev. Lett. 41, 1677 (1978).
13. P.L. Gourley and J.P. Wolfe, Bull. Am. Phys. Soc. 25, 267 (1980).
14. H-h, Chou, G.K. Wong and B.J. Feldman, Phys. Rev. Lett. 39, 959 (1977).
15. J.P. Wolfe and P.L. Goruley, in Physics of Semiconductors, p. 367, Inst. Phys. London, U.K.
16. G. Kirczenow and K.S. Singwi, Phys. Rev. Lett. 41, 326 (1978); Ibid. 42, 1004 (1979); Phys. Rev. B20, 4171 (1979).
17. G. Kirczenow, Phys. Rev. B23, 1902 (1981).
18. J. Bajaj, Fei-Ming Tong and George K. Wong, Phys. Rev. Lett. 46, 61 (1981).
19. T. Timusk and H.G. Zarate, Bull. Am. Phys. Soc. 26, 487 (1981).
20. S. Chakravarty and N.W. Ashcroft, Phys. Rev. B18, 4588 (1978).

KINETIC EQUATIONS AND TWO-PARTICLE CORRELATIONS IN THE HOMOGENEOUS ELECTRON LIQUID

Göran Niklasson

Chalmers University of Technology, Göteborg
Sweden

INTRODUCTION

The homogeneous electron liquid is a hypothetical system which has no true counterpart in the real world. Nevertheless, a great deal of work has been devoted to it. One reason for the persisting interest in this system is of course that the conduction electrons in many real metals can be approximately treated as uniformly distributed in space. Another reason is that an understanding of the homogeneous system is necessary as a starting-point for theories of inhomogeneous systems. A third reason may be that the electron liquid is a sufficiently simple system to be useful as a testing-ground for theoretical ideas about fluids in general and quantum fluids in particular.

In spite of all the work that has been spent on the uniform electron liquid, it can not yet be said that we completely understand its properties. We shall here limit ourselves to the problem of determining the dynamic response to an externally applied electric field. The interesting quantity is then the longitudinal dielectric function $\varepsilon(q,\omega)$. From the dielectric function one can also calculate the dynamic structure factor $S(q,\omega)$, which is the quantity measured in various kinds of scattering experiments. The standard approximation for $\varepsilon(q,\omega)$ and thereby $S(q,\omega)$ has for a long time been the Random Phase Approximation (RPA). However, it is by now well-known that the RPA is unsatisfactory whenever short-range correlations are important. This is clearly manifested by the fact that for electron densities of typical metals the RPA leads to negative values of the pair-distribution function $g(r)$ at small values of r. Furthermore, recent scattering experiments have revealed marked deviations from the RPA in the dynamic behaviour of $S(q,\omega)$ for electrons in a number of different materials. It is generally believed that these deviations are caused by intrinsic properties of the electron liquid and not by the periodic lattice potential.

In order to improve on the RPA a large number of modified approximations have been proposed. There are essentially three different ways to approach the problem. One way is to use the standard perturbation theory, formulated in terms of Green functions and diagrammatic expansions. This is a very systematic procedure, but it becomes quite complicated as soon as one goes beyond the simplest terms in the perturbation expansion. Various sum rules are easily violated, unless one takes specific care to avoid such violations. Furthermore, the physical arguments for keeping certain diagrams and neglecting others are not always clear. An alternative approach is to work with kinetic equations for Wigner distribution functions, which are the quantum-mechanical analogues of the classical phase-space distribution functions. In this way it is easier to get a mental picture of what is going on in the system when it responds to an external disturbance. On the other hand, by relying too much on classical analogies one may lose or misrepresent the specific quantum-mechanical aspects of the problem. The third approach is to use the Mori formulation, in which the central quantity is the so called memory function. It is particularly useful if one wants to study the dynamic behaviour of a system whose static properties are known. By construction, the Mori formalism guarantees that the relevant sum rules in the high-frequency and low-frequency limits are exactly satisfied. The difficult part of the problem is to construct an explicit approximation for the memory function which governs the behaviour at intermediate frequencies.

The main subject of these lectures is the kinetic formalism and in particular the properties of the two-particle correlation function that enters there. However, we shall also take a look at the Mori formalism, which is becoming an increasingly popular tool for studying dynamic properties of condensed matter, although it has so far been used only in a few works on the electron liquid.

1. BASIC DEFINITIONS AND FORMULAS

The purpose of this section is to collect a number of definitions and fundamental relations which form a basis for the following discussions. Much of the material presented here should be well-known. Details can be found in many textbooks, for instance Pines and Nozières [1] .

We consider a uniform system of N fermions of mass m, contained in the volume V and interacting through the pair-potential v(r). The Hamiltonian for such a system can be written in second-quantized form as

$$H_o = -\frac{\hbar^2}{2m}\sum_{\sigma}\int dr\ \psi^{+}(r)\nabla^2\psi_{\sigma}(r) + \frac{1}{2}\sum_{\sigma}\sum_{\sigma'}\int dr\int dr'\ v(r-r')\psi^{+}_{\sigma}(r)\psi^{+}_{\sigma'}(r')\psi_{\sigma'}(r')\psi_{\sigma}(r) \quad (1.1)$$

where $\psi^{+}_{\sigma}(r)$ and $\psi_{\sigma}(r)$ are creation and annihilation operators for a particle with spin σ at position r. Assuming periodic boundary conditions we introduce the Fourier components of $\psi^{+}_{\sigma}(r)$ and $\psi_{\sigma}(r)$ as follows :

$$a^{+}_{k\sigma} = \frac{1}{\sqrt{V}}\int dr\ e^{ik.r}\ \psi^{+}_{\sigma}(r) \quad (1.2)$$

$$a_{k\sigma} = \frac{1}{\sqrt{V}}\int dr\ e^{-ik.r}\ \psi_{\sigma}(r) \quad (1.3)$$

These are the creation and annihilation operators for a particle with momentum $\hbar k$ and spin σ, and they satisfy the following anticommutation rules :

$$[a_{k\sigma},a_{k'\sigma'}]_{+} = [a^{+}_{k\sigma},\ a^{+}_{k'\sigma'}]_{+} = 0 \quad (1.4)$$

$$[a_{k\sigma},a^{+}_{k'\sigma'}]_{+} = \delta^{k}_{k'}\ \delta^{\sigma}_{\sigma'} \quad (1.5)$$

Using eqs. (1.2) and (1.3) we rewrite the Hamiltonian H_o as

$$H_o = \sum_{k\sigma}\frac{\hbar^2}{2m}a^{+}_{k\sigma}\ a_{k\sigma} + \frac{1}{2V}\sum_{k\sigma}\sum_{k'\sigma'}v_q\ a^{+}_{k-\frac{1}{2}q,\sigma}\ a^{+}_{k'+\frac{1}{2}q,\sigma'}\ a_{k'-\frac{1}{2}q,\sigma'}\ a_{k+\frac{1}{2}q,\sigma} \quad (1.6)$$

where v_q is the Fourier-transformed interaction potential

$$v_q = \int dr\ e^{-iq.r}\ v(r) \quad (1.7)$$

In the special case of electrons interacting through the electrostatic Coulomb potential one must also include a uniform positively charged background in order to keep the system together and guarantee overall charge neutrality. The Fourier-transformed Coulomb potential can then be written as

$$v_q = \begin{cases} \dfrac{4\pi e^2}{q^2} & \text{if} \quad q \neq 0 \\ 0 & \text{if } q = 0 \end{cases} \tag{1.8}$$

where the zero in the q = 0 component arises from the interaction with the neutralizing positive background.

The Hamiltonian H_o describes the system in the absence of any external forces. Let us now assume that each particle in the system is acted on by an external time-dependent force field with the potential $\Phi^{ext}(r,t)$. The Hamiltonian can then be written as

$$H(t) = H_o + \sum_\sigma \int dr\, \psi^+_\sigma(r)\, \psi_\sigma(r)\, \Phi^{ext}(r,t) \tag{1.9}$$

or equivalently

$$H(t) = H_o + \frac{1}{V} \sum_q \sum_{k\sigma} a^+_{k-\frac{1}{2}q,\sigma}\, a_{k+\frac{1}{2}q,\sigma}\, \Phi^{ext}_{-q}(t) \tag{1.10}$$

where $\Phi^{ext}_q(t)$ is the Fourier component of the external potential.

The external field induces variations in the local mean density of particles, given by

$$n(r,t) = \sum_\sigma \langle \psi^+_\sigma(r,t)\, \psi_\sigma(r,t) \rangle \tag{1.11}$$

where $\psi^+_\sigma(r,t)$ and $\psi_\sigma(r,t)$ are the creation and annihilation operators in the Heisenberg picture. After Fourier transformation with respect to r this can be written as

$$n_q(t) = \langle \rho_q(t) \rangle = \langle \sum_{k\sigma} a^+_{k-\frac{1}{2}q,\sigma}(t)\, a_{k+\frac{1}{2}q,\sigma}(t) \rangle \tag{1.12}$$

We shall in the following adopt the convention that deviations from equilibrium are denoted by a bar over the symbol. Thus we write

$$\bar{n}(r,t) = n(r,t) - n_o \tag{1.13}$$

$$\bar{n}_q(t) = n_q(t) - N\delta^o_q \tag{1.14}$$

where n_o is the equilibrium density. This is often expressed by means of a dimensionless parameter r_s, defined through

$$r_s = \frac{1}{a_o} \left(\frac{3V}{4\pi N}\right)^{1/2} \tag{1.15}$$

which means that r_s is essentially the interparticle spacing, measured in units of the Bohr radius a_o. One can also interpret r_s as the ratio between the average potential energy and the average kinetic energy. In the high-density limit ($r_s \to 0$), the electrons behave essentially as free particles, whereas the interactions are dominant when $r_s \gg 1$. The region of metallic densities corresponds roughly to $1.8 < r_s < 5.6$, which is an intermediate coupling region.

We shall assume that the applied field is weak enough, so that the standard theory of linear response is applicable. Introducing the spacial and temporal Fourier transforms

$$\bar{n}_q(\omega) = \int_{-\infty}^{\infty} e^{i\omega t}\, \bar{n}_q(t)dt \tag{1.16}$$

$$\Phi_q^{ext}(\omega) = \int_{-\infty}^{\infty} e^{i\omega t}\, \Phi_q^{ext}(t)dt \tag{1.17}$$

one obtains the important result

$$\bar{n}_q(\omega) = \chi(q,\omega)\, \Phi^{ext}(q,\omega) \tag{1.18}$$

where the response function $\chi(q,\omega)$ is given by

$$\chi(q,\omega) = \frac{1}{i\hbar V} \int_{o}^{\infty} dt\, e^{i\omega t} \langle [\rho_q(t), \rho_q^+(o)] \rangle_{eq} \tag{1.19}$$

Here, the symbol $\langle \ldots \rangle_{eq}$ denotes the statistical average in the absence of external disturbances. The response function is the central quantity that one needs to calculate in order to study the behaviour of the electron liquid in an applied electric field. It is closely related to two other functions which give the same information from a different point of view, namely the dielectric function $\varepsilon(q,\omega)$ and the screened response function $\chi^{sc}(q,\omega)$. These two functions are introduced in order to describe the fact that the induced mean density change produces an induced field which screens the external field. The local screened field is

$$\Phi_q^{sc} = \Phi_q^{ext}(\omega) + v_q\, \bar{n}_q(\omega) \tag{1.20}$$

The dielectric function and the screened response function are defined through the relations

$$\Phi_q^{sc}(\omega) = \frac{\Phi_q^{ext}(\omega)}{\varepsilon(q,\omega)} \tag{1.21}$$

$$\bar{n}_q(\omega) = \chi^{sc}(q,\omega)\ \Phi_q^{sc}(\omega) \tag{1.22}$$

Using eq.(1.17) we easily find the following relations between $\chi(q,\omega)$, $\varepsilon(q,\omega)$ and $\chi^{sc}(q,\omega)$:

$$\frac{1}{\varepsilon(q,\omega)} = 1 + v_q\ \chi(q,\omega) \tag{1.23}$$

$$\chi(q,\omega) = \frac{\chi^{sc}(q,\omega)}{1 - v_q\ \chi^{sc}(q,\omega)} \tag{1.24}$$

The screened response function is in the standard Green function theory represented by the series of all proper bubble diagrams, exemplified in figure 1. Keeping only the first of these diagrams one obtains the Random Phase Approximation,

$$\chi_{RPA}(q,\omega) = \frac{\chi_o(q,\omega)}{1 - v_q\ \chi_o(q,\omega)} \tag{1.25}$$

where $\chi_o(q,\omega)$ is the Lindhard free-particle response function, given by

$$\chi_o(q,\omega) = \frac{1}{V}\sum_{k\sigma} \frac{n^o_{k-1/2q,\sigma} - n^o_{k+1/2q,\sigma}}{\hbar\omega - \frac{\hbar^2}{m}\,k.q + i\varepsilon} \tag{1.26}$$

$n^o_{k\sigma}$ being the free-particle occupation number.

Within the Green function formalism the obvious way to improve on the RPA is to include more diagrams in the calculation of $\chi^{sc}(q,\omega)$. The first order corrections, represented by diagrams (2), (3) and (4) in figure 1, have been examined in great detail by Geldart and Taylor in the static limit [2] and more recently by Holas et al in the general case [3]. However, it is not sufficient to include only the first order diagrams. The standard procedure is to select certain classes of diagrams, such as the exchange ladder series represented by diagrams (4) and (8), and sum these to infinite order in some approximative way. These methods will be discussed in the lectures by Professor Devreese, and we shall therefore not go into details here.

There is another way to approach the problem, which is less systematic but perhaps intuitively more appealing. The starting-point is the observation that the RPA implies that the electrons respond as free particles under the influence of the mean field, given in eq.(1.20). One can then argue that there

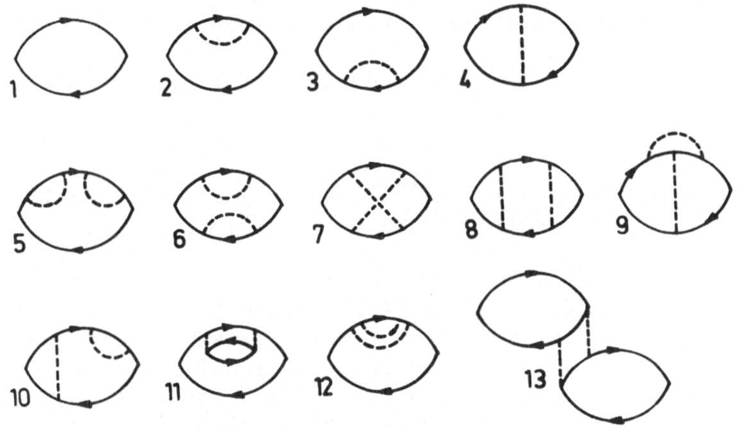

Figure 1

are two defects in the RPA. First of all, the effective field acting on an electron is not equal to the mean field, because it is modified by short-range exchange and correlation effects. Secondly, the electrons do not really respond as free particles. One is therefore led to seek an improved approximation in the form

$$\bar{n}_q(\omega) = \chi^{eff}(q,\omega)\ \Phi_q^{eff}(\omega) \tag{1.27}$$

where $\Phi_q^{eff}(\omega)$ is an effective screened field which includes exchange and correlation effects, and where the effective response function $\chi^{eff}(q,\omega)$ describes the single-particle response to the effective field. It has become conventional to write the effective mean field in the form

$$\Phi_q^{eff}(\omega) = \Phi_q^{ext}(\omega) + v_q[1 - G(q,\omega)]\,\bar{n}_q(\omega) \tag{1.28}$$

where the exchange and correlation effects are contained in the "local field correction" $G(q,\omega)$. Various approximative expressions for $G(q,\omega)$, based on more or less phenomenological considerations, have been suggested in the last decade. Most of these are static approximations, i.e. $G(q,\omega)$ is assumed to be independent of ω. The most well-known approximation of this kind is probably the one proposed by Vashishta et al. [4] . In recent years there has been an increased interest to construct frequency-dependent local field corrections in order to account for the results of the scattering experiments mentioned in the introduction.

Eqs.(1.27) and (1.28) lead to the following expression for the response function :

$$\chi(q,\omega) = \frac{\chi^{eff}(q,\omega)}{1 - v_q[1 - G(q,\omega)]\ \chi^{eff}(q,\omega)} \tag{1.29}$$

Obviously, any approximation for $\chi(q,\omega)$ can be cast in this form. Formally there is also complete freedom to shuffle terms between $\chi^{eff}(q,\omega)$ and $G(q,\omega)$. Intuitively, it would seem that there is a natural distinction between the local field felt by a particle and the response properties of the particle. However, both are manifestations of the same interparticle interactions, and we shall see later that one easily gets into inconsistencies when one tries to separate them. In praxis one usually identifies $\chi^{eff}(q,\omega)$ with the free-particle response function $\chi_o(q,\omega)$, which means that all the deviations from the RPA are lumped into the local field correction $G(q,\omega)$.

There are a number of known exact relations which must be satisfied by the response function $\chi(q,\omega)$ or equivalently by the dielectric function $\varepsilon(q,\omega)$. These relations are very useful as tests of any given approximative theory. Such tests are particularly important for the electron liquid, since there are only limited possibilities to compare the theoretical results with experiments.

One class of exact relations are those which govern the behaviour in the long wave-length limit, i.e. when q tends to zero. For the electron liquid one can show that

$$\varepsilon(q,o) \rightarrow 1 + \frac{4\pi e^2}{q^2} n_o^2 K \tag{1.30}$$

$$\varepsilon(q,\omega) \rightarrow 1 - \frac{\omega_p^2}{\omega^2} \tag{1.31}$$

Here, n_o is the density of particles, ω_p is the plasma frequency, and K is the compressibility. The latter can also be calculated as the second derivative of the total energy with respect to density. The requirement that the value of K obtained from the energy should be equal to the value obtained from eq.(1.30) is known as the compressibility sum rule. It has often been used as a crucial test to judge the validity of various approximative expressions for the static dielectric function.

Another class of exact relations are the so called frequency moment sum rules. These contain information about the short-time behaviour of the system and they therefore govern the assymptotic form of the response function in the limit of large frequencies. Expanding $\chi(q,\omega)$ in inverse powers of ω one obtains

$$\chi(q,\omega) = \frac{M_1(q)}{\omega^2} + \frac{M_3(q)}{\omega^4} + \ldots \tag{1.32}$$

where $M_1(q)$ and $M_3(q)$ are known as the first and third moments respectively. They can be exactly evaluated, and are given by the expressions [5,6]

$$M_1(q) = -\frac{2}{\pi}\int_0^\infty d\omega\, \omega \,\mathrm{Im}\,\chi(q,\omega) = \frac{nq^2}{m} \tag{1.33}$$

$$\begin{aligned} M_3(q) &= -\frac{2}{\pi}\int_0^\infty d\omega\, \omega^3 \,\mathrm{Im}\,\chi(q,\omega) \\ &= \frac{n\omega_p^2}{m} q^2 + \frac{nq^2}{m^2}\left(\frac{\langle p^2\rangle}{m} + \frac{\hbar^2 q^2}{4m}\right) \\ &+ \frac{n^2}{m^2}\int dr[(q.\nabla)^2\, v(r)]\,(1 - e^{iq.r})\,[g(r) - 1] \end{aligned} \tag{1.34}$$

Here, $\langle p^2\rangle$ is the mean square momentum of an electron in the interacting system, and $g(r)$ is the pair-distribution function. Higher moments can also be evaluated, but they are more complicated and not very useful for practical purposes. The first moment sum rule, given in eq.(1.33), is equivalent to the well-known f-sum rule which arises from the conservation of the number of particles.

The imaginary part of $\chi(q,\omega)$, entering in the frequency moment sum rules, describes the dissipative processes through which the system can absorb energy from the applied field. By the famous fluctuation-dissipation theorem this can be related to the dynamic structure factor $S(q,\omega)$, describing the spectrum of density fluctuations in the system. At zero temperature and for $\omega > 0$ the relation takes the form

$$\mathrm{Im}\,\chi(q,\omega) = -\frac{n_o}{\hbar} S(q,\omega) \tag{1.35}$$

where

$$S(q,\omega) = \frac{1}{N}\int_{-}^{\infty} dt\, e^{i\omega t}[\langle\rho_q(t)\,\rho_q^+(o)\rangle - \delta_q^o N^2] \tag{1.36}$$

As mentioned in the introduction, $S(q,\omega)$ can be measured through scattering experiments. For $q \to 0$, it reduces to a δ-function peak at the plasma frequency, as is easily seen from eq.(1.31). For finite q-values $S(q,\omega)$ is composed mainly of two contributions, one being the plasma peak and the other being a broader distribution arising from particle-hole excitations (see fig.2). In the RPA these are the only contributions, and the plasmons are then undamped until the plasma dispersion curve enters the particle-hole region. Approximations that go beyond the RPA lead to changes in the plasmon dispersion curve, and they may also contain mechanisms that contribute to the damping of plasmons. Comparisons with experimental data are unfortunately to a large extent obscured by unknown band structure contributions.

Integrating $S(q,\omega)$ over all frequencies one obtains the static structure factor,

$$S(q) = \int_0^\infty \frac{d\omega}{\pi} S(q,\omega) = -\frac{\hbar}{n_o \pi} \int_0^\infty d\omega \, \mathrm{Im}\, \chi(q,\omega) \qquad (1.37)$$

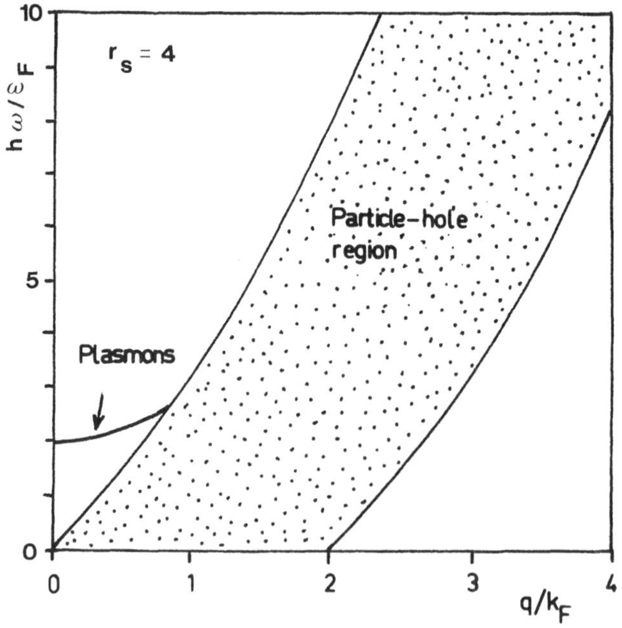

Figure 2. (from ref. [1]).

which is essentially the Fourier transform of the pair-distribution function,

$$g(r) = 1 + \frac{1}{N} \sum_q e^{iq\cdot r} [S(q) - 1] \tag{1.38}$$

The true form of g(r) for the electron liquid is not known, but at least we know that it has to be positive. This criterion is not satisfied by the RPA at metallic densities. Several of the modified approximations found in the literature also tend to give slightly negative values for g(r) near r = 0, particularly in the low-density limit.

2. WIGNER DISTRIBUTION FUNCTIONS AND THE KINETIC EQUATION

In classical non-equilibrium statistical mechanics one often works with kinetic equations for phase-space distribution functions. The one-particle phase-space distribution function $f^{(1)}(p,r;t)$ gives the density of particles with momentum p at position r at time t. Its time-evolution under the influence of an external field with the potential $\Phi^{ext}(r,t)$ is governed by the classical kinetic equation

$$\begin{aligned} &\frac{\partial}{\partial t} f^{(1)}(p,r;t) + \frac{p}{m} \cdot \nabla_r f^{(1)}(p,r;t) \\ &\qquad - \int dr' \nabla_r v(r-r') \cdot \nabla_p f^{(2)}(p,r;p',r';t) \\ &= \nabla_r \Phi^{ext}(r,t) \cdot \nabla_p f^{(1)}(p,r;t) \end{aligned} \tag{2.1}$$

where the two-particle phase-space distribution function $f^{(2)}(p,r;p',r';t)$ enters in the interaction term. This is the first equation in a series known as the BBGKY-hierarchy, in which phase-space distribution functions of higher order enter successively. The central problem is to find good approximative ways to break this hierarchy, so that the one-particle distribution function can be calculated.

When one tries to apply the kinetic approach to quantum fluids one encounters two conceptual difficulties. The first one is that momentum and position are conjugate variables which can not be specified simultaneously. The second one is that the particles are indistinguishable, which means that the two-particle and many-particle distribution functions contain exchange terms that have no classical counterpart. Nevertheless, it is possible to construct kinetic equations which are formally analogous to the classical equations. The central quantities in these equations are the so-called Wigner distribution functions [7]. These are statistical averages involving a "phase-space" density operator $\rho_{k\sigma}(r,t)$, which is defined as

$$\rho_{k\sigma}(r,t) = \frac{1}{V} \int dr' \, e^{-ik.r'} \, \psi^{+}_{\sigma}(r - \frac{r'}{2}, t) \, \psi(r + \frac{r'}{2}, t) \tag{2.2}$$

and has the Fourier components

$$\rho_{k\sigma}(q,t) = \int dr \, e^{-iq.r} \, \rho_{k\sigma}(r,t) = a^{+}_{k-\frac{1}{2}q,\sigma}(t) a_{k-\frac{1}{2}q,\sigma}(t) \tag{2.3}$$

The interpretation of $\rho_{k\sigma}(r,t)$ as the quantum-mechanical analogue of the classical phase-space density is justified by the following two relations :

$$\sum_{k\sigma} \rho_{k\sigma}(r,t) = \sum_{\sigma} \psi^{+}_{\sigma}(r,t) \, \psi_{\sigma}(r,t) \tag{2.4}$$

$$\int dr \, \rho_{k\sigma}(r,t) = a^{+}_{k\sigma}(t) \, a_{k\sigma}(t) \tag{2.5}$$

The first relation gives the particle density operator in r-space and the second relation gives the occupation number operator for particles with momentum $\hbar k$ and spin σ.

The one-particle Wigner distribution function is defined as the expectation value of the density operator $\rho_{k\sigma}(r,t)$

$$f^{(1)}_{k\sigma}(r,t) = \langle \rho_{k\sigma}(r,t) \rangle \tag{2.6}$$

From equation (2.4) it follows that the local particle density n(r,t) is obtained from the Wigner function by summing over k and σ. Similarly, it follows from equation (2.5) that integration over r gives the occupation number $n_{k\sigma}(t)$. The Wigner distribution function has in these respects essentially the same properties as the classical phase-space distribution function. However, the incompatibility of position and momentum forbids a strict interpretation of $f^{(1)}_{k\sigma}(r,t)$ as a probability distribution in phase-space. This is also signalled by the fact that the Wigner function can assume negative values, in contrast to its classical counterpart. From a more intuitive point of view it is clear that the classical interpretation breaks down completely for a short wavelength disturbance such that q is comparable with the Fermi wave-vector. For long wave-length disturbances, on the other hand, it seems quite reasonable to think of $f_{k\sigma}^{(1)}(r,t)$ as a local probability density. At equilibrium, $f_{k\sigma}^{(1)}(r,t)$ reduces to the equilibrium occupation density $n_{k\sigma}/V$. Following our convention, we write the deviation from equilibrium as

$$\bar{f}_{k\sigma}^{(1)}(r,t) = f_{k\sigma}^{(1)}(r,t) - \frac{1}{V} n_{k\sigma} \tag{2.7}$$

or equivalently

$$\bar{f}_{k\sigma}^{(1)}(q,t) = f_{k\sigma}^{(1)}(q,t) - \delta_q^o\, n_{k\sigma} \tag{2.8}$$

Many-particle Wigner density operators are defined as products of one-particle density operators. These products should be ordered with the creation operators to the left and the annihilation operators to the right, and in such a way that the resulting operator is Hermitean. Thus the two-particle Wigner operator is defined as

$$\rho_{k\sigma;k'\sigma'}^{(1)}(r,r';t) = \frac{1}{V^2} \int dr_1 \int dr_2\, e^{-ik.r_1 - ik.r_2}$$
$$\times\, \psi_\sigma^+\left(r - \frac{r_1}{2}, t\right) \psi_{\sigma'}^+\left(r' - \frac{r_2}{2}, t\right) \psi_{\sigma'}\left(r' + \frac{r_2}{2}, t\right) \psi_\sigma\left(r + \frac{r_1}{2}, t\right) \tag{2.9}$$

The corresponding two-particle distribution function is

$$f_{k\sigma,k'\sigma'}^{(2)}(r,r';t) = \langle \rho_{k\sigma;k'\sigma'}^{(2)}(r,r';t) \rangle \tag{2.10}$$

and after Fourier-transformation with respect to r and r' it can be written as

$$f_{k\sigma;k'\sigma'}^{(2)}(q,q';t) =$$
$$\langle a^+_{k-\frac{1}{2}q,\sigma}(t)\, a^+_{k'-\frac{1}{2}q',\sigma'}(t)\, a_{k'+\frac{1}{2}q',\sigma'}(t)\, a_{k+\frac{1}{2}q,\sigma}(t) \tag{2.11}$$

In analogy with eq.(2.7) we write

$$\bar{f}_{k\sigma;k'\sigma'}^{(2)}(r,r';t) = f_{k\sigma;k'\sigma'}^{(2)}(r,r';t) - f_{k\sigma;k'\sigma'}^{(2)}(r-r') \tag{2.12}$$

where $f_{k\sigma;k'\sigma'}^{(2)}(r-r')$ is the equilibrium two-particle distribution function. Because of translational invariance, it can only depend on r-r' but not on r and r' separately. After Fourier transformation we get

$$\bar{f}^{(2)}_{k\sigma;k'\sigma'}(q,q';t) = f^{(2)}_{k\sigma;k'\sigma'}(q,q';t) - \delta^{q}_{-q'}\, f^{(2)}_{k\sigma;k'\sigma'}(q) \tag{2.13}$$

where

$$f^{(2)}_{k\sigma;k'\sigma'}(q) = V \int dr\, e^{-iq.r}\, f^{(2)}_{k\sigma;k'\sigma'}(r)$$

$$= \langle a^{+}_{k-\frac{1}{2}q,\sigma}\, a^{+}_{k'+\frac{1}{2}q,\sigma}\, a_{k'-\frac{1}{2}q,\sigma}\, a_{k+\frac{1}{2}q,\sigma} \rangle \tag{2.14}$$

Summing the equilibrium two-particle distribution over all momenta and spin one obtains the equal-time density-density correlation function, which can be expressed in terms of the pair-distribution function g(r) or the static structure factor S(q). By comparison with eqs.(1.36) - (1.38) it is easy to show that

$$\sum_{k\sigma}\sum_{k'\sigma'} f^{(2)}_{k\sigma;k'\sigma'}(r) = n_o^2\, g(r) \tag{2.15}$$

$$\sum_{k\sigma}\sum_{k'\sigma'} f^{(2)}_{k\sigma;k'\sigma'}(q) = N[S(q) - 1] + N^2\delta^{o}_{q} \tag{2.16}$$

These relations are completely analogous to their classical counterparts. Taking the analogy one step further, one may be tempted to write

$$f^{(2)}_{k\sigma;k'\sigma'}(r) = n_{k\sigma}\, n_{k'\sigma'}\, g(r) \tag{2.17}$$

which differs from the exact classical formula only through the appearance of the quantum-mechanical momentum distributions $n_{k\sigma}$ and $n_{k'\sigma'}$ in place of the Maxwellian distributions. However, the actual form of $f^{(2)}_{k\sigma;k'\sigma'}(r)$ is much more complicated, as can be seen by evaluating a few terms in the perturbation expansion indicated by the diagrams in figures 3 and 4. The lines in the diagrams represent time-ordered propagators with self-energy contributions included. The creation and annihilation points are assigned infinitesimally different times in such a way as to guarantee the correct ordering of the operators in eq.(2.14).
According to figure 3 we can write

$$f^{(2)}_{k\sigma;k'\sigma'}(q) = \delta^{o}_{q}\, n_{k\sigma}\, n_{k'\sigma'} - \delta^{k}_{k'}\, \delta^{\sigma}_{\sigma'}\, n_{k-\frac{1}{2}q,\sigma}\, n_{k+\frac{1}{2}q,\sigma}$$

$$+ H_{k\sigma;k'\sigma'}(q) \tag{2.18}$$

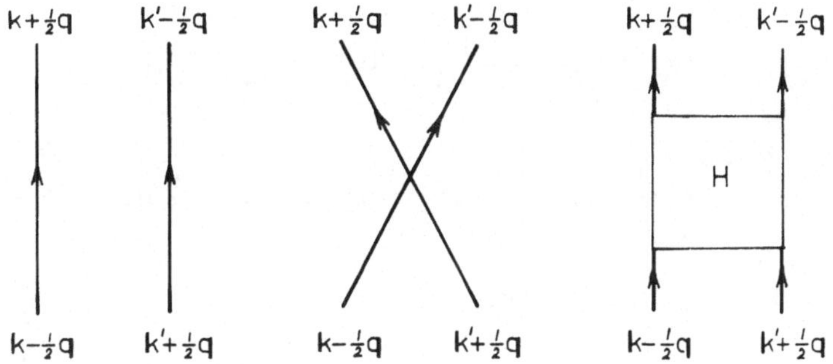

Figure 3

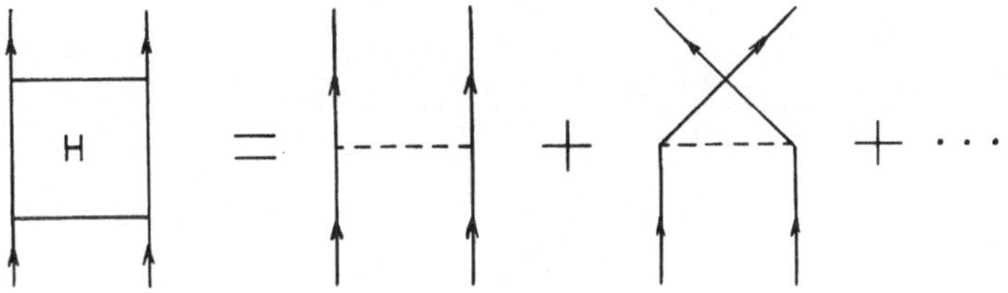

Figure 4

where the first two terms are the uncorrelated direct and exchange terms, respectively, and where $H_{k\sigma;k'\sigma'}(q)$ comes from all connected two-particle diagrams. The lowest order contributions to $H_{k\sigma;k'\sigma'}(q)$, represented by the diagrams in figure 4 can easily be evaluated, if self-energy corrections are neglected. The result is

$$H_{k\sigma;k'\sigma'}(q) = \frac{m}{\hbar^2 V} \frac{1}{(k-k').q} [v_q - v_{k-k'} \delta^{\sigma}_{\sigma'}]$$

$$\times [(1-n^o_{k+\frac{1}{2}q,\sigma})(1-n^o_{k'-\frac{1}{2}q,\sigma'}) n^o_{k'+\frac{1}{2}q,\sigma'} n^o_{k-\frac{1}{2}q,\sigma}$$

$$- n^o_{k+\frac{1}{2}q,\sigma} n^o_{k'-\frac{1}{2}q,\sigma'} (1-n^o_{k'+\frac{1}{2}q,\sigma'})(1-n^o_{k-\frac{1}{2}q,\sigma})] \quad (2.19)$$

where $n^o_{k\sigma}$ is the occupation number for a gas of non-interacting fermions. It is apparent that the approximation in eq.(2.17) is far too simple.

Having defined the Wigner distribution functions, we can now proceed to write down their equations of motion, using the general formula

$$i\hbar \frac{\partial}{\partial t} \langle A(t) \rangle = \langle [A(t),H(t)] \rangle \quad (2.20)$$

The Hamiltonian H was given in eq.(1.9). After some algebra we can write the equation for the one-particle Wigner function in the form

$$\frac{\partial}{\partial t} f^{(1)}_{k\sigma}(r,t) + \frac{\hbar^2}{m} k.\nabla_r f^{(1)}_{k\sigma}(r,t)$$

$$- \frac{1}{i\hbar}\int dr \sum_{k'\sigma'} [f^{(2)}_{k+\frac{1}{2}\nabla_{r'},\sigma;k'\sigma'}(r,r';t) - f^{(2)}_{k-\frac{1}{2}\nabla_{r'},\sigma;k'\sigma'}(r,r';t)]\, v(r-r')$$

$$= \frac{1}{i\hbar} [f^{(1)}_{k+\frac{1}{2}\nabla_{r},\sigma}(r,t) - f^{(1)}_{k-\frac{1}{2}\nabla_{r},\sigma}(r,t)]\, \Phi^{ext}(r,t) \quad (2.21)$$

which should be compared with the corresponding classical equation (2.1). Apart from a minor difference in notation, the classical equation is obtained if we forget about spin, put $k = p/\hbar$ and then let $\hbar \to 0$. Keeping in eq.(2.21) only terms linear in the external field and taking the Fourier transform we get

$$(\hbar\omega - \frac{\hbar^2}{m} k.q)\ \bar{f}^{(1)}_{k\sigma}(q,\omega)$$

$$- \frac{1}{V} \sum_{q'} v_{q'} \sum_{k'\sigma'} [\bar{f}^{(2)}_{k-\frac{1}{2}q',\sigma;k'\sigma'}(q-q',q';\omega)$$

$$- \bar{f}^{(2)}_{k+\frac{1}{2}q',\sigma;k'\sigma'}(q-q',q';\omega)]$$

$$= \frac{1}{V} [n_{k-\frac{1}{2}q,\sigma} - n_{k+\frac{1}{2}q,\sigma}]\, \Phi^{ext}(q,\omega) \tag{2.22}$$

This is the fundamental kinetic equation that forms the basis for our discussion in the following sections.

From the kinetic equation one can extract equations for densities and currents. Of particular importance is the continuity equation which is obtained by summing over all k and σ in eq.(2.21),

$$\frac{\partial}{\partial t} n(r,t) + \nabla.j(r,t) = 0 \tag{2.23}$$

The particle current density is given by

$$j(r,t) = \sum_{k\sigma} \frac{\hbar k}{m} f^{(1)}_{k\sigma}(r,t) \tag{2.24}$$

One can also derive an equation of motion for j(r,t), multiplying both sides of eq.(2.21) by $\hbar k/m$ and then summing over k and σ. This gives

$$\frac{\partial}{\partial t} j(r,t) + \nabla.T(r,t)$$

$$+ \frac{1}{m} \int dr[\nabla v(r-r')] \sum_{k\sigma} \sum_{k'\sigma'} f^{(2)}_{k\sigma;k'\sigma'}(r,r';t)$$

$$= - \frac{1}{m} n(r,t)\ \nabla\Phi^{ext}(r,t) \tag{2.25}$$

where T(r,t) is the kinetic pressure tensor. Similarly, one can proceed to write down equations for the energy density and the energy current. These equations are important in hydrodynamic theories but we shall not discuss them here.

3. APPROXIMATIVE DECOUPLING PROCEDURES FOR THE TWO-PARTICLES WIGNER FUNCTION

A common approach to kinetic equations such as (2.22) is to make an approximative decoupling of the two-particle distribution function into products involving one-particle distribution functions. Such decoupling approximations can be constructed in many different ways on the basis of more or less intuitive physical arguments. Our previous discussion of the equilibrium two-particle Wigner function indicates that one is unlikely to find any simple decoupling scheme that will give a good account of the detailed behaviour for all values of the momentum and position variables. However, as long as one is interested only in the density response $\bar{n}(r,t)$, the detailed dependence on k and k' in $\bar{f}^{(2)}_{k\sigma;k'\sigma'}(r,r';t)$ may not be too important. What one then needs is a decoupling procedure that is good in some average sense. A common but somewhat artificial way to achieve this is to introduce an adjustable parameter (or even an adjustable function) which is determined by means of some suitable sum rule.

The simplest approximation is obviously to neglect correlation and exchange effects completely, which means that the two-particle Wigner function is written as

$$f^{(2)}_{k\sigma;k'\sigma'}(q,q';t) = f^{(1)}_{k\sigma}(q,t)\, f^{(1)}_{k'\sigma'}(q,t) \tag{3.1}$$

Linearizing this we obtain

$$\bar{f}^{(2)}_{k\sigma;k'\sigma'}(q,q';t) = \delta^{o}_{q}\, n_{k\sigma}\, \bar{f}^{(1)}_{k'\sigma'}(q',t) + \delta^{o}_{q'}\, n_{k'\sigma'}\, \bar{f}^{(1)}_{k\sigma}(q,t) \tag{3.2}$$

and after insertion into eq.(2.22) it gives

$$(\hbar\omega - \frac{\hbar^2}{m} k.q)\, \bar{f}^{(1)}_{k\sigma}(q,\omega)$$

$$= \frac{1}{V}\,(n_{k-\frac{1}{2}q,\sigma} - n_{k+\frac{1}{2}q,\sigma}) \times [\Phi^{ext}(q,\omega) + v_q\, \bar{n}(q,\omega)] \tag{3.3}$$

This equation is easily solved, and it leads to the RPA result

$$\bar{n}(q,\omega) = \chi_{RPA}(q,\omega)\, \Phi^{ext}(q,\omega) \tag{3.4}$$

except for one minor modification : the free-particle occupation number $n^{o}_{k\sigma}$ that enters in the Lindhard function is here replaced by the interacting occupation number $n_{k\sigma}$. Numerically, this modification is excepted to be insignificant. However, it illustrates our previous discussion about the distinction between the local field felt by a particle and the response properties of the particle. It would certainly be very artificial to try to incorporate the modification of $n_{k\sigma}$ in a local field correction $G(q,\omega)$.

From a systematic point of view, the first step beyond the RPA should be to include exchange. The two-particle Wigner function is then written as

$$f^{(2)}_{k\sigma;k'\sigma'}(q,q';t) = f^{(1)}_{k\sigma}(q,t)\, f^{(1)}_{k'\sigma'}(q',t)$$

$$- \delta^{o}_{\sigma'} \langle a^{+}_{k-\frac{1}{2}q,\sigma}(t)\, a_{k'+\frac{1}{2}q',\sigma'}(t)\rangle$$

$$\langle a^{+}_{k'-\frac{1}{2}q',\sigma'}(t)\, a_{k+\frac{1}{2}q,\sigma}(t)\rangle \qquad (3.5)$$

which after linearization becomes

$$\bar{f}^{(1)}_{k\sigma;k'\sigma'}(q,q';t) = \delta^{o}_{q}\, n_{k\sigma}\, \bar{f}^{(1)}_{k\sigma}(q',t) + \delta^{o}_{q}\, n_{k'\sigma'}\, \bar{f}^{(1)}_{k\sigma}(q,t)$$

$$- \delta^{k'+\frac{1}{2}q'}_{k-\frac{1}{2}q}\, \delta^{\sigma}_{\sigma'}\, n_{k-\frac{1}{2}q,\sigma}\, \bar{f}^{(1)}_{k-\frac{1}{2}q'\sigma}(q+q',t)$$

$$- \delta^{k'-\frac{1}{2}q'}_{k+\frac{1}{2}q}\, \delta^{\sigma}_{\sigma'}\, n_{k+\frac{1}{2}q,\sigma}\, \bar{f}^{(1)}_{k+\frac{1}{2}q'\sigma}(q+q',t) \qquad (3.6)$$

Inserting this into eq. (2.22) we obtain

$$[\hbar\omega - \frac{\hbar^2}{m} k.q + \Sigma^{HF}_{k+\frac{1}{2}q,\sigma} - \Sigma^{HF}_{k-\frac{1}{2}q,\sigma}]\, \bar{f}^{(1)}_{k\sigma}(q,\omega)$$

$$= \frac{1}{V}(n_{k-\frac{1}{2}q,\sigma} - n_{k+\frac{1}{2}q,\sigma})[\Phi^{ext}(q,\omega) + v_q\, \bar{n}(q,\omega)$$

$$- \sum_{q'} v_{q'}\, \bar{f}^{(1)}_{k-q',\sigma}(q,\omega)] \qquad (3.7)$$

where

$$\Sigma^{HF}_{k\sigma} = \frac{1}{V} \sum_{q'} v_{q'}\, n_{k-q',\sigma} \qquad (3.8)$$

This is just the lowest order self-energy correction. On comparison of eq.(3.7) with eq.(3.3) it appears that there are two distinct modifications caused by exchange effects. One is that the flow term is modified by self-energy corrections, which for small values of q can be incorporated in an effective mass correction. The second modification is represented by the last term on the right hand side, and it can be interpreted as a contribution to a k-dependent "local field" correction. It is important that the self-energy corrections and the "local field" corrections are treated in a consistent way, because one will

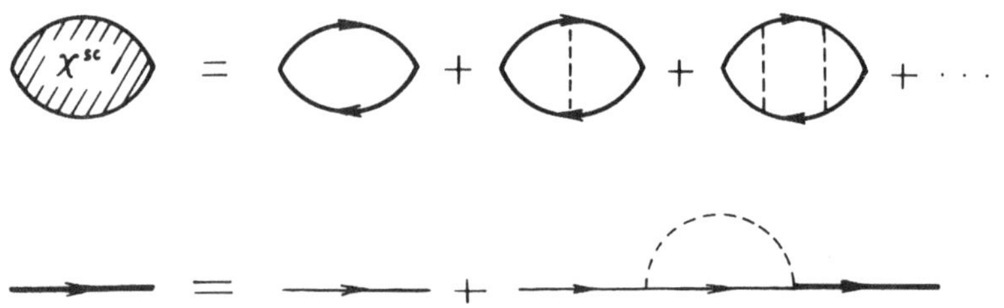

Figure 5

otherwise easily violate the exact continuity equation, given in eqs.(2.23) and (2.24). This is reminiscent of the Landau theory of Fermi liquids, in which the continuity equation imposes an exact relation between the quasi-particle effective mass and the quasi-particle interaction parameters. In fact, by a proper renormalization procedure it should be possible to recover the Landau theory from eq.(2.22) in the limit $q \to 0$, $\omega \to 0$.

The calculation of the dielectric function from eq. (3.7) is equivalent to an infinite sum of ladder diagrams containing Hartree-Fock Green functions, as indicated in figure 5. Several approximative solutions have been proposed, but since the exchange effects will be discussed in detail in the lectures by

Professor Devreese we shall not consider them here. Instead we shall proceed to discuss how one can take into account Coulomb correlation effects which are not included among the diagrams in figure 5. Going back to eq.(3.1) which leads to the RPA, one can argue that the main defect lies in the neglect of short-range correlations. A very simple-minded way to remedy this is to write

$$f^{(2)}_{k\sigma;k'\sigma'}(r,r';t) = f^{(1)}_{k\sigma}(r,t)\ f^{(2)}_{k'\sigma'}(r',t)\ g(r,r';t) \tag{3.9}$$

where $g(r,r';t)$ is a local pair-distribution function which includes exchange as well as Coulomb correlations. We have already seen that this gives a very poor representation of the detailed behaviour of $f^{(2)}_{k\sigma;k'\sigma'}(r,r';t)$, but one may hope that it is acceptable "on the average". In fact, theories based on eq.(3.9) or some modification of it have been remarkably successful. The basic idea was originally put forward by Singwi et al. [8] They constructed a semi-classical model, starting from eq.(3.9) with the further approximation that the equilibrium pair-distribution function $g(|r-r'|)$ was inserted in place of $g(r,r';t)$. We shall here derive their result in a way that is somewhat different from their own. After inserting $g(|r-r'|)$ into eq.(3.9) we linearize the equation in terms of the external disturbance and make a Fourier transformation with respect to r, r' and t. This gives

$$\begin{aligned}\bar{f}^{(2)}_{k\sigma,k'\sigma'}(q,q';t) &= \delta^{o}_{q}\ n_{k\sigma}\ \bar{f}^{(1)}_{k'\sigma'}(q+q';\omega)\\ &+ \delta^{o}_{q'}\ n_{k'\sigma'}\ \bar{f}^{(1)}_{k'\sigma'}\ (q+q';\omega)\\ &+ \frac{n_o}{V}\ n_{k\sigma}[S(q)-1]\ \bar{f}^{(1)}_{k'\sigma'}(q+q';\omega)\\ &+ \frac{n_o}{V}\ n_{k'\sigma'}[S(q')-1]\ \bar{f}^{(1)}_{k\sigma}(q+q';\omega)\end{aligned} \tag{3.10}$$

Wit this approximation, the kinetic equation (2.22) can be written as

$$\begin{aligned}(\hbar\omega - \frac{\hbar^2}{m}k.q)\ \bar{f}^{(1)}_{k\sigma}(q,\omega) &= \frac{1}{V}\ (n_{k-\frac{1}{2}q,\sigma} - n_{k+\frac{1}{2}q,\sigma})\\ &\times \{\Phi^{ext}(q,\omega) + v_q[1-G_k(q)]\,\bar{n}(q,\omega)\}\end{aligned} \tag{3.11}$$

where

$$G_k(q) = -\frac{1}{N}\sum_{q'} \frac{q^2}{q'^2} \frac{n_{k-\frac{1}{2}q',\sigma} - n_{k+\frac{1}{2}q',\sigma}}{n_{k-\frac{1}{2}q,\sigma} - n_{k+\frac{1}{2}q,\sigma}} [S(q-q') - 1] \quad (3.12)$$

The function $G_k(q)$ can be interpreted as a k-dependent local field correction (compare eq.(1.28)). The semi-classical approach used by Singwi et al. had the effect of removing the k-dependence. We can arrive at their result by going to the classical limit in the above expression for $G_k(q)$. This means that we replace k by $p/\hbar$ and then let $\hbar \to 0$, which gives

$$G_k(q) = -\frac{1}{N}\sum_{q'} \frac{q^2}{q'^2} \frac{p\cdot q'}{p\cdot q} [S(q-q') - 1] \quad (3.13)$$

Performing the angular part of the summation one sees that this function is actually independent of p and is equal to

$$G(q) = -\frac{1}{N}\sum_{q'} \frac{q\cdot q'}{q'^2} [S(q-q') - 1] \quad (3.14)$$

which is the local field correction obtained by Singwi et al. In order to evaluate it they determined the structure factor self-consistently by means of eq.(1.37). In this way they obtained a pair-distribution function g(r) that was positive for all r over almost the whole range of metallic densities. The compressibility sum rule, on the other hand, was not satisfied.

In a later work, Hasegawa and Shimizu [9] solved eq. (3.11) without any classical approximation. Their result can be written in the conventional form

$$\chi(q,\omega) = \frac{\chi_o(q,\omega)}{1 - v_q[1 - G(q,\omega)]\chi_o(q,\omega)} \quad (3.15)$$

with

$$G(q,\omega) =$$

$$-\frac{1}{\chi_o(q,\omega)}\frac{1}{N}\sum_{q'}\sum_{k\sigma} \frac{n_{k-\frac{1}{2}q',\sigma} - n_{k+\frac{1}{2}q',\sigma}}{\hbar\omega - \frac{\hbar^2}{m} k.q + i\varepsilon} \frac{q^2}{q'^2} [S(q-q') - 1] \quad (3.16)$$

Thus, the k-dependence of $G_k(q)$ in eq.(3.11) shows up as a frequency dependence in the final local field correction. The numerical results of Hasegawa and Shimizu were very similar to those of Singwi et al, but slightly better in the sense that the pair-distribution function stayed positive for lower densities. The compressibility sum rule was not satisfied.

Vashishta and Singwi [4] improved the model of Singwi et al by considering $g(r,r';t)$ in eq.(3.9) to be a function of the local density. Originally, this idea was proposed by Schneider et al.[10]. We shall here describe the latest calculation along these lines which was made by Hayashi and Shimizu [11]. Their formulation is more straightforward than Vashishta-Singwi's, and they also obtain slightly better numerical results. The starting-point is eq.(3.9) which after linearization becomes

$$\bar{f}^{(1)}_{k\sigma;k'\sigma'}(r,r';t) = \bar{f}^{(1)}_{k\sigma}(r)\ g(|r-r'|)\ n_{k'\sigma'} + n_{k\sigma}\ g(|r-r'|)\ \bar{f}^{(1)}_{k'\sigma'}(r') + n_{k\sigma}\ \bar{g}(r,r';t)\ n_{k'\sigma'} \tag{3.17}$$

Here $g(|r-r'|)$ is the equilibrium pair-distribution function, which is a function of the equilibrium density n_o. The non-equilibrium pair-distribution function $g(r,r';t)$ is assumed to be the same as in equilibrium, except that it depends on the local density instead of the equilibrium density. The local density is taken to be the average of the densities at r and r'. Thus one gets

$$\bar{g}(r,r';t) = \frac{1}{2}[\bar{n}(r,t) + \bar{n}(r',t)]\ \frac{\partial g(|r-r'|)}{\partial n} \tag{3.18}$$

Using this approximation in eq.(2.22), it is straightforward to obtain the following expression for the local field correction :

$$G(q,\omega) = -\frac{1}{\chi_o(q,\omega)}\frac{1}{2N}\sum_{q'}\sum_{k\sigma}\frac{n_{k-\frac{1}{2}q',\sigma} - n_{k+\frac{1}{2}q',\sigma}}{\hbar\omega - \frac{\hbar^2}{m}k.q + i\varepsilon}\ \frac{q^2}{q'^2} \times \left(1 + n_o\frac{\partial}{\partial n_o}\right)[S(q-q') - 1] \tag{3.19}$$

Just as in the theory of Singwi et al, the pair-distribution function should be determined self-consistently. Figures 6 and 7 show a comparison between the four different calculations of g(r) discussed here. Figure 8 shows a comparison between the compressibilities obtained from these four theories through eq.(1.30). The results of Vashishta-Singwi and of Hayashi-Shimizu agree with each other, and they also agree with the result obtained from the ground state energy, which is not the case for the two other calculations. Thus the last term in eq.(3.17) is crucial for satisfying the compressibility sum rule. On the other hand, it does not improve the pair-distribution function, but instead it makes it more negative at the origin for large values of r_s. The compressibility reflects the behaviour of the

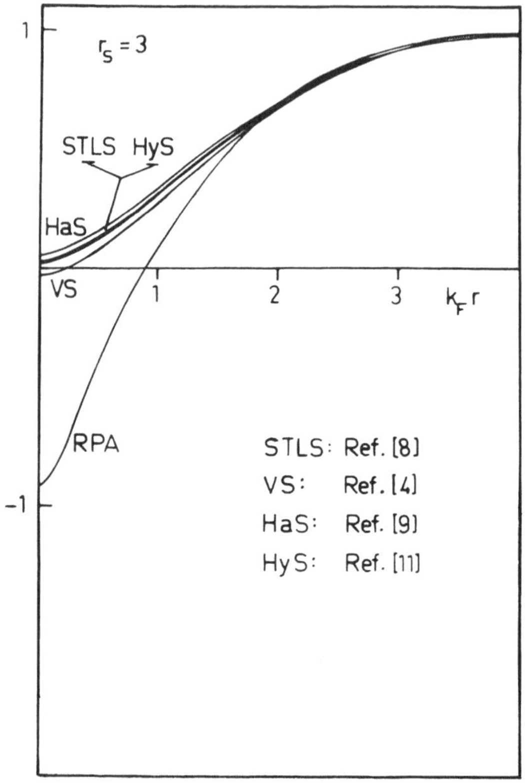

Figure 6

response function in the static long wave-length limit. It is quite clear that the local pair-distribution function in that limit should depend on the local density, and therefore the crucial role of the last term in eq.(3.17) is understandable. On the other hand, in the calculation of g(r) through eqs.(1.37) and (1.38) it is the short wave-length and large frequency behaviour of the response function that is most important. For such rapid changes in the external field it is not really meaningful to introduce a local modification of $g(|r-r'|)$, and one should not expect any improvements due to the last term in eq.(3.17).

The behaviour of the dielectric function for large frequencies is determined by the frequency moment sum rules, given in eqs.(1.32) - (1.34). All approximations based on eq. (3.9) satisfy the first moment sum rule, arising from the conservation of the number of particles, but they violate the third moment sum rule. This is usually satisfied only by approximations which are specifically designed to give the

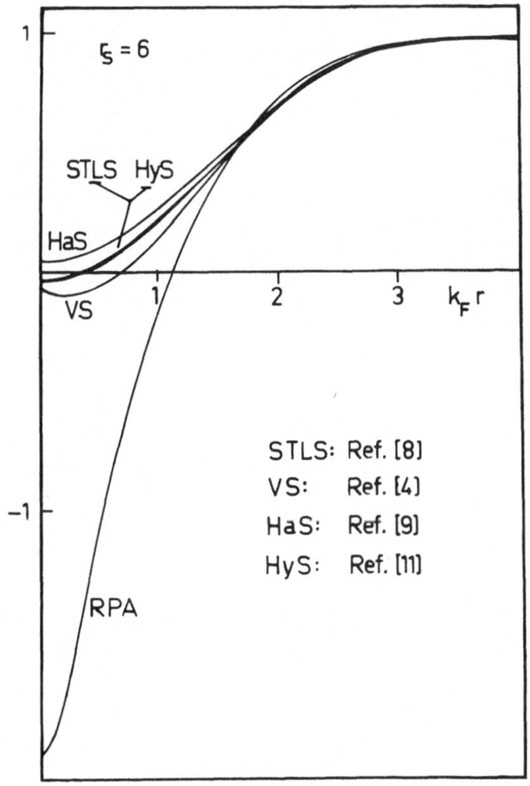

Figure 7

correct high-frequency behaviour. One such approximation was proposed by Toigo and Woodruff [12] , who decoupled the interaction term in eq.(2.22) by writing it as

$$\frac{1}{V} \sum_{q'} v_{q'} \sum_{k'\sigma'} [\bar{f}^{(2)}_{k-\frac{1}{2}q',\sigma;k'\sigma'}(q-q',q';\omega)$$

$$- \bar{f}^{(2)}_{k+\frac{1}{2}q',\sigma;k'\sigma'}(q-q',q';\omega)]$$

$$= A_{k\sigma}(q)\, \bar{n}(q,\omega) \tag{3.20}$$

where the coefficients $A_{k\sigma}(q)$ were determined by expanding both sides in powers of $1/\omega$ (or equivalently powers of t) and keeping only the leading term. Toigo and Woodruff made the further approximation of neglecting all contributions to $A_{k\sigma}(q)$ of higher than linear order in the interaction potential v_q. Their procedure was therefore an approximative calculation of the

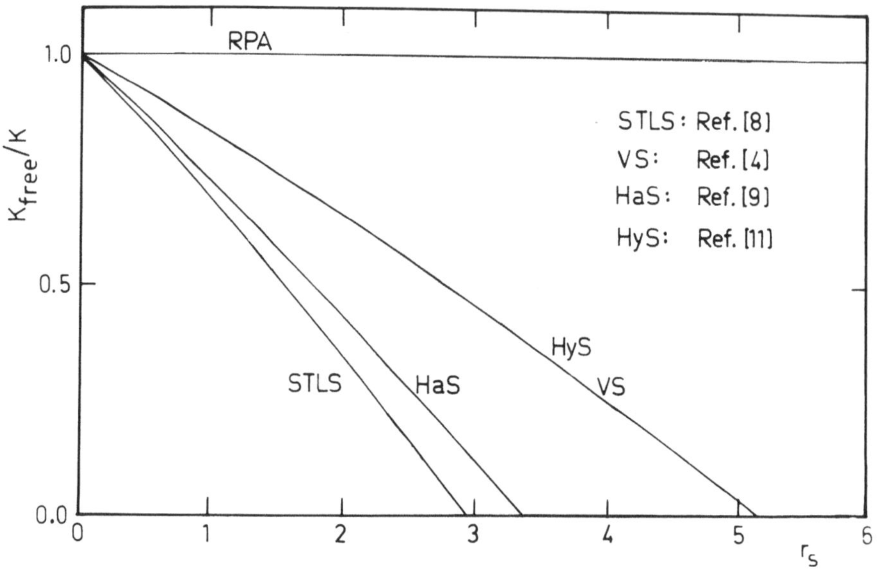

Figure 8

dynamic-exchange correction to the RPA. They obtained a pair-distribution function that was somewhat more negative than the one obtained by Vashishta and Singwi. The compressibility sum rule was satisfied in the sense that the compressibility obtained from the static limit of the dielectric function agreed with the one obtained from the Hartree-Fock ground state energy.

Tripathy and Mandal [13] improved the approximation of Toigo and Woodruff by "conserving frequency moments to infinite order". This means that they replaced the coefficient $A_{k\sigma}(q)$ in equation (3.20) by a frequency-dependent coefficient $A_{k\sigma}(q,\omega)$, chosen in such a way that all the terms in an $1/\omega$-expansion were given correctly. Like Toigo and Woodruff they worked within the dynamic-exchange approximation, keeping $A_{k\sigma}(q,\omega)$ only to terms linear in the interaction potential. Their result turns out to be identical with the one obtained by Brosens et al. [14-16] through a completely different method, as will be discussed in the lectures of Professor Devreese. The result can be written in the conventional form with a local field correction given by

$$G(q,\omega) = \frac{1}{[\chi_o(q,\omega)]^2} \sum_{k\sigma} \sum_{k'} \frac{q^2}{|k-k'|^2} \frac{n_{k-\frac{1}{2}q,\sigma} - n_{k+\frac{1}{2}q,\sigma}}{\hbar\omega - \frac{\hbar^2}{m} k.q + i\varepsilon}$$

$$\times (n_{k'-\frac{1}{2}q,\sigma} - n_{k'+\frac{1}{2}q,\sigma})\left[\frac{1}{\hbar\omega - \frac{\hbar^2}{m} k'.q + i\varepsilon} - \frac{1}{\hbar\omega - \frac{\hbar^2}{m} k.q + i\varepsilon}\right] \tag{3.21}$$

We shall see in the next section that there is still another way to obtain this expression.

In addition to the decoupling procedures described here, one finds in the literature several others, based on more or less similar ideas. We shall only mention the recent work by Utsumi and Ichimaru [17], who took great care to obtain correct behaviour in the high-frequency as well as low-frequency limit. They introduced a static local field correction G(q) similar to the one used by Singwi et al, but they also took into account relaxation effects in such a way that the third frequency moment sum rule was satisfied. However, their model is somewhat too complicated to be presented in any detail here.

4. THE KINETIC EQUATION FOR THE TWO-PARTICLE WIGNER FUNCTION AND SOME EXACT ASYMPTOTIC FORMULAS

In the preceding section we discussed various ways to approximate the interaction term in the kinetic equation for the one-particle Wigner function. An alternative procedure is to go one step further in the hierarchy of kinetic equations and write down the equation of motion for the two-particle function $\bar{f}^{(2)}_{k\sigma;k'\sigma'}(q,q';\omega)$. The derivation of this equation is straight-forward, although somewhat tedious. It has the form

$$(\hbar\omega - \frac{\hbar^2}{m} k.q - \frac{\hbar^2}{m} k'.q')\ \bar{f}^{(2)}_{k\sigma;k'\sigma'}(q,q';\omega)$$

$$+ \frac{1}{V} \sum_{q''} v_{q''} [\bar{f}^{(2)}_{k-\frac{1}{2}q'',\sigma;k'+\frac{1}{2}q'',\sigma'}(q+q'',q'-q'';\omega)$$

$$- \bar{f}^{(2)}_{k+\frac{1}{2}q'',\sigma;k'-\frac{1}{2}q'',\sigma'}(q+q'',q'-q'';\omega)]$$

$$= \frac{1}{V}[\bar{f}^{(2)}_{k-\frac{1}{2}(q+q'),\sigma;k'\sigma'}(-q') - f^{(2)}_{k+\frac{1}{2}(q+q'),\sigma;k'\sigma'}(-q')$$

$$+ f^{(2)}_{k\sigma;k'-\frac{1}{2}(q+q'),\sigma'}(q) - f^{(2)}_{k\sigma;k'+\frac{1}{2}(q+q'),\sigma'}(q)]\,\Phi^{ext}(q,\omega)$$

$$+ \frac{1}{V}\sum_{q''} v_{q''} \sum_{k''\sigma''} [\bar{f}^{(3)}_{k-\frac{1}{2}q'',\sigma;k'\sigma';k''\sigma''}(q-q'',q',q'';\omega)$$

$$- \bar{f}^{(3)}_{k+\frac{1}{2}q'',\sigma;k'\sigma';k''\sigma''}(q-q'',q'';\omega)$$

$$+ \bar{f}^{(3)}_{k\sigma;k'-\frac{1}{2}q'',\sigma';k''\sigma''}(q,q'-q'',q'';\omega)$$

$$- \bar{f}^{(3)}_{k\sigma;k'+\frac{1}{2}q'',\sigma';k''\sigma''}(q,q'-q'',q'';\omega)] \qquad (4.1)$$

where $\bar{f}^{(3)}_{k\sigma;k'\sigma';k''\sigma''}(q,q',q'';\omega)$ is the non-equilibrium part of the three-particle Wigner function, defined in analogy with the one- and two-particle functions. This equation has been used by Niklasson [18] to study the high-frequency and short wave-length behaviour of the dielectric function, and it has also been used as the starting-point for a more extensive calculation by Aravind et al.[19]. It should be noted that the notation here is somewhat different from the one used in ref.[17] and [18]. There the uncorrelated part of $f^{(2)}_{k\sigma;k'\sigma'}(q,q';\omega)$ was extracted separately, and a similar decomposition was made in the three-particle correlation function.

It is obvious that one has to make rather drastic simplifications in order to obtain an explicit solution of eq. (4.1). The difficulties arise from the interaction terms, containing the Coulomb potential v_q. A very simple-minded approximation is to replace these terms with an effective local field, added to the external field $\Phi^{ext}(q,\omega)$. Thus one may write

$$(\hbar\omega - \frac{\hbar^2}{m} k.q - \frac{\hbar^2}{m} k'.q')\, \bar{f}^{(2)}_{k\sigma;k'\sigma'}(q-q',q';\omega)$$

$$= \frac{1}{V}\,[f^{(2)}_{k-\frac{1}{2}(q+q'),\sigma;k'\sigma'}(-q') - f^{(2)}_{k+\frac{1}{2}(q+q'),\sigma;k'\sigma'}(-q')$$

$$+ f^{(2)}_{k\sigma;k'-\frac{1}{2}(q+q'),\sigma'}(q) - f^{(2)}_{k\sigma;k'+\frac{1}{2}(q+q'),\sigma'}(q)]$$

$$\times \{\Phi^{ext}(q,\omega) + v_q[1 - G(q,\omega)]\bar{n}(q,\omega)\} \tag{4.2}$$

where $G(q,\omega)$ is the same local field correction as the one appearing in the density response. This means that $G(q,\omega)$ is defined by the relation

$$\bar{n}(q,\omega) = \chi_o(q,\omega)\{\Phi^{ext}(q,\omega) + v_q[1 - G(q,\omega)]\bar{n}(q,\omega)\} \tag{4.3}$$

where $\chi_o(q,\omega)$ is the modified Lindhard free-particle response function, given by eq.(1.26) but containing the true occupation number $n_{k\sigma}$ in place of the free-particle Fermi step function. Combining eqs.(4.2) and (4.3) with eq.(2.22) it is straightforward to extract an expression for $G(q,\omega)$. The result is

$$G(q,\omega) =$$

$$1 - \frac{1}{v_q[\chi_o(q,\omega)]^2}\frac{1}{V^2}\sum_{q'}\sum_{k\sigma}\sum_{k'\sigma'}\frac{v_{q'}}{\hbar\omega - \frac{\hbar^2}{m}k.(q-q') - \frac{\hbar^2}{m}k'.q'}$$

$$\times \left[\frac{1}{\hbar\omega - \frac{\hbar^2}{m}(k+\frac{1}{2}q').q} - \frac{1}{\hbar\omega - \frac{\hbar^2}{m}(k-\frac{1}{2}q').q}\right]$$

$$\times [f^{(2)}_{k-\frac{1}{2}q,\sigma;k'\sigma'}(-q') - f^{(2)}_{k+\frac{1}{2}q,\sigma;k'\sigma'}(-q')$$

$$+ f^{(2)}_{k\sigma;k'-\frac{1}{2}q,\sigma'}(q-q') - f^{(2)}_{k\sigma;k'+\frac{1}{2}q,\sigma'}(q-q')] \tag{4.4}$$

This is slightly different from the result obtained by Aravind et al. The difference arises because they kept only the Hartree mean field $v_q\bar{n}(q,\omega)$ in eq.(4.2); neglecting there the local field correction $G(q,\omega)$.

It should be noted that $f^{(2)}_{k\sigma;k'\sigma'}(q)$, appearing in eq. (4.4), should in principle be the exact two-particle correlation function. Obviously, in order to calculate $G(q,\omega)$ explicitly one needs some approximation for this function. The simplest approximation is to keep only the Hartree-Fock contributions,

$$f^{(2)HF}_{k\sigma;k'\sigma'}(q) = \delta^o_q\, n_{k\sigma}\, n_{k'\sigma'} - \delta^k_{k'}\,\delta^\sigma_{\sigma'}\, n_{k-\frac{1}{2}q,\sigma}\, n_{k+\frac{1}{2}q,\sigma} \tag{4.5}$$

and one then finds that eq.(4.4) reduces to the expression given previously in eq.(3.21). This is quite interesting, because eq. (3.21) was obtained by Brosens et al and by Tripathy et al through completely different methods. The derivation given here may perhaps contribute to shed some further light on its physical content. It also seems to provide a natural way to go beyond the dynamical-exchange approximation by including higher order Coulomb correlations in $f^{(2)}_{k\sigma;k'\sigma'}(q)$. Aravind et al., suggested two alternative approximations for $f^{(2)}_{k\sigma;k'\sigma'}(q)$, namely

$$f^{(2)A}_{k\sigma;k'\sigma'}(q) = f^{(2)HF}_{k\sigma;k'\sigma'}(q) + \frac{1}{N} n_{k\sigma} n_{k'\sigma'} [S(q) - S^{HF}(q)] \tag{4.6}$$

and

$$f^{(2)B}_{k\sigma;k'\sigma'}(q) = f^{(2)HF}_{k\sigma;k'\sigma'}(q) + \frac{1}{2N} (n_{k-\frac{1}{2}q,\sigma} n_{k'+\frac{1}{2}q,\sigma'} + n_{k+\frac{1}{2}q,\sigma} n_{k'-\frac{1}{2}q',\sigma'}) [S(q) - S^{HF}(q)] \tag{4.7}$$

where the superscript HF refers to the Hartree-Fock approximation. Both alternatives satisfy the exact equation

$$\sum_{k\sigma} \sum_{k'\sigma'} f^{(2)}_{k\sigma;k'\sigma'}(q) = N[S(q) - 1] + N^2\delta^o_q \tag{4.8}$$

but it is questionable if any of them gives a fair representation of the true dependence on k and k'. From the numerical results of Aravind et al., who studied the dispersion and damping of plasmons, it appears that the two alternative forms lead to quite different results, particularly for the plasmon damping. The conclusion is therefore that the dependence of $f^{(2)}_{k\sigma;k'\sigma'}(q)$ on k and k' must be studied carefully before one tries to use eq.(4.4) for further calculations.

One interesting property of the expression in eq.(4.4) is that it is exact in the limit of large ω and also in the limit of large q. If the external disturbance varies rapidly in time, the response of the system is determined by the motion of the particles for short times. This is essentially a free-particle behaviour, since it takes a certain time before the motion of a particle is changed by the forces from other particles. The response to a disturbance that varies rapidly in space is also determined by the free-particle behaviour, because a particle must travel a certain distance before it is affected by the presence of other particles. Mathematically, this behaviour shows up in the kinetic equations through the fact that the flow term contains terms proportional to ω or q which dominate over the interaction terms for sufficiently large q or ω. Thus in eq. (4.1) only the flow term and the driving term proportional to $\Phi^{ext}(q,\omega)$ are important in these limits. Both these terms were exactly taken into account in the derivation of eq.(4.4). Therefore, we can from eq.(4.4) obtain exact expressions for the leading terms when expanding in powers of $1/\omega$ or $1/q$. The result of such an expansion is [18]

$$G(q,\omega) = -\frac{1}{N}\sum_{q'}\frac{(q\cdot q')^2}{q^2 q'^2}[S(q'-q)-1] + \frac{1}{2}\left\{\left[\frac{\hbar\omega + \frac{\hbar^2}{2m}q^2}{\hbar\omega - \frac{\hbar^2}{2m}q^2}\right]^2 + \left[\frac{\hbar\omega - \frac{\hbar^2}{2m}q^2}{\hbar\omega + \frac{\hbar^2}{2m}q^2}\right]^2\right\}\frac{1}{N}\sum_{q'}\frac{(q\cdot q')^2}{q^2 q'^2}[S(q')-1] \tag{4.9}$$

where the two-particle correlations enter only through the static structure factor S(q). Eq.(4.9) is expected to be valid for values of q and ω far away from the particle-hole region in the q-ω-plane,

$$\left|\hbar\omega - \frac{\hbar^2}{2m}q^2\right| \gg \varepsilon_F \tag{4.10}$$

Specializing to the case when q is finite and ω tends to infinity one obtains

$$G_\infty(q) = \lim_{\omega\to\infty} G(q,\omega) = \frac{1}{N}\sum_{q'}\frac{(q\cdot q')^2}{q^2 q'^2}[S(q') - S(q'-q)] \tag{4.11}$$

which is identical with the result obtained by Pathak and Vashishta by means of the third moment sum rule [6] . Keeping ω finite and letting q tend to infinity one obtains

$$\lim_{q\to\infty} G(q,\omega) = \frac{2}{3}[1 - g(0)] \tag{4.12}$$

where g(0) is the value of the pair-distribution function at r = 0. In particular, the Hartree-Fock approximation leads to g(0) = 1/2 and thus

$$\lim_{q\to\infty} G^{HF}(q,\omega) = \frac{1}{3} \tag{4.13}$$

By means of the fluctuation-dissipation theorem one can derive another exact relation concerning the behaviour of the pair-distribution function at small r. Inserting the assymptotic expression from eq.(4.9) into eq.(1.37) one can show that [18]

$$-\lim_{q\to\infty} q^4 [S(q) - 1] = \frac{8\pi e^2 m n_o}{\hbar^2} g(0) \tag{4.14}$$

On the other hand it follows from the general properties of Fourier transforms that

$$\lim_{q\to\infty} q^4 [S(q) - 1] = -8\pi n_o g'(0) \tag{4.15}$$

where g'(0) is the derivative of g(r) at r = 0. Comparing these two equations one concludes that

$$\frac{g'(0)}{g(0)} = \frac{me^2}{\hbar^2} = \frac{1}{a_o} \tag{4.16}$$

where a_o is the Bohr radius. This exact relation was first derived by Kimball [20] , in a different and much simpler way. His argument was that when two electrons are very close together their mutual interaction dominates over the interactions with the surrounding particles. Therefore, one only needs to solve the two-particle problem in order to obtain the correct correlations at small distances. Introducing the "pair wave function" $\psi(r)$, Kimball writes down the Schrödinger equation in the form

$$\left(-\frac{\hbar^2}{2\mu}\frac{\partial^2}{\partial r^2} - \frac{\hbar^2}{\mu r}\frac{\partial}{\partial r} + \frac{e^2}{r}\right)\psi(r) = \mathcal{E}\psi(r) \tag{4.17}$$

where $\mathcal{E}$ is a bounded operator which contains all the many-body contributions. The reduced mass μ is half the electron mass m. It is assumed that the two electrons are in a relative s state, since higher angular momentum states do not contribute to g(r) or g'(r) at r = 0. Expanding $\psi(r)$ in powers of r,

$$\psi(r) = a + br + \ldots \tag{4.18}$$

and inserting into eq.(4.17) one obtains

$$-\frac{\hbar^2}{\mu} b + e^2 a = 0 \tag{4.19}$$

which leads to

$$\psi(r) = a\left(1 + \frac{1}{2a_o} r + \ldots\right) \tag{4.20}$$

where a_o is the Bohr radius $\hbar^2/me^2$. Eq.(4.16) now follows from the observation that g(r) is proportional to $|\psi(r)|^2$.

5. DYNAMIC PROPERTIES AND THE MORI FORMALISM

We have seen in the previous sections that the kinetic equations for Wigner distribution functions can be quite useful for studying the density response in an electron liquid. However, there are certain aspects of the problem which are difficult to discuss within this formalism. In the conventional Green's function theory one encounters concepts such as self-energy corrections, quasi-particle life-time, particle-hole interactions, etc. In the kinetic theory all these things tend to be hidden in a smeared-out local field correction. Any detailed structure, arising for instance from life-time effects, would easily be lost. Particularly in the particle-hole region of the q-ω-plane one may expect complications of this kind. In fact, the experimental measurements of S(q,ω) [21-24] show in this region a characteristic structure which is generally absent in the theoretical calculations, as exemplified in figure 9. The experimental dynamic structure factor obtained by Platzman and Eisenberger [21] is there compared with

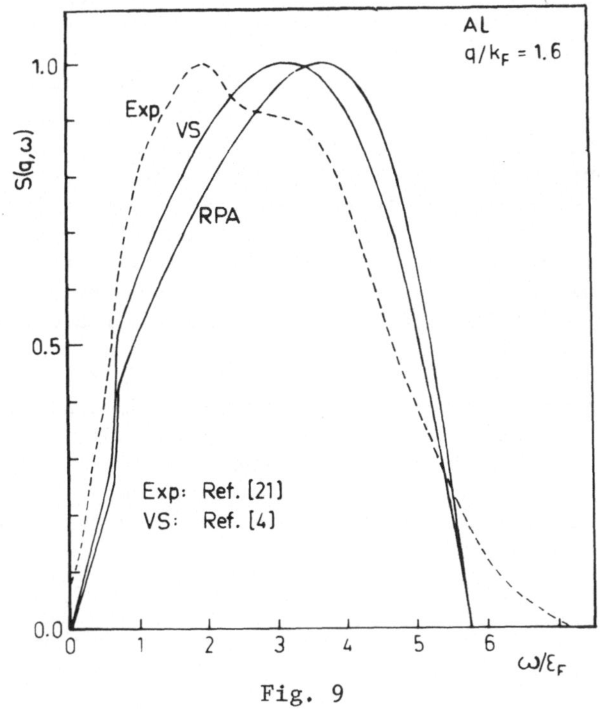

Fig. 9

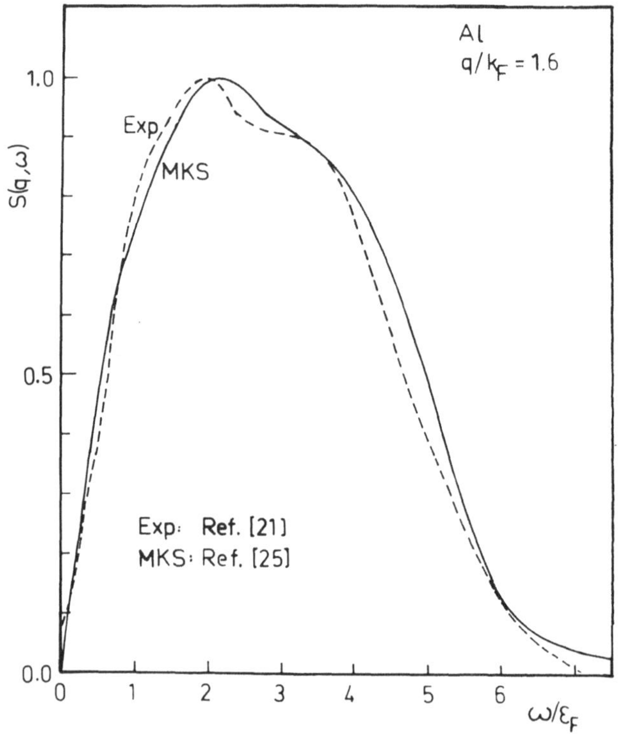

Figure 10

the theoretical results obtained by Vashishta and Singwi. Almost all theoretical models show a similar behaviour in the sense that they fail to reproduce the tendency towards a two-peak structure that is apparent in the experimental curve. This structure has been observed in several different materials, and it is generally believed to arise from intrinsic properties of the electron liquid and not from band-structure effects. An obvious interpretation of the experimental results is that a plasmon-like excitation exists within the region of particle-hole excitations. The question is then how it can survive long enough to be observable. Normally, one would expect it to decay very rapidly into particle-hole excitations (Landau damping). In recent years several attempts have been made to explain the anomaly through modifications of the existing local-field theories. Mukhopadhyay et al [25] made a simple modification of the Vashishta-Singwi model by including life-time effects in the effective free-particle response function. Thus they replaced the Lindhard function by

$$\chi^{eff}(q,\omega) = \frac{1}{V} \sum_{k\sigma} \frac{n_{k-\frac{1}{2}q,\sigma} - n_{k+\frac{1}{2}q,\sigma}}{\hbar\omega - \frac{\hbar^2}{m} k.q + \frac{i}{\tau_{k-\frac{1}{2}q}} + \frac{i}{\tau_{k+\frac{1}{2}q}}} \qquad (5.1)$$

where τ_k is the life-time of a quasi-particle (or quasi-hole) with wave-vector k. The life-time was determined through an approximative calculation of the imaginary part of the self-energy. As is illustrated in figure 10, Mukhopadhyay et al managed in this way to reproduce the characteristic features of the experimental results. Recently, Awa et al [26] have obtained very similar results by including life-time effects in a calculation based on conventional Green function techniques. However, if one traces the approximation of Mukhopadhyay et al back to the kinetic equation for $f^{(1)}_{k\sigma}(r,t)$ one finds that it violates the continuity equation, and this has created some scepticism against the results. In the Green function formalism used by Awa et al the continuity equation does not appear explicitly, but their approximations seem to be largely equivalent to the ones made by Mukhopadhyay et al. Therefore, the role of the continuity equation in this context needs to be clarified further before one is to be convinced by the agreement between theory and experiment. One would also like to understand in more intuitive terms why the introduction of finite life-times for particle-hole excitations has the effect of enhancing the plasma oscillations. A possible argument may be as follows. If a plasmon decays into an undamped particle-hole pair, the plasmon energy is quickly carried away from the region where the decay took place. If, on the other hand, the particle-hole pair is scattered against other excitations, the energy stays around for a longer time and some of it may be back-scattered into the plasmon mode. Therefore, the Landau damping becomes less efficient and the plasmon can survive for a while, even inside the particle-hole continuum.

A number of recent studies of the dynamic structure in $S(q,\omega)$ have been based on the Mori formalism. This makes it possible to include life-time effects in a consistent way without violation of sum rules or conservation laws. The remaining part of these lectures is devoted to a short discussion of the Mori formalism and its application to the electron liquid.

In all many-body systems there are fluctuations of various kinds. Some of these fluctuations decay slowly, because they are associated with conservation laws such as conservation of the number of particles or conservation of energy. Most of the fluctuations are, however, such that they decay very rapidly. Usually, one wants to eliminate all rapidly decaying variables in order to obtain kinetic equations for slow variables like the particle density or the current. In the Mori formalism one achieves this elimination by incorporating the effects of the rapid fluctuations in a so called memory function $\Gamma(t)$. We shall illustrate the basic idea by first studying a very simple mathematical example. Let us consider two variables x and y which obey the following coupled kinetic equations,

$$\frac{dx(t)}{dt} + k_{11}\ x(t) + k_{12}\ y(t) = 0 \tag{5.2}$$

$$\frac{dy(t)}{dt} + k_{21}\ x(t) + k_{22}\ y(t) = 0 \tag{5.3}$$

Let us now assume that the initial conditions are such that x(0) is finite, whereas y(0) is zero. We wish to study the time evolution of x(t), and we achieve this by eliminating y(t) from the equations. The formal solution of eq.(5.3) is

$$y(t) = -k_{21} \int_0^t e^{-k_{22}(t-t')}\ x(t')\ dt' \tag{5.4}$$

and after insertion into eq.(5.2) this gives

$$\frac{dx(t)}{dt} + k_{11}\ x(t) + \int_0^t \Gamma(t-t')\ x(t')\ dt' \tag{5.5}$$

where we have introduced the "memory function"

$$\Gamma(t) = k_{12}\ k_{21}\ e^{-k_{22}t} \tag{5.6}$$

Eq.(5.5) is the Mori equation for the variable x. We conclude that a memory function enters in a very natural way when one eliminates some of the variables from a complete set of kinetic equations. In the illustrating example above we could explicitly evaluate the memory function, but in practical applications this is usually very difficult. The Mori theory leads to an exact formal expression for the memory function, but in most cases one prefers to construct an approximative memory function on the basis of physical arguments, intuition or pure guesswork. This is facilitated by the fact that the limiting values $\Gamma(0)$ and $\Gamma(\infty)$ can be found exactly by means of sum rules and conservation laws. These are then satisfied exactly, independently of the particular form chosen for $\Gamma(t)$.

We shall first present the Mori formalism for the special case that there is only one slow variable, represented by an operator A(t) in the Heisenberg picture. We wish to obtain a kinetic equation for this variable by means of the Mori formalism. The central quantity is then the so called relaxation function F(t), defined by

$$F(t) = \frac{1}{i\hbar} \int_{-\infty}^{0} dt' \langle [A(t), A^+(t')]_- \rangle \tag{5.7}$$

The physical significance of the relaxation function can be explained as follows. Suppose that the system for $t < 0$ is subject to a static external disturbance which couples to the variable A and therefore generates a deviation $\delta\langle A\rangle$ from the equilibrium mean value. The relaxation function F(t) describes how $\delta\langle A\rangle$ decays back to zero, if the disturbance is suddenly turned off at $t = 0$. Mori showed that the kinetic equation governing the relaxation can be written in the form

$$\frac{d}{dt} F(t) - i\Omega F(t) + \int_0^t dt'\Gamma(t-t')F(t') = 0 \tag{5.8}$$

where Ω is a certain characteristic frequency and $\Gamma(t)$ is the memory function. The formal expressions for Ω and $\Gamma(t)$ were given by Mori in terms of certain "scalar products" in an abstract vector space, spanned by the observables of the system. The scalar product between two operators A(t) and B(t') is defined as

$$(A(t),B(t')) = \frac{1}{i\hbar} \int_{-\infty}^{0} dt'' \langle [A(t),B^{+}(t'+t'')] \rangle \tag{5.9}$$

We notice that the relaxation function F(t) is just the scalar product between A(t) and A(0). Mori further introduced a projector operator P onto the subspace spanned by A(0). This is defined by

$$PB = \frac{(B,A(0))}{(A(0),A(0))} A(0) \tag{5.10}$$

The time evolution of the various observables is governed by the Liouville operator L, defined through

$$\frac{d}{dt} A(t) = \frac{i}{\hbar} [H_0,A(t)] = iLA(t) \tag{5.11}$$

where H_0 is the Hamiltonian. The expressions obtained by Mori for Ω and $\Gamma(t)$ can now be written as

$$i\Omega = - \frac{(A(0),\dot{A}(0))}{(A(0),A(0))} = \frac{\dot{F}(0)}{F(0)} \tag{5.12}$$

$$\Gamma(t) = \frac{(\dot{A}(0)(1-P),e^{i(1-P)L(1-P)t}(1-P)\dot{A}(0))}{(A(0),A(0))} \tag{5.13}$$

We notice that the memory function is determined by the time-evolution of operators in the subspace spanned by the projection operator 1-P. This is the space of rapidly decaying variables, orthogonal to the slow variable A.

The above equations are easily generalized to cases where there are several slow variables, denoted by $A_k(t)$ with

$k = 1,2,3,\ldots$. One then introduces a matrix $\mathbb{F}(t)$ of relaxation functions,

$$F_{k\ell}(t) = \frac{1}{i\hbar} \int_{-\infty}^{0} dt' \langle [A_k(t), A_\ell^+(t')] \rangle \tag{5.14}$$

and the kinetic equation takes the form

$$\frac{d}{dt} \mathbb{F}(t) - i\Omega . \mathbb{F}(t) + \int_0^t dt' \Gamma(t-t') . \mathbb{F}(t') = 0 \tag{5.15}$$

where the matrices Ω and $\mathbb{F}(t)$ are given by

$$i\Omega = \dot{\mathbb{F}}(0) . \mathbb{F}^{-1}(0) \tag{5.16}$$

$$\Gamma_{k\ell}(t) = \sum_m (\dot{A}_k(0)(1-P), e^{i(1-P)L(1-P)t}(1-P)\dot{A}_m(0)) F^{-1}_{m\ell}(0) \tag{5.17}$$

The operator P is now the projector onto the subspace spanned by all the variables $A_k(t)$,

$$PB = \sum_k \frac{(B, A_k(0))}{(A_k(0), A_k(0))} A_k(0) \tag{5.18}$$

We have here assumed that the operators $A_k(t)$ are orthogonal to each other in the sense that their equal-time scalar products are zero. This is always possible to achieve by a suitable orthogonalization procedure.

The first problem one encounters in applications of the Mori formalism to specific physical problems is the selection of the set $A_k(t)$. We have argued that these should represent the slowly decaying variables of the system, but the distinction between slow and rapid variables is not always very sharp. The Mori formalism as such remains valid for any set of selected variables, and the selection is therefore a matter of choice rather than necessity.

Let us now try to apply the Mori formalism to the problem of calculating the density response in an electron liquid. It is then obvious that one of the selected slow variables has to be the local density. The longitudinal current should also be treated as a slow variable, because it is connected to the density through the continuity equation. For simplicity we shall limit ourselves to these two variables. We write the corresponding operators as

$$A_1(q,t) = \rho_q(t) = \sum_{k\sigma} a^+_{k-\frac{1}{2}q,\sigma}(t)\, a_{k+\frac{1}{2}q,\sigma}(t) \tag{5.19}$$

$$A_2(q,t) = -iq.j_q(t)$$

$$= -i \frac{\hbar}{m} \sum_{k\sigma} q.k \, a^+_{k-\frac{1}{2}q,\sigma}(t) \, a_{k+\frac{1}{2}q,\sigma}(t) \qquad (5.20)$$

From the continuity equation it follows that $A_2(q,t)$ is the time derivative of $A_1(q,t)$. The definition of the scalar product in eq.(5.9) then implies that $A_1(q,t)$ and $A_2(q,t)$ are orthogonal to each other. In order to write down the matrix equation (5.15) we need to calculate the initial values of the 2×2 matrices $\mathbb{F}(0)$ and $\dot{\mathbb{F}}(0)$. Using the definition in eq.(5.14) together with some commutator algebra one finds that

$$\mathbb{F}(0) = \begin{pmatrix} V\chi(q) & 0 \\ 0 & -\frac{Nq^2}{m} \end{pmatrix} \qquad (5.21)$$

$$\dot{\mathbb{F}}(0) = \begin{pmatrix} 0 & -\frac{Nq^2}{m} \\ \frac{Nq^2}{m} & 0 \end{pmatrix} \qquad (5.22)$$

where $\chi(q)$ is the static response function, obtained by putting $\omega = 0$ in eq.(1.19). It is characteristic for the Mori formalism that the static response functions enter as initial values in the theory. The values of these static quantities can not be obtained from the Mori equations themselves. Therefore this formalism is essentially a technique for studying the dynamic behaviour of a system whose static properties are known. From eqs.(5.21) and (5.22) we can obtain the matrix Ω, given by eq.(5.16). The only problem that remains is to find the matrix of memory functions from eq.(5.17). Actually, there is only one memory function, namely $\Gamma_{22}(t)$. The other three elements of the matrix $\mathbb{\Gamma}$ vanish because of the orthogonality of $A_1(q,t)$ and $A_2(q,t)$ together with the fact that $A_2(q,t) = \dot{A}_1(q,t)$. We can now write down the four components of the matrix equation (5.15) in the following way

$$\frac{d}{dt} F_{11}(q,t) - F_{21}(q,t) = 0 \qquad (5.23)$$

$$\frac{d}{dt} F_{12}(q,t) - F_{22}(q,t) = 0 \qquad (5.24)$$

$$\frac{d}{dt} F_{21}(q,t) + \omega_o^2(q) \, F_{11}(q,t) + \int_0^t \Gamma_{22}(q,t-t') \, F_{21}(q,t') = 0 \qquad (5.25)$$

$$\frac{d}{dt} F_{22}(q,t) + \omega_o^2(q)\ F_{12}(q,t) + \int_0^t \Gamma_{22}(q,t-t')\ F_{22}(q,t') = 0 \tag{5.26}$$

where we have introduced the symbol $\omega_o(q)$, defined by

$$\omega_o^2(q) = -\frac{n_o q^2}{m\chi(q)} \tag{5.27}$$

From the equations above we may extract a closed equation for the current-current relaxation function $F_{22}(q,t)$ in the form

$$\frac{d^2}{dt^2} F_{22}(q,t) + [\omega_o^2(q) + \Gamma_{22}(q,0)]\, F_{22}(q,t) + \int_0^t \dot{\Gamma}_{22}(q,t-t')\ F_{22}(q,t')dt' = 0 \tag{5.28}$$

The initial value $\Gamma_{22}(q,0)$ of the memory function can be determined exactly from the short-time behaviour of $F_{22}(q,t)$. This is in fact related to the third moment of the density response function, as given in eq.(1.34). From eq.(5.28) it follows that

$$\Gamma_{22}(q,0) + \omega_o^2(q) = \left[\frac{\frac{d^2}{dt^2} F_{22}(q,t)}{F_{22}(q,t)}\right]_{t=0} \tag{5.29}$$

which leads to the result

$$\Gamma_{22}(q,0) = \omega_\infty^2(q) - \omega_o^2(q) \tag{5.30}$$

where ω_∞ is given by

$$\omega_\infty^2(q) = \frac{m}{n_o q^2} M_3(q) \tag{5.31}$$

Equation (5.26) can now be written in the form

$$\frac{d^2}{dt^2} F_{22}(q,t) + \omega_\infty^2(q)\ F_{22}(q,t) + [\omega_\infty^2(q) - \omega_o^2(q)] \int_0^t \dot{\Gamma}_o(q,t-t')\ F_{22}(q,t') = 0 \tag{5.32}$$

where $\Gamma_o(q,t)$ is a new memory function normalized in such a way that $\Gamma_o(q,0) = 1$ and $\Gamma_o(q,\infty) = 0$. The behaviour at intermediate times is not known. A common procedure is to chose some simple form, for instance a gaussian or an experimentally decaying function. The characteristic decay time for the memory function is either estimated on the basis of some physical arguments or used as an adjustable parameter to fit experimental data.

Once the memory function is given, one can solve equations (5.23 - 5.26) and then obtain the response function $\chi(q,\omega)$ through the relation

$$\chi(q,\omega) = -\frac{1}{V}\int_0^\infty e^{i\omega t}\frac{d}{dt}F_{11}(q,t)\,dt \tag{5.33}$$

which follows from eqs.(1.19) and (5.14). The result can be written in the form

$$\chi(q,\omega) = \frac{\chi(q)}{1-\frac{\omega^2}{\omega_o^2}+i\omega\left(1-\frac{\omega_\infty^2}{\omega_o^2}\right)\Gamma_o(q,\omega)} \tag{5.34}$$

where

$$\Gamma_o(q,\omega) = \int_0^\infty e^{i\omega t}\,\Gamma_o(q,t)\,dt \tag{5.35}$$

Eq.(5.34) is by construction such that it goes over to the exact static response function in the limit $\omega \to 0$ and it also behaves correctly in the high-frequency limit, since it satisfies both the first moment sum rule and the third moment sum rule.

The method presented here was used by Mukhopadhyay and Sjölander [28] to study the behaviour of the dynamic structure factor. They determined the static response function from the Vashishta-Singwi model, and then they tried a number of simple choices for the memory function. None of these led to the two-peak structure seen in the experiments. Essentially the same conclusion was reached by Yoshida et al [29], who used a Gaussian memory function in a model containing a couple of adjustable parameters. It is of course always possible to reproduce the experimental results by choosing a sufficiently complicated memory function, but this is uninteresting unless one can motivate the form of the memory function on physical grounds.

De Raedt and De Raedt [30] developed the theory presented here one step further by also including the energy density among the slow variables. They obtained good agreement with experiments, but their theory contained some adjustable parameters and the physical significance of their results is therefore not yet quite clear.

REFERENCES

1. D. Pines and P. Nozières : The Theory of Quantum Liquids, Vol.1 (Benjamin, New York, 1966).

2. D.J.W. Geldart and R. Taylor : Can. J. Phys. 48, 155 (1970).

3. A. Holas, P.K. Aravind and K.S. Singwi : Phys. Rev. B 20, 4912 (1979).

4. P. Vashishta and K.S. Singwi : Phys. Rev. B 6, 875 (1972).

5. R.D. Puff : Phys. Rev. 137, A 406 (1965).

6. K.N. Pathak and P. Vashishta : Phys. Rev. B 7, 3649 (1973).

7. E.P. Wigner : Phys. Rev. 40, 749 (1932).

8. K.S. Singwi, M.P. Tosi, R.H. Land and A. Sjölander : Phys. Rev. 176, 589 (1968).

9. T. Hasegawa and M. Shimizu : J. Phys. Soc. Japan 38, 965 (1975).

10. T. Schneider, R. Brout, H. Thomas and J. Feder : Phys. Rev. Letters 25, 1423 (1970).

11. H. Hayashi and M. Shimizu : J. Phys. Soc. Japan 48, 16 (1980).

12. F. Toigo and T.O. Woodruff : Phys. Rev. B 2, 3958 (1970); B 4, 371 (1971); B 4, 4312 (1971).

13. D.N. Tripathy and S.S. Mandal : Phys. Rev. B 16, 231 (1977).

14. F. Brosens, L.F. Lemmens and J.T. Devreese : Phys. Stat. Sol. (b) 74, 45 (1976).

15. J.T. Devreese, F. Brosens and L.F. Lemmens : Phys. Rev. B 21, 1349 (1980).

16. F. Brosens, J.T. Devreese and L.F. Lemmens : Phys. Rev. B 21, 1363 (1980).

17. K. Utsumi and S. Ichimaru : Phys. Rev. B 22, 1522 (1980); 22, 5203 (1980); 23, 3291 (1981).

18. G. Niklasson : Phys. Rev. B 10, 3052 (1974).

19. P.K. Aravind, A. Holas and K.S. Singwi : submitted to Phys. Rev. B.

20. J.C. Kimball : Phys. Rev. A 7, 1648 (1973).

21. P.M. Platzman and P. Eisenberger : Phys. Rev. Letters 33, 152 (1974).

22. P. Eisenberger, P.M. Platzman and P. Schmidt : Phys. Rev. Letters 34, 18 (1975).

23. P. Eisenberger and P.M. Platzman : Phys. Rev. B 13, 934 (1976).

24. G.D. Priftis, J. Boviatsis and A. Vradis : Phys. Letters 68 A, 482 (1978).

25. G. Mukhopadhyay, R.K. Kalia and K.S. Singwi : Phys. Rev. Lett. 34, 950 (1975).

26. K. Awa, H. Yasuhara and T. Asai : Solid State Communications (in press).

27. H. Mori : Progr. Theor. Phys. 33, 423 (1965).

28. G. Mukhopadhyay and A. Sjölander : Phys. Rev B 17, 3589 (1978).

29. F. Yoshida, S. Takeno and H. Yasuhara : Progr. Theoret. Phys. 64, 40 (1980).

30. H. De Raedt and B. De Raedt : Phys. Rev. B 18, 2039 (1978).

DYNAMICAL EXCHANGE EFFECTS IN THE DIELECTRIC FUNCTION OF THE ELECTRON GAS°

J.T. Devreese* and F. Brosens°°

Physics Department, Universitaire Instelling Antwerpen, Universiteitsplein 1, B-2610 Wilrijk, Belgium

ABSTRACT

This set of lectures is subdivided into 3 parts.

In part I, some elementary concepts related to the dielectric function of the electron gas are summarized, and related to the single-particle picture and the Hohenberg-Kohn formalism, in order to elucidate the meaning of the local field correction $G(q,\omega)$, describing exchange- and correlation effects in the dielectric function.

In part II, the dynamical exchange decoupling is applied on the equation of motion for the Wigner distribution function. A variational approach is used to treat the resulting integro-differential equation. An explicit expression for $G(q,\omega)$ is obtained as a sixfold integral, which is reduced to a double integral by analytical methods. Given the interest of this type of integral in related problems, the analytical methods are presented in quite some detail.

° The formalism, presented in part II and part III, has been developed in collaboration with L.F. Lemmens.

* And : Institute of Applied Mathematics, R.U.C.A. Groenenborgerlaan 171, B-2020 Antwerpen
And : University of Technology, Eindhoven (The Netherlands).

°° Research Associate of the National Fund for Scientific Research, (Belgium).

Finally, in part III, the numerical results obtained for the dielectric function and the dynamical structure factor are discussed. It is shown that dynamical exchange effects drastically lower the slope of the plasmon dispersion. Several sum rules and consistency requirements are examined. The variational result we derived is finally compared with the proper polarizability to first order in the electron-electron interaction.

PART I - THE JELLIUM MODEL : ELEMENTARY CONCEPTS

A. GROUND STATE ENERGY

Consider a system of N electrons in an arbitrarily large box of volume Ω, and imagine the total charge neutralized by a uniform positive background. The single parameter characterizing the system is then the averaged electron density n : N/Ω. In practice, one often uses the Wigner-Seitz radius r_s, which is the radius (measured in units of the Bohr radius a_o) of a sphere which has a volume equal to the averaged volume per electron : $\frac{1}{n} = \frac{4\pi}{3} r_s^3 a_o^3$. Some values of r_s for real metals, taking the valence electrons into account, are listed in table I.

Table I : Values of r_s and valence z

Metal	z	r_s	Metal	z	r_s	Metal	z	r_s
Li	1	3.24	Be	2	1.87	Cd	2	2.60
Na	1	3.96	Mg	2	2.66	Ra	2	3.70
K	1	4.96	Ca	2	3.26	Al	3	2.07
Rb	1	5.23	Zn	2	2.30	Ga	3	2.18
Cs	1	5.63	Sr	2	3.54	In	3	2.41

If one assumes a uniform electron density, the simplest possible approximation would be to neglect the Coulomb forces. Indeed, if the electrons were not fermions, the Coulomb forces from the electrons and the uniform background would exactly cancel each other. The individual electrons are then described by plane waves, on which periodic boundary conditions are imposed in order to eliminate surface effects.

However, the electrons are fermions, and the exclusion principle prevents two electrons to occupy the same state. Therefore, a given wave vector $\vec{k}$ cannot be occupied by more than two electrons of opposite spin. The maximum occupancy of the states with lowest possible kinetic energy defines the Fermi sphere in wave vector space. Its radius k_F is given by

$$k_F^3 = 3\pi^2 n = \frac{9\pi}{4} \frac{1}{r_s^3 a_o^3}$$

This condition follows by imposing that the number of occupied states equals the number of electrons.

However, due to the exclusion principle this free-particle description breaks down because in this approximation the total electron wave function is assumed to be a product of single-particle states, which is not antisymmetric for interchanging two particles. The exclusion principle is automatically built in if one considers a Slater determinant of single-particle wave functions as a trial wave function. On applying the variational principle of quantum mechanics, one then obtains the Hartree-Fock equation for the single-particle wave functions. In a uniform compensating background, plane waves still are a solution of the Hartree-Fock equations. For the total energy however one now obtains an exchange contribution, apart from the kinetic terms. The single-particle energy for an electron with wave vector $\vec{k}$ then becomes :

$$E_{\vec{k}} = \frac{\hbar^2 k^2}{2m} - \frac{e^2 k_F}{\pi} \left(1 + \frac{1-\eta^2}{2\eta} \ln \left|\frac{1+\eta}{1-\eta}\right|\right)$$

with

$$\eta = k/k_F$$

and the total Hartree-Fock energy of the system equals :

$$E_{HF} = N\left\{\frac{3}{10}\frac{\hbar^2 k_F^2}{m} - \frac{3}{4\pi} e^2 k_F\right\}$$

$$= N\left\{\frac{2.2099}{r_s^2} - \frac{0.9163}{r_s}\right\}\frac{e^2}{2a_o}$$

Instead of considering a Slater determinant of plane waves, where each wave vector is occupied by two electrons of opposite spins, one can examine whether a different spin occupation produces a lower Hartree-Fock energy. This question was already raised in 1929 by Bloch [1] . He found that depending on the density either the above mentioned paramagnetic state, or the ferromagnetic state (i.e. all electrons with parallel spin) has the lowest Hartree-Fock energy. For the ferromagnetic case, one obtains :

$$(E_{HF})_{ferro} = N\left\{\frac{3.508}{r_s^2} - \frac{1.155}{r_s}\right\}\frac{e^2}{2a_o}$$

which is smaller than the paramagnetic Hartree-Fock energy for

$r_s \gtrsim 5.47$. As was shown in table I, most simple metals have a smaller r_s.

Already some 50 years ago, [2] attempts have been made to estimate the correlation energy, i.e. the difference between the exact ground state energy and the Hartree-Fock energy. A well-known result was obtained by Gell'Mann and Brückner [3] :

$$E = \{\frac{2.2099}{r_s^2} - \frac{0.9163}{r_s} - 0.094 + 0.0622 \ln(r_s) + \ldots\} \frac{e^2}{2a_o}$$

which was derived by a diagrammatic expansion. A discussion of this correction and subsequent alternative approaches, lies beyond the scope of the present set of lectures. A concise review can e.g. be found in [4] .

Apart from these studies of the homogeneous electron distribution, one also has examined the possibility of a non-uniform electron gas (in a uniform background). At extremely low density, Wigner [5] has shown that a Hartree-Fock state with electrons localized near lattice positions, has lower energy than the paramagnetic Hartree-Fock state. The result which he obtained served as a basis for interpolating between the low- and high-density regime.

The existence of the Wigner lattice in three-dimensional solids remained a rather academic problem. However, recently experimental evidence was obtained, showing the occurrence of Wigner crystallization in a two-dimensional electron gas [6] . In contrast to a three-dimensional electron gas, two-dimensional electron systems form a Wigner lattice at high areal densities [7] .

A second important deviation from the paramagnetic Hartree-Fock ground state has been found by Overhauser [8,9] . The problem of the occurrence of spin- and charge-density waves is discussed in the lectures of Overhauser in the present volume.

Thus in a simple model like jellium, with well-defined interactions, even the problem of the ground state energy is far from solved. The dynamical properties of this system are still more complex, and no exact solution has been found. We will try to show some of the more recent developments under the assumption of a homogeneous paramagnetic ground state. But first of all we will briefly sketch what can be learned about the response properties of the jellium from a simple perturbative treatment.

B. COLLECTIVE AND SINGLE-PARTICLE EXCITATIONS

Although a detailed calculation of the dielectric response properties of the electron gas requires a many-body treatment, the single-particle picture is physically very transparent in discussing some of the important concepts in linear-response theory. This treatment has also the advantage to be closely connected to the standard pseudopotential theory of simple metals (see e.g. [10,11]).

Starting from a free-electron gas, with the electrons occupying free-electron states $|\vec{k}>$ with energy $E_{\vec{k}} = \hbar^2 k^2/2m$, the interaction between the electrons, with the background charges, and with possible external perturbations is described by some time-dependent operator W, which in general is non-local. This means that the interaction potential depends on the electron state under consideration. For simplicity, we will consider a local interaction, and treat the non-locality in the many-body treatment of part II. Also the spin dependence is not taken into account in the present simple illustrative treatment.

The potential W, seen by the electrons, can be expanded in a Fourier series :

$$W(\vec{r},t) = \sum_{\vec{q},\omega} W_{\vec{q}\omega} \, e^{i(\vec{q}.\vec{r}-\omega^+ t)} \tag{1}$$

where $\omega^+ = \omega + i\delta$ accounts for switching on the interaction adiabatically at $t = -\infty$. Considering then the time-evolution of the electron wave function :

$$|\psi_{\vec{k}}(t)> = |\vec{k}>e^{-iE_{\vec{k}}t/\hbar} + \sum_{\vec{q}}{}' a_{\vec{q}}(\vec{k},t)|\vec{k}+\vec{q}>e^{-iE_{\vec{k}+\vec{q}}t/\hbar} \tag{2}$$

the expansion coefficients $a_{\vec{q}}(\vec{k},t)$ can easily be determined to first order in the perturbation W, from the Schrödinger equation :

$$i\hbar \frac{\partial}{\partial t} |\psi_{\vec{k}}(t)> = (T+W)|\psi_{\vec{k}}(t)> \tag{3}$$

where T is the kinetic energy operator, whose eigenstates are the initial plane waves $|k>$:

$$T|\vec{k}> = E_{\vec{k}}|\vec{k}> \; ; \qquad E_{\vec{k}} = \frac{\hbar^2 k^2}{2m} \tag{4}$$

One readily obtains :

$$\sum_{\vec{q}}{}' \left[i\hbar \frac{\partial}{\partial t} a_{\vec{q}}(\vec{k},t)\right] |\vec{k}+\vec{q}\rangle e^{-iE_{\vec{k}+\vec{q}}t/\hbar} =$$

$$W|k\rangle e^{-iE_{\vec{k}}t/\hbar} + \sum_{\vec{q}}{}' a_{\vec{q}}(\vec{k},t) W |\vec{k}+\vec{q}\rangle e^{-iE_{\vec{k}+\vec{q}}t/\hbar} \tag{5}$$

Multiplication from the left with $\langle\vec{k}+\vec{q}'|$, and omitting the last term in (5) which is of second order, one finds

$$i\hbar \left[\frac{\partial}{\partial t} a_{\vec{q}}(\vec{k},t)\right] e^{-iE_{\vec{k}+\vec{q}}t/\hbar} = \langle\vec{k}+\vec{q}|W|\vec{k}\rangle e^{-iE_{\vec{k}}t/\hbar} \tag{6}$$

Given the boundary condition $a_{\vec{q}}(\vec{k},t = -\infty) = 0$, one obtains

$$i\hbar\, a_{\vec{q}}(\vec{k},t) = \int_{-\infty}^{t} dt\, \langle\vec{k}+\vec{q}|W|\vec{k}\rangle e^{-i(E_{\vec{k}}-E_{\vec{k}+\vec{q}})t/\hbar} \tag{7}$$

With the local approximation (1), the matrix element $\langle\vec{k}+\vec{q}|W|\vec{k}\rangle$ becomes

$$\langle\vec{k}+\vec{q}|W|\vec{k}\rangle = \sum_{\omega} W_{\vec{q}\omega}\, e^{-i\omega^{+}t} \tag{8}$$

and the expansion coefficient $a_{\vec{q}}(\vec{k},t)$ of the wave function to first order in $W_{\vec{q}\omega}$ is given by

$$a_{\vec{q}}(\vec{k},t) = \sum_{\omega} W_{\vec{q}\omega} \frac{e^{-it(\hbar\omega^{+} + E_{\vec{k}} - E_{\vec{k}+\vec{q}})/\hbar}}{\hbar\omega^{+} + E_{\vec{k}} - E_{\vec{k}+\vec{q}}} \tag{9}$$

The electron density $n(\vec{r},t)$ to first order is then readily calculated :

$$n(\vec{r},t) = \sum_{|\vec{k}|<k_F} |\psi_{\vec{k}}(\vec{r},t)|^2 \tag{10}$$

$$= n + \frac{1}{\Omega} \sum_{\vec{q}}{}' \left(e^{i\vec{q}.\vec{r}} \sum_{|k|<k_F} a_{\vec{q}}(\vec{k},t)\, e^{-i(E_{\vec{k}+\vec{q}} - E_{\vec{k}})t/\hbar} + c.c.\right)$$

where n is the averaged density N/Ω. Inserting (9), and performing the Fourier expansion of the density :

$$n(\vec{r},t) = n + \sum_{q\omega}{}' n_{\vec{q}\omega}\, e^{i(\vec{q}.\vec{r}-\omega^{+}t)} \tag{11}$$

the density fluctuations $n_{\vec{q}\omega}$ are given by

$$n_{\vec{q}\omega} = \frac{1}{\Omega} \sum_{|\vec{k}|<k_F} \left[\frac{W_{\vec{q}\omega}}{\hbar\omega^+ + E_{\vec{k}} - E_{\vec{k}+\vec{q}}} + \frac{W^{\star}_{-\vec{q},-\omega}}{-\hbar\omega^+ + E_{\vec{k}} - E_{\vec{k}+\vec{q}}} \right] \tag{12}$$

Because the interaction potential has to be hermitian, it follows that $W^{*}_{-\vec{q},-\omega} = W_{\vec{q}\omega}$. If furthermore in the last term $\vec{k}$ is replaced by $-\vec{k}$, (12) becomes :

$$\frac{4\pi e^2}{q^2} n_{\vec{q}\omega} = -Q_o(q,\omega)\, W_{\vec{q}\omega} \tag{13}$$

where

$$Q_o(q,\omega) = -\frac{4\pi e^2}{q^2} \sum_{|\vec{k}|<k_F} \left\{ \frac{1}{\hbar\omega^+ + E_{\vec{k}} - E_{\vec{k}+\vec{q}}} - \frac{1}{\hbar\omega^+ - E_{\vec{k}} + E_{\vec{k}+\vec{q}}} \right\} \tag{14}$$

The response function $Q_o(q,\omega)$ thus relates the induced Coulomb potential to the averaged potential W felt by the electrons. Replacing the summation by an integral ($\sum_{|\vec{k}|} \rightarrow \frac{2\,\Omega}{(2\pi)^3} \int d^3k$ where a factor of 2 accounts for the spins) the integral can be done analytically, giving

$$\mathrm{Re}\, Q_o(q,\omega) = \frac{1}{2} \frac{k_{TF}^2}{q^2} \left\{ 1 + \frac{m^2}{2k_F q^3 \hbar^2} [4E_F E_q - (E_q + \hbar\omega)^2] \ln \left| \frac{E_q + \hbar q v_F + \hbar\omega}{E_q - \hbar q v_F + \hbar\omega} \right| \right.$$
$$\left. + \frac{m^2}{2k_F q^3 \hbar^2} [4E_F E_q - (E_q - \hbar\omega)^2] \ln \left| \frac{E_q + \hbar q v_F - \hbar\omega}{E_q - \hbar q v_F - \hbar\omega} \right| \right\} \tag{15}$$

$$\mathrm{Im}\, Q_o(q,\omega) = \frac{\pi}{2} \frac{\omega}{q v_F} \frac{k_{FT}^2}{q^2} \qquad \text{if } |E_q - \hbar q v_F| > \hbar\omega > 0$$

$$= \frac{\pi}{4} \frac{k_F}{q} \left[1 - \left(\frac{\hbar\omega - E_q}{\hbar q v_F}\right)^2\right] \frac{k_{FT}^2}{q^2} \qquad \text{if } E_q + \hbar q v_F > \hbar\omega > |E_q - \hbar q v_F|$$

$$= 0 \qquad \text{elsewhere}$$

and

$$\mathrm{Im}\ Q_o(q,-|\omega|) = -\mathrm{Im}\ Q_o(q,|\omega|) \tag{16}$$

In these expressions, k_{FT} is the Thomas-Fermi wave vector, defined by

$$k_{FT}^2 = \frac{4mk_F e^2}{\pi\hbar^2} \tag{17}$$

The imaginary part of this response function $Q_o(q,\omega)$ results from the resonances which are apparent in (14), and which are due to the excitation of electron-hole pairs if the exciting energy $\hbar\omega$ equals the difference in energy between states inside and outside the Fermi sphere. Two possibilities for resonance are possible. The first term in (14) gives rise to the absorption process :

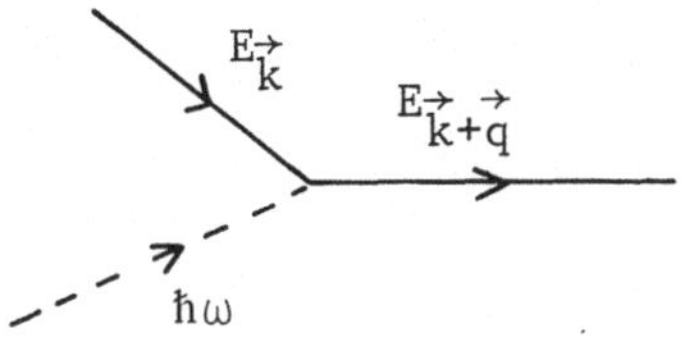

(a)

which implies the condition $\hbar\omega = E_{\vec{k}+\vec{q}} - E_{\vec{k}}$. For $|k| < k_F$ this condition implies

$$-qk_F + \frac{q^2}{2} \leqslant \frac{m\omega}{\hbar} \leqslant qk_F + \frac{q^2}{2}$$

The second term involves the emission process

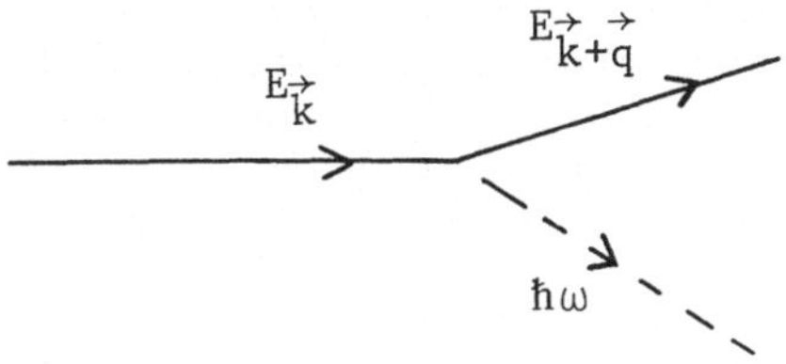

(e)

with the condition $E_{\vec{k}} = \hbar\omega + E_{\vec{k}+\vec{q}}$, or :

$$-qk_F - \frac{q^2}{2} \leqslant \frac{m\omega}{\hbar} \leqslant qk_F - \frac{q^2}{2}$$

These resonances determine the continuum of single particle excitations in the (ω,q) plane.

The function $Q_o(q,\omega)$ forms the basic ingredient of the Lindhard dielectric function [12], which is commonly called the RPA dielectric function, from a famous attempt to describe the electron gas in terms of collective coordinates [13].

This dielectric function is given by

$$\varepsilon^{RPA}(q,\omega) = 1 + Q_o(q,\omega) \tag{18}$$

and is readily derived if one makes the Hartree approximation for the effective potential, i.e. if one assumes that the Coulomb interaction between the electrons governs their dynamical behaviour. In this approximation, the potential W felt by the electrons consists of the induced Coulomb interaction $\frac{4\pi e^2}{q^2} n_{\vec{q}\omega}$, and the applied external potential $W^{ext}_{\vec{q}\omega}$:

$$W_{\vec{q}\omega\,\mathrm{Hartree}} \simeq \frac{4\pi e^2}{q^2} n_{\vec{q}\omega} + W^{ext}_{\vec{q}\omega} \tag{19}$$

Inserting this approximation (19) into the response equation (13), and using the definition of the dielectric function

$$V^{tot}_{\vec{q}\omega} = W^{ext}_{\vec{q}\omega} + \frac{4\pi e^2}{q^2} n_{\vec{q}\omega} = \frac{W^{ext}_{\vec{q}\omega}}{\varepsilon(q,\omega)} \tag{20}$$

one obtains the RPA approximation (18) for the dielectric function. Incidentally, the RPA approximation (19) assumes that the potential W felt by the electrons equals the potential $V^{tot}_{\vec{q}\omega}$ which a test probe outside the medium would see.

However, because the electrons are fermions, also an exchange and correlation interaction has to be introduced to account for the exclusion principle and correlation effects. If we formally represent the exchange and correlation potential to first order in the induced electron density by :

$$V^{xc}_{\vec{q}\omega} = -\frac{4\pi e^2}{q^2} G(q,\omega)\, n_{\vec{q}\omega} \tag{21a}$$

the potential acting upon the electrons becomes :

$$W_{\vec{q}\omega} = \frac{4\pi e^2}{q^2} (1 - G(q,\omega))\, n_{\vec{q}\omega} + W^{ext}_{\vec{q}\omega} \tag{21b}$$

Solving then the response equation (13) combined with (21) for the density, one obtains :

$$\frac{4\pi e^2}{q^2} n_{\vec{q}\omega} = - \frac{Q_o(q,\omega)}{1 + Q_o(q,\omega)[1 - G(q,\omega)]} W^{ext}_{\vec{q}\omega} \tag{22}$$

whereas the potential W becomes :

$$W_{\vec{q}\omega} = \frac{1}{1 + Q_o(q,\omega)[1 - G(q,\omega)]} W^{ext}_{\vec{q}\omega} \tag{23}$$

Inserting (22) into the defining equation (20) for the dielectric function, an expression for the dielectric function is obtained which is commonly used in the electron gas theory :

$$\varepsilon(q,\omega) = 1 + \frac{Q_o(q,\omega)}{1 - G(q,\omega)Q_o(q,\omega)} \tag{24}$$

The function $G(q,\omega)$ in (24) describes the exchange and correlation effects in the dielectric function. Its calculation is one of the fundamental problems in many-body theory.

It should be noted that the potential $V^{tot}_{\vec{q}\omega}$, which defines the dielectric function, is different from the potential acting upon the electrons. The dielectric function $\varepsilon(q,\omega)$ in fact is a test-charge - test-charge dielectric function. In analogy with (20), one could consider (23) as the defining equation for the electron - test-charge dielectric function

$$W_{\vec{q}\omega} = \frac{W^{ext}_{\vec{q}}}{\varepsilon^{et}(q,\omega)} \tag{25}$$

which describes how the potential from an external test charge upon an electron is screened by the surrounding electrons, and which is given by

$$\varepsilon^{et}(q,\omega) = 1 + Q_o(q,\omega)[1 - G(q,\omega)] \tag{26}$$

The distinction between both dielectric functions has been emphasized in the past [15-18] . Recently, a further development has been presented [19] to account for electron-electron screening (taking also the spin dependence into account), and which is discussed by Overhauser in a special lecture.

In the RPA, the function $G(q,\omega)$, describing the exchange and correlation hole, is supposed to be zero, so that no distinction is made between the electron - test-charge and the test-charge - test-charge dielectric functions (26) and (24). Despite the neglect of exchange and correlation, the RPA dielectric function has been widely used because of its simple conceptual basis, and also because of some remarkable results, especially in the longe-wave-length limit, where the long-range Coulomb forces are the dominant interactions.

In the long-wave length limit ($E_q \ll h\omega$; $qv_F \ll \omega$), i.e. at sufficiently high frequency, a standard Taylor series expansion of $Q_o(q,\omega)$ from (15) gives

$$Q_o(q,\omega) \underset{\substack{q\to 0 \\ \omega \gg qv_F}}{\rightarrow} -\frac{\omega_p^2}{\omega^2}\{1+\frac{3}{5}\frac{q^2v_F^2}{\omega^2}+\ldots\} \tag{27}$$

where ω_p is the plasma frequency :

$$\omega_p^2 = \frac{4\pi ne^2}{m} \tag{28}$$

In this limit the RPA dielectric function has a zero for

$$\omega(q) = \omega_p \{1+\frac{3}{10}\frac{q^2v_F^2}{\omega_p^2}+\ldots\} \tag{29}$$

From the definition (20) of the dielectric function, it is intuitively clear that a zero in the dielectric function implies the possibility of density fluctuations in the electron gas, without an external field. In general, as will be discussed later, these collective excitations will be described as well-defined peaks in the dynamical structure factor

$$S(q,\omega) = -\frac{q^2}{4\pi ne^2}\,\mathrm{Im}\,(\frac{1}{\varepsilon(q,\omega)}) \tag{30}$$

but for the RPA, the dispersion of the maxima in $S(q,\omega)$ is given by (29) in the long-wave-length limit. One of the reasons why the RPA is so widely used is precisely the prediction of the plasmon dispersion $\omega_{q=o} = \omega_p$ for $q \to 0$, which is surprisingly accurate. With increasing wave vector however, the Coulomb interaction becomes less dominant, and the plasmon dispersion (29) in RPA increasingly deviates from the experimental observations, as discussed in detail by Schnatterly.

Another interesting relation which derives from RPA, is its connection with the Thomas-Fermi screening [20-21]

$$\varepsilon^{RPA}(q,o) \underset{q\to o}{\to} \varepsilon^{TF}(q) = 1 + \frac{k_{FT}^2}{q^2} \tag{31}$$

which is precisely the result from RPA in the static limit ($\omega = 0$) for sufficiently small wave vector.
The RPA-screened potential from a point charge (ze) is given by

$$\frac{4\pi ze^2}{q^2\varepsilon^{RPA}(q,o)} \underset{q\to o}{\longrightarrow} \frac{4\pi ze^2}{q^2\varepsilon^{TF}(q)} = \frac{4\pi ze^2}{q^2 + k_{FT}^2} \tag{32}$$

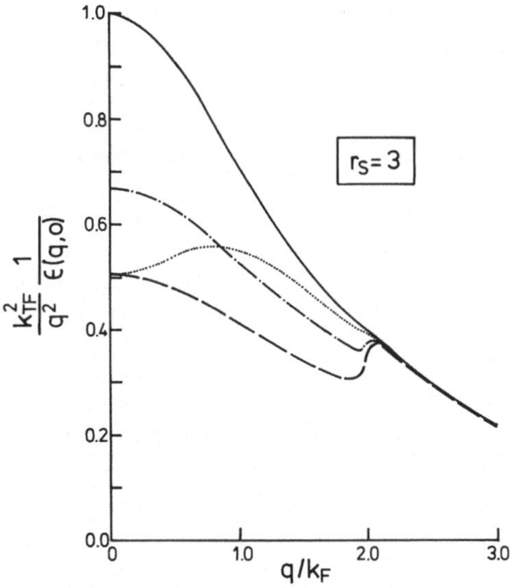

Figure 1. The screened Coulomb potential, screened by the test-charge - test-charge dielectric function of Eq.(24) for r_s = 3 and with G(q,o) from ref.[22] (...), ref. [24] (-·-) and ref. [23](---). The full line is with RPA screening.

which is the Fourier transform of the Yukawa potential $\frac{e^{-k_{FT}r}}{r}$.
The screening, due to the induced density in the electron gas, therefore results in an exponentially decreasing effective potential at large distances (typically in the scale of a unit cell).

The inclusion of exchange effects via the function $G(q,\omega)$ has appreciable effects on the screened interaction potential. This is illustrated in figure 1, where we compare the static effective potential $\frac{4\pi e^2}{q^2\varepsilon(q,o)}$ as obtained from RPA, and from three different approximations for the function $G(q,\omega)$ in the static limit [22-24] for $r_s = 3$. The Hubbard approximation leads to the form

$$G(q,\omega) = G^H(q) = \frac{1}{2}\frac{q^2}{q^2+k_s^2} \tag{33}$$

and was historically one of the first attempts to take exchange effects into account. The approximations of ref. [23] and [24] will be discussed in Part II of these lectures.

For comparison, the effective potential $\frac{4\pi e^2}{q^2\varepsilon^{et}(q,o)}$ which would be seen by an electron (26) is given in figure 2. This potential is the screened pseudopotential resulting from a bare Coulomb potential. It should be noted that both types of screening are quite different if exchange is taken into account.

Finally, in figure 3, we plot the effective electron-electron potential, which is obtained if, instead of an external test charge, a perturbation from an unpolarized electronic distribution $\delta\rho_{q\omega}$ is considered. Because of the exchange and correlation potential, the external pseudopotential $W^{ext}_{\vec{q}\omega}$ in eq. (21) has to be replaced by

$$W^{ext}_{\vec{q}\omega} \rightarrow (1 - G(q,\omega))\,\frac{4\pi e^2}{q^2}\,\delta\rho_{\vec{q}\omega} \tag{34}$$

The induced density $n_{\vec{q}\omega}$ of eq.(13) then becomes :

$$\frac{4\pi e^2}{q^2}\,n_{\vec{q}\omega} = -\,Q_o(q,\omega)\,[1 - G(q,\omega)]\,\frac{4\pi e^2}{q^2}\,(n_{\vec{q}\omega} + \delta\rho_{\vec{q}\omega}) \tag{35}$$

which is readily solved for $n_{\vec{q}\omega}$:

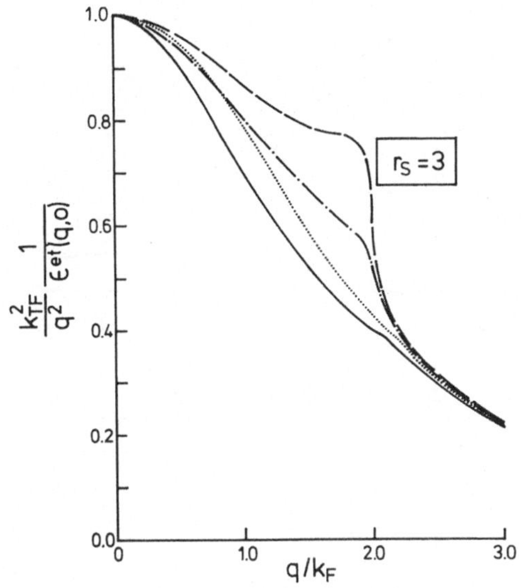

Figure 2. Similar as figure 1, but screened by the electron - test-charge dielectric function of eq.(26).

$$n_{\vec{q}\omega} = \frac{-Q_o(q,\omega)[1-G(q,\omega)]}{1+Q_o(q,\omega)[1-G(q,\omega)]}\,\delta\rho_{\vec{q}\omega} \tag{36}$$

Inserting this induced density, together with the replacement (34) in the effective potential (21), one obtains the effective electron-electron potential :

$$W^{ee}_{q\omega} = \frac{1-G(q,\omega)}{1+Q_o(q,\omega)[1-G(q,\omega)]}\,\frac{4\pi e^2}{q^2}\,\delta\rho_{\vec{q}\omega} \tag{37}$$

In analogy with the defining equations (20) and (25) for the test-charge - test-charge and the electron - test-charge dielectric functions respectively, (37) can be considered as the definition of an electron-electron dielectric function

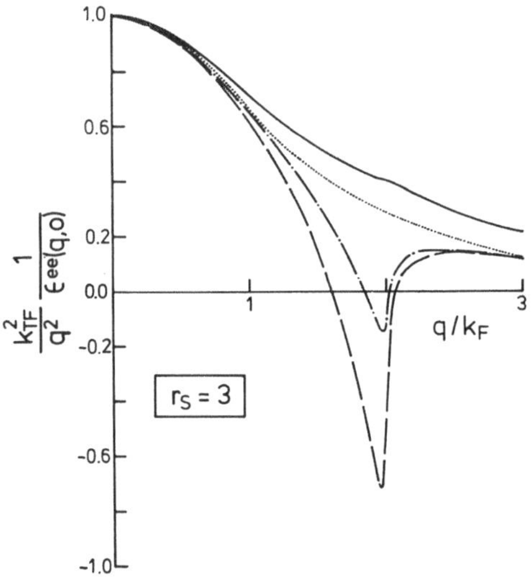

Figure 3. Similar as in figure 1, but screened by the electron-electron dielctric function of eq.(38).

$$\varepsilon^{ee}(q,\omega) = \frac{1 + Q_o(q,\omega)\,[1 - G(q,\omega)]}{1 - G(q,\omega)} = Q_o(q,\omega) + \frac{1}{1 - G(q,\omega)} \tag{38}$$

The electron-electron potential (37) is half the sum of the potentials $\tilde{V}_{\uparrow\uparrow}$ and $V_{\uparrow\downarrow}$ obtained by Kukkonen and Overhauser [19] by treating the spins separetely. This is to be expected since we consider an unpolarized electronic perturbation.

The most remarkable fact of the screening via (38) is that it leads to an attractive static interaction near $q \simeq 2k_F$ for both the approximations [24] of Holas, Aravind and Singwi, and the approximation [23] which we developed. The reason for this behaviour is that the function $G(q,\omega)$ exceeds unity in these approximations near $q \simeq 2k_F$.

The different potentials sketched here will be discussed in a forthcoming paper [25], whereas the function $G(q,\omega)$ will be studied in part II of these lectures.

C. EFFECTIVE POTENTIAL AND DENSITY-FUNCTIONAL FORMALISM

In the static limit, the single-particle formulation of the preceding section can be studied in the context of an alternative formulation of the electron gas problem. In a famous paper, Hohenberg and Kohn [26] considered an interacting electron gas under the influence of an external potential $W^{ext}(\vec{r})$ and with a non-degenerate ground state. They proved that the potential $W^{ext}(\vec{r})$ is a unique functional (to within a constant) of the density $n(r)$, and therefore that the full many-particle ground state is a unique functional of the density.

Defining then the energy functional

$$E[n] \equiv \int d^3r\, W^{ext}(\vec{r})n(\vec{r}) + F[n] \tag{39}$$

where $F[n]$ is a universal functional for the kinetic and interaction energy, valid for any number of particles and any external potential, they proved that $E[n]$ has as its minimum value the correct ground state energy E for the correct density $n(\vec{r})$. (As noted by Hohenberg and Kohn, they could not prove whether an arbitrary positive density distribution can be realized by some external potential).

The problem of determining the ground-state energy and density in a given external potential therefore requires the minimization of a functional, which is proven to exist, but is not known. Hohenberg and Kohn could express this function $F[n]$ in terms of the correlation energy and the polarizability of a uniform electron gas in the two situations of an almost constant density and of a slowly varying density.

In a quasi-uniform electron gas :

$$n(\vec{r}) \simeq n_o + \tilde{n}(\vec{r}) \qquad \tilde{n}(\vec{r}) \ll n_o \tag{40}$$

and under the influence of a weak external potential $W^{ext}(\vec{r})$, a formal expansion of the energy functional with respect to the density gives :

$$E[n(r)] = E_o[n_o] + E_W + \int d^3r\, W^{ext}(\vec{r})\, \tilde{n}(\vec{r}) + \int d^3r \int d^3r' \left[\frac{1}{2}\frac{e^2}{|r-r'|} + K_{n_o}(\vec{r}-\vec{r}')\right] \tilde{n}(\vec{r})\, \tilde{n}(\vec{r}') + \ldots \tag{41}$$

where $K_{n_o}(\vec{r}-\vec{r}')$ stands for the kinetic, exchange and correlation contributions from the inhomogeneity, $E_o[n_o]$ is the ground state energy of the homogeneous electron gas, and E_W represents possible energy contributions not related to the electron system. The Fourier expansion

$$\tilde{n}(\vec{r}) = \sum_{\vec{q}} n_{\vec{q}}\, e^{i\vec{q}.\vec{r}} \tag{42}$$

and similarly for the external potential $W^{ext}(\vec{r})$ and the kernel $K_{n_o}(\vec{r}-\vec{r}')$, transforms (41) into :

$$E[n(\vec{r})] = E_o[n_o] + E_W + \Omega \sum_{\vec{q}} W^{ext}_{\vec{q}}\, n_{\vec{q}} + \frac{1}{2}\Omega \sum_{\vec{q}} \frac{4\pi e^2}{q^2} |n_{\vec{q}}|^2 + \Omega \sum_{\vec{q}}{}' K_q |n_{\vec{q}}|^2 + \ldots \tag{43}$$

As the functional $E[n(\vec{r})]$ has to assume its minimum value for the correct density, one obtains to first order in the density fluctuations :

$$W^{ext}_{\vec{q}} + \frac{4\pi e^2}{q^2} n_{\vec{q}} + 2\, K_{\vec{q}} n_{\vec{q}} = 0 \tag{44}$$

Using then the defining equation (20) for the dielectric function to express the induced density in terms of the dielectric function and the external potential, (44) implies :

$$K_{\vec{q}} = \frac{2\pi e^2}{q^2} \frac{1}{\varepsilon(q) - 1} \tag{45}$$

as derived by Hohenberg and Kohn from first-order perturbation theory. The exchange and correlation contribution to the kernel K_q is obtained by subtracting the RPA contribution from (45) :

$$K_q^{xc} = \frac{2\pi e^2}{q^2} \{\frac{1}{\varepsilon(q) - 1} - \frac{1}{\varepsilon^{RPA}(q) - 1}\} \tag{46}$$

because in the RPA the exchange and correlation interaction is neglected between the density fluctuations.

Using then the explicit expressions (18) and (24) for $\varepsilon^{RPA}(q)$ and $\varepsilon(q)$, K^{xc} is given by :

$$K_{\vec{q}}^{xc} = - \frac{2\pi e^2}{q^2} G(q) \tag{47}$$

where $G(q)$ denotes the static limit of $G(q,\omega)$, and similarly for $\varepsilon(q)$ and $n_{\vec{q}}$. Inserting (47) in (43), the exchange and correlation energy, due to the density fluctuations, becomes :

$$E^{xc}[\tilde{n}(\vec{r})] = - \Omega \sum_{\vec{q}}{}' \frac{2\pi e^2}{q^2} G(q)|n_{\vec{q}}|^2 \tag{48}$$

implying an exchange and correlation interaction

$$V_{\vec{q}}^{xc} = - \frac{4\pi e^2}{q^2} G(q)\, n_{\vec{q}} \tag{49}$$

This is precisely the static limit of the exchange and correlation potential (21) introduced ad hoc in the time-dependent perturbation theory of the previous section. Thus for a quasi-uniform electron gas, the single-particle treatment with the local exchange and correlation potential (21) is consistent with the Hohenberg-Kohn formulation for the quasi-homogeneous many-electron system.

It should be noted that the total energy, as given in (43) to second order in the density fluctuations, can be written in terms of the external potential W^{ext}. Using (45) for $K_{\vec{q}}$ and (20) to eliminate the density fluctuations $n_{\vec{q}}$, one obtains after some algebra :

$$E[n] = E_o[n_o] + E_W - \frac{\Omega}{2} \sum_{\vec{q}} |W_{\vec{q}}^{ext}|^2 \frac{q^2}{4\pi e^2} \frac{\varepsilon(q) - 1}{\varepsilon(q)} \tag{50}$$

If for $W^{ext}(\vec{r})$ one considers a local unscreened pseudopotential from the pseudo-ions at positions $\vec{R}_j$:

$$W^{ext}(\vec{r}) = \sum_j w^\circ(\vec{r}-\vec{R}_j) \tag{51}$$

eq.(50) can be rewritten as :

$$E[n] = E_o[n_o] + E_W + \frac{\Omega}{2}\sum_{\vec{q}} |S(\vec{q})|^2 F(q) \tag{52}$$

where F(q) is the energy-wavenumber characteristic :

$$F(q) = |w^\rho_q|^2 \frac{q^2}{4\pi e^2}\{\frac{1}{\varepsilon(q)} - 1\} \tag{53}$$

and where $S(\vec{q})$ is the geometrical structure factor of the lattice :

$$S(\vec{q}) = \frac{1}{N_i}\sum_j e^{-i\vec{q}.\vec{R}_j} \tag{54}$$

with N_i denoting the number of pseudo-ions, and $\vec{R}_j$ their positions.

The energy expression (52) is a standard result of local pseudopotential theory [10-11], but exchange and correlation is accounted for via ε(q), if G(q) would be known. This form (53), including exchange and correlation, has also been obtained by Hedin and Lundqvist [27] in an alternative context. They applied the Kohn-Sham [28] one-electron scheme, as derived from the Hohenberg-Kohn theorem, with the effective electron potential (23), i.e. related to the electron - test-charge dielectric function (26).

Thus the exchange and correlation potential (21), as seen by an electron in the presence of a (weak) density fluctuation $n_{\vec{q}}$, leads to a consistent description if used in a one-electron picture or in the framework of the density-functional formalism.

D. FUNDAMENTAL PROPERTIES OF THE DIELECTRIC FUNCTION

In the previous sections, we essentially concentrated on the concept of the dielectric function and its relation with the exchange and correlation interaction in the single-particle picture, emphasizing the concept of the effective interaction.

But the knowledge of the dielectric function is important for the study of many physical properties of simple metals (see e.g. ref.[29]). Without attempting to be exhaustive, we briefly mention some of them.

The compressibility sum rule relates the long wave-length limit of the static dielectric function to the compressibility of the interacting electron gas :

$$\varepsilon(q,o) \underset{q\to 0}{\to} 1 + \frac{k_{FT}^2}{q^2}\frac{\kappa}{\kappa_o} \tag{55}$$

where κ_o is the compressibility of the free-particle model. As pointed out in [30], the compressibility sum rule implies that :

$$G(q,o) \underset{q\to 0}{\to} \frac{q^2}{4\pi e^2 n^2}\left(\frac{1}{\kappa_o} - \frac{1}{\kappa}\right) \tag{56}$$

If the energy is calculated in the Hartree-Fock approximation, the resulting compressibility can be inserted in (56), and one obtains for the function G(q,o) in the Hartree-Fock approximation :

$$G_{HF}(q,o) \underset{q\to 0}{\to} \frac{1}{4}\left(\frac{q}{k_F}\right)^2 \tag{57}$$

Another interesting property has been derived by Niklasson [31] :

$$\lim_{q\to\infty} G(q,o) = \frac{2}{3}\,(1 - g(o)) \tag{58}$$

where g(r) is the pair correlation function [32], defined as :

$$g(r) = \frac{1}{nN}\sum_{i\neq j}^{N} \delta(\vec{r}_i - \vec{r}_j - \vec{r}) \tag{59}$$

where $\vec{r}_i$ and $\vec{r}_j$ denote the electron positions.

The static structure factor S(q), essentially defined as the Fourier transform of the pair-correlation function [29] :

$$g(r) - 1 = \frac{1}{n}\frac{1}{(2\pi)^3}\int d^3q\,[S(q) - 1]\; e^{i\vec{q}.\vec{r}} \tag{60}$$

is related to the dynamical dielectric function via the fluctuation-dissipation theorem [33]

$$S(q) = -\frac{q^2}{4\pi e^2}\frac{\hbar}{\pi n}\int_0^\infty d\omega \, \mathrm{Im}\frac{1}{\varepsilon(q,\omega)} \tag{61}$$

From general convergence considerations for the Fourier transform (60), it was shown in [34] that the derivative g'(r) of the pair correlation function at the origin has to satisfy the relation :

$$g'(o) = -\frac{1}{8\pi n}\lim_{q\to\infty} q^4 [S(q) - 1] \tag{62}$$

Because the Coulomb interaction between two electrons becomes dominant if two electrons come sufficiently close to each other, the probability distribution at short distance is determined by this two-particle problem. On the basis of this physical argument, it was shown [34] that

$$g'(o) = \frac{1}{a_o} g(o) \tag{63}$$

Niklasson [31] derived this relation on a mathematical basis. Eliminating then g'(o) from (62) and (63), one obtains :

$$g(o) = -\frac{a_o}{8\pi n}\lim_{q\to\infty} q^4 [S(q) - 1] \tag{64}$$

The relation (64) requires thus the dynamical dielectric function, taking (61) into account, whereas the relation (58) only requires the static limit.

In order to fulfill the conditions (64) and (58) simultaneously the exchange and correlation effects in $G(q,\omega)$ have to be included dynamically. Any static approximation $G_s(q)$ for $G(q,\omega)$ necessarily violates one of both criteria, since with a static approximation (64) would lead to [34] :

$$\lim_{q\to\infty} G_s(q) = 1 - g(o) \tag{65}$$

which contradicts (58).

Apart from the compressibility sum rule and the pair correlation function, we also mention some relations involving the frequency moments of the inverse dielectric function :

$$M_n(q) \equiv \int_{-\infty}^{\infty} \frac{d\omega}{\pi} \omega^n \, \mathrm{Im} \frac{1}{\varepsilon(q,\omega)} \tag{66}$$

It is well known that Im $\varepsilon^{-1}(q,\omega)$ is an odd function of frequency from causality arguments [33,35] . Therefore all even moments are identically zero. The causality mathematically means that the response function $[\varepsilon^{-1}(q,\omega) - 1]$ is an analytical function of the complex variable ω for Im $\omega > 0$, and consequently has to satisfy the Kramers-Kronig relation :

$$\frac{1}{\varepsilon(q,\omega)} - 1 = \int_{-\infty}^{\infty} \frac{d\omega}{\pi} \frac{1}{\omega' - \omega - io^+} \, \mathrm{Im} \frac{1}{\varepsilon(q,\omega')} \tag{67}$$

At sufficiently high frequency, one thus obtains the expansion :

$$\frac{1}{\varepsilon(q,\omega)} - 1 = - \frac{M_1(q)}{\omega^2} - \frac{M_3(q)}{\omega^4} - 0\left(\frac{1}{\omega^6}\right) \tag{68}$$

Both the moments $M_1(q)$ and $M_3(q)$ have been evaluated. The first moment is given by [36] :

$$M_1(q) = -\omega_p^2 \tag{69}$$

The third frequency-moment is more complicated [37] :

$$M_3(q) = -\omega_p^2 \left\{ \left(\frac{\hbar q^2}{2m}\right)^2 + \frac{2q^2}{m} \langle KE \rangle + \omega_p^2 + I(q) \right\} \tag{70a}$$

$$I(q) = \frac{4\pi e^2}{m} \int \frac{d^3 q'}{(2\pi)^3} \left[\frac{|\hat{q}\cdot(\vec{q}+\vec{q}')|^2}{|q+q'|^2} - \frac{|\hat{q}\cdot\vec{q}'|^2}{q^2} \, [S(q') - 1] \right] \tag{70b}$$

where <KE> denotes the kinetic energy per electron of the interacting electron gas. From the sum rule (70), the following relations have been derived [38] for the high-frequency limit of $G(q\,\omega)$:

$$G(q,\infty) \underset{q\to\infty}{\to} - \frac{2q^2}{m\omega_p^2} \{\langle KE \rangle - \langle KE \rangle_o + \frac{2}{3}(1 - g(o)\} \tag{71}$$

$$G(q,\infty) \underset{q\to o}{\to} - \frac{2q^2}{m\omega_p^2} \{\langle KE \rangle - \langle KE \rangle_o + \frac{2}{15} \langle V \rangle\} \tag{72}$$

where <V> represents the potential energy per particle and $\langle\ \rangle_o$ denotes expectation values for the non-interacting electron gas.

Finally, we mention that the interaction energy for electrons in the ground state can be written as :

$$E^{int} = -\frac{2\hbar^2}{me^4 r_s^2} \int_o^{r_s} r_s dr_s \sum_{\vec{q}} \{\frac{2\pi e^2}{q^2 \Omega} + \frac{1}{N} \int_o^{\infty} \frac{d\omega}{2\pi} \operatorname{Im} \varepsilon^{-1}(q,\omega)\} \tag{73}$$

as first derived by Hubbard [22] using diagrammatic techniques, and by Nozières and Pines from linear response theory.

E. DYNAMICAL EXCHANGE EFFECTS IN THE DIELECTRIC FUNCTION

In the previous sections, the dielectric function remained a rather formal quantity, in the sense that the function $G(q,\omega)$, which accounts for exchange and correlation, was not specified. In part II and III we show in quite some detail our actual calculation of $G(q,\omega)$, in which we attempted to include exchange effects dynamically [23].

The derivation was based on the equation of motion for the Wigner distribution function, which was linearized in the externally applied field, and where the two-particle density operator was decoupled in the equation of motion according to the Hartree-Fock prescription. The resulting integro-differential equation for the Wigner distribution function was approximately solved via a variational procedure, which leads to a complicated sixfold integral for $G(q,\omega)$. This integral was reduced to a tractable double integral by two independent analytical methods, which are shown in detail. A summary of the method and the results has been published previously [23].

It should be emphasized that Holas, Aravind and Singwi [24] obtained a strongly related expression $G^p(q,\omega)$ by diagrammatic techniques. They also developed a third method to evaluate the same sixfold integral. If $G^{var}(q,\omega)$ denotes our variational result, it is related to the approach of [24] via :

$$G^{var}(q,\omega) = \frac{G^p(q,\omega)}{1 - G^p(q,\omega) Q_o(q,\omega)} \tag{74}$$

The function $G^p(q,\omega)$ can be obtained from the integro-differential equation for the Wigner distribution function by an iteration to first order in the deviation of the Lindhard solution. Except for the compressibility sum rule (from which $G^{p}(q,\omega)$ does not give the Hartree-Fock compressibility) both the functions G^{p} and G^{var} satisfy the same tested sum rules and consistency relations. Qualitatively, both functions are very similar in the range of metallic densities.

A detailed comparison between our variational approach and the perturbative result of ref.[24] will be given in part III, after the explicit derivation of $G(q,\omega)$ in the framework of the dynamical exchange decoupling.

REFERENCES

[1] F. Bloch, Z. Phys. 55, 545 (1929).

[2] E.P. Wigner and F. Seitz, Phys. Rev. 43, 804 (1933); 46, 509 (1934).

[3] M. Gell'Mann and K. Brueckner, Phys. Rev. 106, 364 (1957).

[4] G.D. Mahan, Many-particle physics, Plenum Press, New York, 1981.

[5] E.P. Wigner, Trans. Faraday Soc. 34, 678 (1938).

[6] C.C. Grimes and G. Adams, Phys. Rev. Lett. 42, 795 (1979).

[7] R.S. Crandall and R. Williams, Phys. Lett. 34A, 404 (1971).

[8] A.W. Overhauser, Phys. Rev. Lett. 4, 462 (1960).

[9] A.W. Overhauser, Adv. Phys. 27, 343 (1978) and references cited therein.

[10] W.A. Harrison, Pseudopotentials in the theory of simple metals, Benjamin Inc., New York, 1966.

[11] V. Heine, M.L. Cohen and D. Weaire, Solid State Phys. 24, Eds. F. Seitz and D. Turnbull, Academic Press, New York, 1964.

[12] J. Lindhard, Kgl. Dan. Videnskab. Selsk. Mat. Fys. Medd. 28 (8) (1954).

[13] D. Pines and D. Bohm, Phys. Rev. 85, 338 (1952); 92, 609 (1953).

[14] P. Nozières and D. Pines, Phys. Rev. 111, 442 (1958).

[15] L. Kleinman, Phys. Rev. 160, 585 (1967); 172, 383 (1968).

[16] R.W. Shaw, Jr., J. Phys. C : Solid State Phys. 3, 1140 (1970).

[17] R. Lobo, Phys. Rev. B8, 5348 (1973).

[18] L. Hedin and B.I. Lundqvist, J. Phys. C : Solid State Phys. 4, 2064 (1971).

[19] C.A. Kukkonen and A.W. Overhauser, Phys. Rev. B20, 550 (1979).

[20] L.H. Thomas, Proc. Cambridge Philos. Soc. 23, 542 (1927).

[21] E. Fermi, Z. Phys. 48, 73 (1928).

[22] J. Hubbard, Proc. Roy. Soc. (London) A243, 336 (1957).

[23] F. Brosens, L.F. Lemmens and J.T. Devreese, Phys. Stat. Sol. (b) 74, 45 (1976); 81, 551 (1977).
F. Brosens, J.T. Devreese and L.F. Lemmens, Phys. Stat. Sol. (b) 80, 99 (1979); Phys. Rev. B21, 1363 (1980).
J.T. Devreese, F. Brosens, L.F. Lemmens, Phys. Stat. Sol. (b) 91, 349 (1979); Phys. Rev. B21, 1349 (1980).

[24] A. Holas, P.K. Aravind and K.S. Singwi, Phys. Rev. B20, 4912 (1979).

[25] F. Brosens and J.T. Devreese, to be published.

[26] P. Hohenberg and W. Kohn, Phys.Rev. 136, B864 (1964).

[27] L. Hedin and B.I. Lundqvist, J. Phys. C : Solid State Phys. 4, 2064 (1971).

[28] W. Kohn and L.J. Sham, Phys. Rev. 140, A1133 (1965).

[29] D. Pines, Elementary Excitations in Solids, W.A. Benjamin, New York, 1963.
P. Nozières, Theory of Interacting Fermi Systems, W.A. Benjamin, New York, 1963.
P. Nozières and D. Pines, Theory of Quantum Liquids, W.A. Benjamin, New York, 1966.
L. Hedin and S. Lundqvist, Solid State Physics 23, 1969.

[30] D.J.W. Geldart and S.H. Vosko, Canad. J. Phys. 44, 2137 (1966).

[31] G. Niklasson, Phys. Rev. B10, 3052 (1974).

[32] A.J. Glick and R.A. Ferrell, Ann.of Phys. 11, 359 (1960).

[33] P. Nozières and D. Pines, Phys. Rev. 109, 1009 (1958); 111, 442 (1958); Il Nuovo Cimento 9, 470 (1958).

[34] J.C. Kimball, Phys. Rev. A7, 1648 (1973); J. Phys. A : Math. Gen. 8, 1513 (1975); Phys. Rev. B14, 2371 (1976).

[35] L.P. Kadanoff and P.C. Martin, Ann. Phys. (N.Y.) 24, 419 (1963).

[36] R.A. Ferrell, Phys. Rev. 107, 450 (1957).

[37] R.D. Puff, Phys. Rev. 137, A406 (1967).

[38] A.A. Kugler, J. Stat. Phys. 12, 35 (1975).

PART II - DIELECTRIC FUNCTION OF THE ELECTRON GAS WITH DYNAMICAL EXCHANGE DECOUPLING

A. INTRODUCTION

Many properties of simple metals can be described and calculated from the dielectric function. In studying the effects, that are essentially due to the electron-electron interactions, the jellium model is widely used. In this model, the discrete ion lattice is supposed not to have essential influence on the dielectric function $\varepsilon(q,\omega)$, and is replaced by a uniform positive background.

In the well-known random-phase approximation (RPA) [1], $\varepsilon(q,\omega)$ was first calculated by Lindhard [2], who studied the motion of the electrons in the presence of an electromagnetic field, under the assumption that this motion is governed by classical laws, and only restricted by the Pauli exclusion principle in the initial state.

Because the RPA only takes into account the long range interaction of the classical Hartree potential, a satisfactory description of the long wavelength collective excitations is obtained. However, due to the neglect of the exchange and correlation interactions, the RPA insufficiently describes short range effects, which e.g. is reflected in a negative pair correlation function from RPA for small interparticle distances [3].

By summing up several exchange diagrams, Hubbard [4] introduced a first correction to the RPA, in the form of a frequency independent function $G(q)$. Several improvements on this local field correction, going beyond Hubbard's expansion, are proposed [5-19].

Several attempts have also been made to include the exchange effects dynamically [20-28], leading to a frequency dependent local field correction $G(q,\omega)$, and it has been shown [29] that an internally consistent theory of the electron gas cannot be obtained by neglecting this frequency dependence. However, most explicit calculations of $G(q,\omega)$ were restricted to the static limit $\omega = 0$ and to a few limiting cases. But even this fragmentary information from approximate exchange treatments, already indicates appreciable effects of the exchange interaction on the dielectric response properties of the electron gas. These effects are indeed to be expected from the interesting general properties of $G(q,\omega)$, derived in [26].

It therefore is interesting to evaluate the exchange influence on the dielectric function explicitly, without further approximations, and including the full wave vector and frequency dependence. The outline of such an attempt has been published previously [27], where the present authors derived general expressions for the transverse and longitudinal dielectric function of jellium, including exchange. The solution obtained was still rather formal, because $G(q,\omega)$ was given as a sixfold integral, that only could be solved analytically in a few limiting cases.

This formal solution was obtained by considering the equation of motion for the Wigner distribution function, and where dynamical exchange effects were included by making the exchange decoupling in the equation of motion. By a variational procedure an approximate solution of the resulting integro-differential equation was obtained, that rigorously satisfies the equation of motion for the charge and current density. Various limits were also studied.

The formal expression for $G(q,\omega)$ turned out to be the same as the one obtained in [23], by variationally solving the integral equation for the irreducible vertex function with linear exchange processes.

The same formal expression for $G(q,\omega)$ was afterwards also obtained in [28] from the equation of motion for the double-time retarded commutator of the charge density fluctuation operators, and where it was shown that all frequency moments are conserved to infinite order by the decoupling technique used.

The static limit $G(q,o)$ has been evaluated, and shows a sharp peak near $q = 2\ k_F$ [28,30], in contrast to earlier theories. In [31], we reported on some preliminary results for $\varepsilon(q,\omega)$ including dynamical exchange effects from $G(q,\omega)$, and in [32] we have shown that the high frequency behaviour of $G(q,\omega)$ is consistent with its static limit, in the sense that both limits lead to the same value 1/2 of the pair correlation function at the origin. Furthermore, we have shown [33] that the excitation spectrum of the electron gas, obtained with the dynamical exchange decoupling, agrees rather well with the experimental data in Al [34].

In section B of the present paper we give an extensive derivation of the dielectric function with dynamical exchange decoupling. The starting point is the equation of motion for the Wigner distribution function, which is decoupled according to the Hartree-Fock prescription, and in which only terms to first order in the applied external potential are retained. The resulting integro-differential equation is treated with a variational principle, which results in an explicit expression for $G(q,\omega)$ as a sixfold integral. Our expression for $G(q,\omega)$ as a function of q/k_F and $\hbar\omega/E_F$ turns out to be independent of the averaged electron density. The comparison with other approximations is briefly discussed, but for a more detailed comparison with the related derivation of Holas, Aravind and Singwi [40], we refer to part III of this paper.

The last section of part II is devoted to the explicit calculation of $G(q,\omega)$. For most of the mathematical details however, we refer to [41-42], where one of the methods to express $G(q,\omega)$ as a sixfold integral is given rather extensively. In Appendix A, we present an alternative evaluation by elementary methods.

B. DERIVATION OF G(q,ω) VIA DYNAMICAL EXCHANGE DECOUPLING

§ 1. Dynamical Exchange Decoupling in the Equation of Motion for the Wigner Distribution Function

Consider a gas of electrons, interacting with some scalar potential $\varphi(\vec{r},t)$, which in the absence of any magnetic or spin-spin interaction is described by the Hamiltonian :

$$H = \frac{\hbar^2}{2m} \sum_{\sigma} \int d^3r\, \overline{\nabla}\psi^+(\vec{r}).\overline{\nabla}\psi(\vec{r}) + \sum_{\sigma} \int d^3r\, \psi^+_{\sigma}(\vec{r})\, e\varphi(\vec{r},t)\psi_{\sigma}(\vec{r})$$
$$+ \frac{1}{2} \sum_{\sigma\sigma'} \int d^3r \int d^3r'\, \psi^+_{\sigma'}(\vec{r}')\, \psi^+_{\sigma}(\vec{r})\, \frac{e^2}{|\vec{r}-\vec{r}'|}\psi_{\sigma}(\vec{r})\psi_{\sigma'}(\vec{r}') \tag{1}$$

where $\psi^+_{\sigma}(\vec{r})$ and $\psi_{\sigma}(\vec{r})$ are the creation and annihilation field operators for particles of spin σ, which obey the standard anti-commutation rules for fermion fields.

Defining the Wigner distribution function :

$$f_{\sigma}(\vec{p},\vec{R},t) = \frac{1}{(2\pi\hbar)^3} \int d^3r\, e^{-i\vec{p}.\vec{r}/\hbar} \langle\psi^+_{\sigma}(\vec{R}-\frac{\vec{r}}{2})\, \psi_{\sigma}(\vec{R}+\frac{\vec{r}}{2})\rangle_t \tag{2}$$

where the brackets $\langle\ \rangle_t$ denote the expectation value at time t, its equation of motion can be obtained from the Liouville equation, by calculating the commutator of the operator $\psi^+_{\sigma}(\vec{R}-\frac{\vec{r}}{2})\, \psi_{\sigma}(\vec{R}+\frac{\vec{r}}{2})$ with the Hamiltonian (1). This equation of motion is the quantum mechanical analogue of the classical Boltzmann equation :

$$\frac{d}{dt} f_\sigma(\vec{p},\vec{R},t) = -\frac{\vec{p}.\vec{\nabla}}{m} f_\sigma(\vec{p},\vec{R},t) + \frac{i}{\hbar}\int d^3k\, e^{i.\vec{k}.\vec{R}}\, U_{\vec{k}}(t)$$

$$[f_\sigma(\vec{p} + \frac{\hbar\,\vec{k}}{2}, \vec{R},t) - f_\sigma(\vec{p} - \frac{\hbar\,\vec{k}}{2}, \vec{R},t)] - \frac{i}{\hbar} X_\sigma^{tot}(\vec{p},\vec{R},t) \tag{3a}$$

where $U_{\vec{k}}(t)$ is the spatial Fourier transform of the Hartree potential :

$$U_{\vec{k}}(t) = \frac{1}{(2\pi)^3}\int d^3r\, e^{i\vec{k}.\vec{r}}[e\varphi(\vec{r},t) + \sum_\sigma \int d^3x \langle\psi_\sigma^+(\vec{x})\psi_\sigma(\vec{x})\rangle_t \frac{e^2}{|\vec{r}-\vec{x}|}] \tag{3b}$$

and where $X_\sigma^{tot}(\vec{p},\vec{R},t)$ is a term arising from exchange and correlation effects :

$$X_\sigma^{tot}(\vec{p},\vec{R},t) = \frac{1}{(2\pi\hbar)^3}\int d^3r\, e^{-i\vec{p}.\vec{r}/\hbar}\int d^3r'\, \frac{1}{2}[\frac{e^2}{|\vec{R}+\vec{r}/2-\vec{r}'|} - \frac{e^2}{|\vec{R}-\vec{r}/2-\vec{r}'|}]$$

$$\langle\{\psi_\sigma^+(\vec{R}-\vec{r}/2)\psi_\sigma(\vec{R}+\vec{r}/2), \sum_{\sigma'}\psi_{\sigma'}^+(r')\psi_{\sigma'}(\vec{r}) - \sum_{\sigma'}\langle\psi_{\sigma'}^+(\vec{r}')\psi_{\sigma'}(\vec{r}')\rangle_t\}\rangle \tag{3c}$$

where $\{A,B\}$ denotes the anticommutator AB + BA.

The equation of motion (3) in phase space is analogous to the classical form of the collisonless Boltzmann equation. This relation to classical statistical mechanics has been elaborated in [35] . A very familiar classical result can be obtained by integrating (3) over all momenta. Noting that the electron density n(R,t) and the current density $\vec{j}(\vec{R},t)$ in the absence of a magnetic field are given by :

$$n(\vec{R},t) \equiv \sum_\sigma \langle\psi_\sigma^+(\vec{R})\psi_\sigma(\vec{R})\rangle_t = \sum_\sigma \int d^3p\, f_\sigma(\vec{p},\vec{R},t) \tag{4}$$

$$\vec{j}(\vec{R},t) = \sum_\sigma \int d^3p\, \frac{\vec{p}}{m} f_\sigma(\vec{p},\vec{R},t) \tag{5}$$

the continuity equation follows directly from (3), because both the Hartree and the exchange and correlation contribution in (3) yield zero by integration over all momenta. Various other relationships are discussed in [35] . In fact, eq. (3) has precisely the form of the first member of the BBKGY hierarchy, because the exchange and correlation term X_σ^{tot} can be expressed in terms of the two-particle distribution function. In a similar manner, the time dependence of the (N-1)-particle distribution function is shifted by the equation of motion to the N-particle distribution function. Therefore, in order to include all exchange and correlation effects, one has to solve an infinite set of differential equations.

However, the exchange contribution $X_\sigma(\vec{p},\vec{R},t)$ to the exchange and correlation term $X_\sigma^{tot}(\vec{p},\vec{R},t)$ can be separated out by the decoupling :

$$<\psi_1^+ \psi_2^+ \psi_3 \psi_4> \approx <\psi_1^+ \psi_4><\psi_2^+\psi_3> - <\psi_1^+ \psi_3><\psi_2^+ \psi_4> \tag{5}$$

This decoupling in the equation of motion dynamically preserves the Pauli exclusion principle for fermions, but neglects correlation effects, arising from the rapid oscillations, superimposed on the mean movement of the particles.

Applying this decoupling in (3c), and expressing the resulting expectation values of the form $<\psi_\sigma^+(\vec{r}_1)\psi_\sigma(\vec{r}_2)>_t$ in their Fourier transform $f_\sigma(p, \frac{\vec{r}_1+\vec{r}_2}{2}, t)$ according to (2), one obtains :

$$X_\sigma(\vec{p},\vec{R},t) = -\frac{1}{(2\pi\hbar)^3}\int d^3q \frac{4\pi e^2}{q^2}\int d^3p_1 \int d^3p_2 \int d^3r_1 \int d^3r_2$$

$$(e^{i\vec{q}.\vec{r}_1/2} - e^{-i\vec{q}.\vec{r}_1/2})\, e^{-i(\hbar\vec{q}-\vec{p}_1+\vec{p}_2).\vec{r}_2/\hbar}$$

$$e^{-i(\vec{p}-\vec{p}_2/2-\vec{p}_1/2).\vec{r}_1/\hbar} \quad f_\sigma(\vec{p}_1,\vec{R}+\frac{\vec{r}_2-\vec{r}_1/2}{2}, t)$$

$$f_\sigma(\vec{p}_2,\vec{R}+\frac{\vec{r}_2+\vec{r}_1/2}{2}, t) \tag{7}$$

The dynamical influence of exchange in the equation of motion for $f_\sigma(\vec{p},\vec{R},t)$ is thus described by making the approximation :

$$X_\sigma^{tot}(\vec{p},\vec{R},t) \approx X_\sigma(\vec{p},\vec{R},t) \tag{8}$$

in eq.(3), leading to a non-linear integro-differential equation for the Wigner distribution function.

§ 2. Linearization in the external field

In order to study dynamical exchange effects on the dielectric function, one has to examine the deviation from the equilibrium distribution, induced by a weak externally applied potential.

In the dynamical exchange decoupling approximation (8), this equilibrium distribution should also satisfy (3) combined with (8), but $e\varphi(\vec{r},t)$ then only describes the interaction of the electrons with the homogeneous neutralizing background, and therefore is independent of position. This means that, without external field, the Coulomb term (3b) is identically zero if the equilibrium density is supposed to be homogeneous in space.

Furthermore, one easily checks that $X_\sigma(\vec{p},\vec{R},t)$, given by (7), is identically zero if $f_\sigma(\vec{p},\vec{R},t)$ is independent of position.

Therefore, if no external fields are applied, a homogeneous equilibrium distribution function satisfies the equation of motion (3) under the dynamical exchange decoupling approximation (8), because the r.h.s. of (3) then becomes zero, and thus this distribution remains constant in time.

If the Coulomb interaction and the interaction with the neutralizing background are applied adiabatically, the equilibrium distribution at $t = -\infty$ is given by the well-known Fermi distribution of the non-interacting electron gas :

$$f^o_\sigma(p) = \begin{cases} (2\pi\hbar)^{-3} & p \leqslant p_F \\ 0 & p > p_F \end{cases} \tag{9}$$

If the distribution is supposed to remain homogeneous in space, i.e. if no charge or spin density waves induce a state of lower energy, the dynamical exchange decoupling (8) in the equation of motion (3) thus implies that the distribution remains unaffected by switching on the Coulomb interaction adiabatically. It should be emphasized that these considerations do not rely on any assumption about the weakness of the Coulomb potential, but only involve the dynamical exchange decoupling.

Thus including dynamical exchange effects by (8), and neglecting all other correlations, the only possible homogeneous equilibrium distribution is given by (9), although it can not be excluded that spin or charge density waves might yield a state of lower energy [36] .

In the subsequent derivations, the homogeneous distribution $f^o_\sigma(p)$ is always supposed to describe the equilibrium state of the electron gas, and is separated out :

$$f_\sigma(\vec{p},\vec{R},t) = f^o_\sigma(p) + f^1_\sigma(\vec{p},\vec{R},t) \tag{10}$$

The dielectric response properties have to be derived from the deviations $f^1_\sigma(\vec{p},\vec{R},t)$ from equilibrium, due to an adiabatic external perturbation. If this external perturbation is supposed to be small enough to allow for neglecting terms of second order in $f^1_\sigma(\vec{p},\vec{R},t)$, the Fourier transform in space and time of (3) becomes :

$$\tilde{f}_\sigma(\vec{p},\vec{q},\omega) = \frac{-1/2\; U_{\vec{q},\omega} N_{\vec{q}}(\vec{p}) + 1/\hbar\; X_\sigma(\vec{p},\vec{q},\omega)}{\omega + i\varepsilon - \vec{p}.\vec{q}/m} \tag{11}$$

where $\tilde{f}_\sigma(\vec{p},\vec{q},\omega)$ is the Fourier transform of $f^1_\sigma(\vec{p},\vec{R},t)$ and :

$$U_{\vec{q},\omega} = e\varphi_{\vec{q},\omega} + \frac{4\pi e^2}{q^2} \sum_\sigma \int d^3p\; \tilde{f}_\sigma(\vec{p},\vec{q},\omega) \tag{12}$$

$$X_\sigma(\vec{p},\vec{q},\omega) = \frac{\hbar}{2} \int d^3p' \frac{4\pi e^2\hbar^2}{|\vec{p}-\vec{p}'|^2} [N_{\vec{q}}(\vec{p})\tilde{f}_\sigma(\vec{p}',\vec{q},\omega) - N_{\vec{q}}(\vec{p}')\tilde{f}_\sigma(\vec{p},\vec{q},\omega)] \tag{13}$$

$$N_{\vec{q}}(\vec{p}) = \frac{1}{\hbar} \sum_\sigma [f^o_\sigma(|\vec{p}+\hbar\vec{q}/2|) - f^o_\sigma(|\vec{p}-\hbar\vec{q}/2|)] \tag{14}$$

§ 3. Variational procedure

The linearized equation of motion (11-14), describing the deviations from equilibrium due to a weak external field, includes dynamical exchange effects by the function $X_\sigma(\vec{p},\vec{q},\omega)$. Due to this term, we know no method for solving the integral equation exactly. However, neglecting this term, equation (11) combined with (12) can be solved [2], yielding :

$$\bar{f}^L_\sigma(\vec{p},\vec{q},\omega) = -\frac{1}{2} e\varphi_{\vec{q},\omega} \frac{1}{1+Q_o(q,\omega)} \frac{N_{\vec{q}}(\vec{p})}{\omega + i\varepsilon - \vec{p}.\vec{q}/m} \tag{15}$$

where $Q_o(q,\omega)$ is the Lindhard polarizability

$$Q_o(q,\omega) = \frac{4\pi e^2}{q^2} \int d^3p \frac{N_{\vec{q}}(\vec{p})}{\omega + i\varepsilon - \vec{p}.\vec{q}/m} \tag{16}$$

It should be noted that the integral equation (11) can also be solved if one only takes the second term of $X_\sigma(\vec{p},\vec{q},\omega)$ into account. One then only has to replace the denominator $\omega + i\varepsilon - \vec{p}.\vec{q}/m$ in (15) and (16) by $\omega + i\varepsilon - \vec{p}.\vec{q}/m + \frac{1}{2} \int d^3p' \, 4\pi e^2\hbar^2 N_{\vec{q}}(\vec{p}')/|\vec{p} - \vec{p}'|^2$. However, the distribution function with self-energy terms obtained this way violates the continuity equation. A recent exchange approximation by Brener and Fry [37] is based on this separation of both exchange terms, but can otherwise essentially be derived along the lines outlined here. In order to treat exchange effects consistently, it is rather important to treat both terms in $X_\sigma(\vec{p},\vec{q},\omega)$ on the same footing.

From the Lindhard distribution function (15), one obtains the dielectric function in the well-known random phase approximation (RPA) : it only takes the Hartree potential into account in the equation of motion (11). However, although dynamical exchange effects are neglected by putting $X_\sigma(\vec{p},\vec{q},\omega) = 0$, the ground state energy obtained from this RPA dielectric function (see Eq.(73) of Part I) already includes some correlation terms [38] . The Hartree-Fock ground state energy is obtained if not only the exchange term, but also the mutual Coulomb interactions between the electrons, are neglected [39] . Therefore, the inclusion of dynamical exchange effects does not necessarily yield the Hartree-Fock ground state energy, and in order to avoid confusion, the dynamical exchange decoupling (8) should not be referred to as the Hartree-Fock approximation.

The Lindhard distribution function (15) can be used as a starting point for studying the dynamical exchange effects in the integral equation (11). Writing the Wigner distribution function as :

$$\tilde{f}_\sigma(\vec{p},\vec{q},\omega) = \bar{f}^L_\sigma(\vec{p},\vec{q},\omega) \, \gamma_{\vec{q},\omega}(\vec{p}) \tag{17}$$

reduces (11) to an integral equation for $\gamma_{\vec{q}\omega}(\vec{p})$:

$$\gamma_{\vec{q},\omega}(\vec{p}) = 1 + Q_o(q,\omega) - \frac{4\pi e^2}{q^2} \int d^3p' \, \gamma_{\vec{q},\omega}(p') \frac{N_{\vec{q}}(\vec{p}')}{\omega + i\varepsilon - \vec{p}'.\vec{q}/m}$$

$$+ \frac{1}{2} \int d^3p' \, \frac{4\pi e^2 \hbar^2}{|\vec{p} - \vec{p}'|^2} N_{\vec{q}}(\vec{p}') \left[\frac{\gamma_{q,\omega}(\vec{p}')}{\omega + i\varepsilon - \vec{p}'.\vec{q}/m} - \frac{\gamma_{q,\omega}(\vec{p})}{\omega + i\varepsilon - \vec{p}.\vec{q}/m} \right] \tag{18}$$

We know of no way to solve (18) exactly. But a variational solution can be found by observing that this integral equation can be derived from some functional, in a similar manner as the classical equations of motions are obtained from the principle of least action.

By inspection, one finds that for fixed $\vec{q}$ and ω, (18) follows from $\frac{\delta F[\gamma]}{\delta \gamma} = 0$, where $F[\gamma_{\vec{q},\omega}]$ is given by :

$$F[\gamma_{\vec{q},\omega}] = \int d^3p \, \frac{N_{\vec{q}}(\vec{p}) \gamma_{\vec{q},\omega}(\vec{p})}{\omega + i\varepsilon - \vec{p}.\vec{q}/m} \{ \gamma_{\vec{q},\omega}(\vec{p}) - 2[1 + Q_o(q,\omega)]$$

$$+ \frac{4\pi e^2}{q^2} \int d^3p' \, \frac{N_{\vec{q}}(\vec{p}') \, \gamma_{\vec{q},\omega}(\vec{p}')}{\omega + i\varepsilon - \vec{p}'.\vec{q}/m} \tag{19}$$

$$- \frac{1}{2} \int d^3p' \, \frac{4\pi e^2 \hbar^2}{|\vec{p} - \vec{p}'|^2} N_{\vec{q}}(\vec{p}') \left[\frac{\gamma_{\vec{q},\omega}(\vec{p}')}{\omega + i\varepsilon - \vec{p}'.\vec{q}/m} - \frac{\gamma_{\vec{q},\omega}(\vec{p})}{\omega + i\varepsilon - \vec{p}.\vec{q}/m} \right] \}$$

It is important to realize that the dielectric response properties of the electron gas, are determined by the induced density, which is an integral of the Wigner distribution function over all momenta. One thus is mainly interested in the momentum dependence of $f_\sigma(\vec{p},\vec{q},\omega)$ in as far as the integral over all momenta is concerned. In looking for an approximate solution of (18), it thus seems reasonable to neglect the momentum dependence as much as possible, but under the condition that (18) is fulfilled as close as possible with a momentum independent $\gamma_{\vec{q}\omega}$. The fact that (18) derives variationally from the functional (19) provides a criterion for the best possible $\gamma_{\vec{q}\omega}$, which is constant in $\vec{p}$. Imposing then $\delta F[\gamma] / \delta\gamma = 0$ yields

$$\gamma_{q,\omega} = \frac{1 + Q_o(q,\omega)}{1 + Q_o(q,\omega) - G(q,\omega) \, Q_o(q,\omega)} \tag{20}$$

where :

$$G(q,\omega) = \frac{4\pi e^2}{q^2} \frac{2\pi e^2\hbar^2}{Q_o^2(q,\omega)} \int d^3p \int d^3p' \frac{1}{|\vec{p}-\vec{p}'|^2} \frac{N_{\vec{q}}(\vec{p}) N_{\vec{q}}(\vec{p}')}{\omega + i\varepsilon - \vec{p}.\vec{q}/m}$$

$$\left[\frac{1}{\omega + i\varepsilon - \vec{p}'.\vec{q}/m} - \frac{1}{\omega + i\varepsilon - \vec{p}.\vec{q}/m}\right] \qquad (21)$$

In this approximation, the trial form for $\tilde{f}_\sigma(\vec{p},\vec{q},\omega)$ thus becomes :

$$\tilde{f}_\sigma(\vec{p},\vec{q},\omega) \simeq -\frac{1}{2} e\varphi_{\vec{q},\omega} \frac{1}{1 + Q_o(q,\omega) - G(q,\omega)\; Q_o(q,\omega)} \frac{N_{\vec{q}}(\vec{p})}{\omega + i\varepsilon - \vec{p}.\vec{q}/m} \qquad (22)$$

The trial solution (22) for $\tilde{f}_\sigma(\vec{p},\vec{q},\omega)$ clearly does not longer satisfy the integral equation rigorously. But, as mentioned above the detailed dependence on the momentum $\vec{p}$ is not very important in the study of the dielectric response of the electron gas, where the quantity of interest is the Fourier transform of the induced charge density :

$$n_{\vec{q},\omega} = \sum_\sigma \int d^3p\; \tilde{f}_\sigma(\vec{p},\vec{q},\omega) \qquad (23)$$

from which the dielectric function can be obtained, using its definition in terms of the total and the applied potential :

$$\varphi_{\vec{q},\omega} + \sum_\sigma \frac{4\pi e}{q^2} \int d^3p\; \tilde{f}_\sigma(\vec{p},\vec{q},\omega) = \frac{\varphi_{\vec{q},\omega}}{\varepsilon(\vec{q},\omega)} \qquad (24)$$

The point to be checked for the dielectric properties of the electron gas to be described by the trial solution (22) is thus if the integrated equation (11) is satisfied :

$$n_{\vec{q},\omega} = -\frac{1}{2} \sum_\sigma \int d^3p \frac{U_{\vec{q}\omega}\, N_{\vec{q}}(\vec{p}) + 1/\hbar\; X_\sigma(\vec{p},\vec{q},\omega)}{\omega + i\varepsilon - \vec{p}.\vec{q}/m} \qquad (25a)$$

Using (22) in (23), it immediately follows from the definition (16) of $Q_o(q,\omega)$ that

$$n_{\vec{q},\omega} = -e\varphi_{\vec{q},\omega} \frac{q^2}{4\pi e^2} \frac{Q_o(q,\omega)}{1 + Q_o(q,\omega) - G(q,\omega)\, Q_o(q,\omega)} \tag{25b}$$

and thus from (12)

$$U_{\vec{q},\omega} = e\varphi_{\vec{q},\omega} \frac{1 - G(q,\omega)\, Q_o(q,\omega)}{1 + Q_o(q,\omega) - G(q,\omega)\, Q_o(q,\omega)} \tag{25c}$$

Furthermore, from the definition (21) of $G(q,\omega)$, and the definition (12) of $X_\sigma(\vec{p},\vec{q},\omega)$, one obtains :

$$-\frac{1}{2}\sum_\sigma \frac{1}{\hbar}\int d^3p \frac{X_\sigma(\vec{p},\vec{q},\omega)}{\omega + i\varepsilon - \vec{p}.\vec{q}/m}$$

$$= -\frac{q^2}{4\pi e^2}\, e\varphi_{\vec{q},\omega} \frac{G(q,\omega)\, Q_o^2(q,\omega)}{1 + Q_o(q,\omega) - G(q,\omega)\, Q_o(q,\omega)} \tag{25d}$$

Combining terms, some elementary algebra then shows that the trial solution (22) indeed indentically satisfies the integrated equation of motion (25a), and therefore we expect that (22) yields an adequate description of the bare exchange effects in the electron gas as far as the dielectric response properties are concerned. In this respect, we mention that the first-frequency moment conserving approximation of Toigo and Woodruff [21] can easily be obtained from the hydrodynamic equation, using the approximation (17) with $\gamma_{\vec{q},\omega}$ independent of $\vec{p}$. This derivation is discussed elsewhere [32] but it is not variational, and does not satisfy the integrated equation of motion (25a).

From the defining equation (24) for the dielectric function, one readily obtains by using (23) and (25b) :

$$\varepsilon(q,\omega) = 1 + \frac{Q_o(q,\omega)}{1 - G(q,\omega)\, Q_o(q,\omega)} \tag{26}$$

which is of the form currently used since Hubbard [4] to some extent corrected the RPA dielectric function for local field effects, by approximating several exchange diagrams by a static exchange field. However, the function $G(q,\omega)$ in (26) not only depends on the wave vector q, but includes exchange effects dynamically by its frequency dependence.

§ 4. Theorem

In the units

$$k = q/k_F \qquad \nu = \frac{\hbar\omega}{2E_F} \tag{27}$$

the integral (21) for $G(q,\omega)$ reduces to [31]

$$G(kk_F, 2\nu E_F/\hbar) = f(k,\nu) \int d^3r \int d^3r' \frac{1}{|\vec{r}-\vec{r}'|^2} [N(\vec{r}+\vec{k}/2) - N(\vec{r}-\vec{k}/2)]$$
$$[N(\vec{r}' + \vec{k}/2) - N(\vec{r}' - \vec{k}/2)] \frac{1}{\nu + i\varepsilon - \vec{r}.\vec{k}} \left[\frac{1}{\nu + i\varepsilon - \vec{r}'.\vec{k}} - \frac{1}{\nu + i\varepsilon - \vec{r}.\vec{k}}\right] \tag{28}$$

where k_F is the Fermi wave vector, E_F is the Fermi energy, and the functions $N(\vec{r})$ and $f(k,\nu)$ are given by

$$N(\vec{r}) = \begin{cases} 1 & |\vec{r}| \leqslant 1 \\ 0 & |\vec{r}| > 1 \end{cases} \tag{29}$$

$$f(k,\nu) = \frac{1}{2} \frac{m^2 e^4}{\pi^4 \hbar^4 k^2 k_F^2 Q_o^2(kk_F, 2\nu E_F/\hbar)} \tag{30}$$

From the definition (16) of the Lindhard polarizability $Q_o(q,\omega)$ it then follows that $f(k,\nu)$ is independent of the density and therefore $G(kk_F, 2\nu E_F/\hbar)$ is a <u>universal function of k and ν for all densities</u>.

§ 5. Comparison with other dynamical approximations

The equation of motion (11) for the Wigner distribution, including the exchange effects from the dynamical exchange decoupling (8) and linearized in the external field, has been deduced previously in [20] where also an iteration to first order was performed, starting from the Lindhard distribution function as the zeroth order solution. In the static limit, this iterative result was also obtained from a diagrammatic expansion [12]. In a previous paper, we have shown that this iteration to first order provides the terms to order e^4 in the geometric progression of the variational solution in powers of e^2.

The variational procedure was also proposed previously to treat integral equations of the same type [10, 22, 23] where a similar trial solution, neglecting the momentum dependence was used.

But the application given here on the equation of motion for the Wigner distribution function clearly indicates that only the bare exchange effects are taken into account, neglecting all other correlations, and that these exchange effects are treated dynamically and to the full extent. Furthermore, because the trial solution rigorously satisfies the integrated equation of motion (25a), we expect that not only the equation of motion, but also the trial solution describes these dynamical exchange effects rather accurately in the dielectric response properties of the electron gas.

It should also be emphasized that the derivation of the dielectric function by Toigo and Woodruff [21] is a first attempt to include dynamical exchange effects. In a previous paper [32], we have shown that this dielectric function can be obtained by making the approximation (17) with $\gamma_{\vec{q},\omega}$ independent of momentum in $\omega^2 n_{\vec{q},\omega} = \omega\vec{q}.\vec{j}_{\vec{q},\omega}$. However, this procedure is not unique and is not variational, and in [21] a first-frequency moment conserving solution is proposed, as is discussed in detail in [32]. Along the same lines as in [21], Tripathy and Mandal derived a dielectric function, conserving frequency moments to infinite order [28]. The resulting expression for $G(q,\omega)$ happens to be identical to the expression (21), which we already derived in [27]. The conservation of the frequency moments in the equation of motion for the double-time retarded commutator of the charge density fluctuation operators, thus seems to be equivalent to satisfying the integrated equation of motion (25a) which we achieved to obtain by variational technique.

Although the formal expression (21) for $G(q,\omega)$ can thus be found by several methods in a more or less transparant way, the evaluation of the integrals involved has not been done, to the best of our knowledge, before we published the first frequency dependent results [30]. (We reported the first plots of Re $G(q,\omega)$ in Bull. Am. Phys. Soc. 22, 438 (1977)).

A very elaborate derivation and evaluation of a frequency dependent exchange correction has recently been published by

Holas, Aravind and Singwi [40], who calculated the diagrams for the proper polarizability to first order in the electron-electron interaction. The interrelation between both approaches will be examined in detail in Part III, in connection with the results.

C. ANALYTICAL TREATMENT OF G(q,ω)

In the previous section, we derived that the influence of exchange effects on the dielectric function is described by the function G(q,ω), given in (28) as a sixfold integral. A direct numerical evaluation of this integral is quite complicated.

Because G(q,ω) is still a function of two variables, its evaluation in a large (q,ω) domain would be very time consuming, since each (q,ω) point would require a sixfold numerical integration. However, a far more serious problem concerns the numerical accuracy.

A minor difficulty in this respect is presented by the fact that in (28) each three-dimensional integration variable is restricted to two displaced Fermi spheres. Even if the integration in one Fermi sphere would be reasonably accurate, the substraction of these integrals could produce a rather inaccurate final result in a large (q,ω) domain.

The main accuracy problem however arises from the fact that a pole of second order in one variable, has to be substracted from a product of first order poles in two variables. Because a sixfold numerical integration containing first or second order poles is already rather inaccurate, this supplementary substraction could introduce unacceptable errors.

Finally, the integrand in (28) contains the rather singular factor $|\vec{r}-\vec{r}'|^2$. For analytical purposes its occurence is not dramatic, because from symmetry considerations

$$\frac{1}{|\vec{r}-\vec{r}'|^2}\,\frac{1}{\nu+i\varepsilon-\vec{r}.\vec{k}}\left[\frac{1}{\nu+i\varepsilon-\vec{r}'.\vec{k}}-\frac{1}{\nu+i\varepsilon-\vec{r}.\vec{k}}\right]$$

can be replaced by

$$-\frac{1}{2}\,\frac{[(\vec{r}-\vec{r}').\vec{k}]^2}{|\vec{r}-\vec{r}'|^2}\,\frac{1}{(\nu+i\varepsilon-\vec{r}.\vec{k})^2}\,\frac{1}{(\nu+i\varepsilon-\vec{r}'.\vec{k})^2}$$

But the latter form now contains a product of poles of second order, and therefore is not easier to treat numerically.

Essentially, two independent methods were worked out in order to achieve the twofold goal of reducing the computation time, and simultaneously solving the accuracy problem. A method based on Fourier transforms has been presented rather extensively in [41]. The final result is :

$$G(kk_F, 2\nu E_F/\hbar) = \tag{31}$$

$$\frac{2\pi^2}{k} f(k,\nu) \int_{-1}^{1} dz\, \mathcal{T}(z,k) \left[\frac{1}{\nu + i\varepsilon - k^2/2 - kz} - \frac{1}{\nu + i\varepsilon + k^2/2 + kz}\right]$$

where

$$\mathcal{T}(z,k) = [\mathcal{H}(z,o) - \mathcal{H}(z,k)] - [\mathcal{F}(z,o) - \mathcal{F}(z,k)] \tag{32}$$

In this expression, $\mathcal{F}(z,k)$ is an elementary function :

$$\mathcal{F}(z,k) = -k \frac{1 - (1 + k^2 + 2zk)}{\sqrt{1 + k^2 + 2zk}} \ln \left|\frac{1 + \sqrt{1 + k^2 + 2zk}}{1 - \sqrt{1 + k^2 + 2zk}}\right|$$
$$+ (z + k) \frac{1 - (z + k)^2}{\sqrt{(z + k)^2}} \ln \left|\frac{1 + \sqrt{(z + k)^2}}{1 - \sqrt{(z + k)^2}}\right| \tag{33}$$

The function $\mathcal{H}(z,k)$ requires a numerical integration :

$$\mathcal{H}(z,k) = -8\sqrt{1 - z^2} \int_{o}^{\pi/2} d\theta\, \mathrm{sh}^{-1}(\mathrm{ctg}\,\theta)\, F((z+k)\cos\theta, \sqrt{1-z^2}\sin\theta) \tag{34a}$$

where

$$F(b,a) = \frac{1-b^2}{4a} \{\Theta(a - |1+b|) \sqrt{1 - (\tfrac{a}{1+b})^2}$$
$$- \Theta(a - |1-b|) \sqrt{1 - (\tfrac{a}{1-b})^2}\}$$
$$+ \frac{a}{4} \Theta(a - |1-b|) \ln \left|\frac{|1 + b| + \sqrt{(1+b)^2 - a^2}}{|1 - b| + \sqrt{(1-b)^2 - a^2}}\right|$$

$$+ \frac{a}{4} \Theta(a - |1+b|)\, \Theta(|1-b| - a) \ln \left| \frac{|1+b| + \sqrt{(1+b)^2 - a^2}}{a} \right| \qquad (34b)$$

The numerical evaluation of the function $H(z,k)$ presents no difficulties, and the result is plotted in figure 1 as a function of z for several values of k. In figure 2, the total function $T(z,k)$, given by (32), is shown.

Apart from the result (31-34), obtained by the Fourier transformation of all the functions appearing in the sixfold integral (28) for $G(q,\omega)$ and described in ref.[41], we developed a second method. In this procedure, (28) is obtained as a double integral from a straightforward evaluation, as described in Appendix A. With this second method however, the integrand in the final result still shows a product of factors containing first order poles. Therefore, it requires more computation time and gives slightly less accurate results than the evaluation of (31-34). Nevertheless, it provides an independent check of the results obtained with the method of Fourier transforms. Furthermore, it turned out that this alternative form for $H(z,k)$ is more transparant for the evaluation of some limiting cases, as will be discussed in part III.

Given the function $T(z,k)$, which is well behaved, the evaluation of $G(q,\omega)$ from (31) can be performed by standard numerical techniques, except if one of the first order poles approaches an integration limit, or when the denominators tend to cancel each other. But these two problems have been handled analytically.

Under the condition $|\nu| >> |\frac{k^2}{2} \pm k|$, the denominators in (31) almost equal each other, which might introduce numerical inaccuracies. This problem was previously considered in the limit $k \to 0$ [27]. For arbitrary k the derivation is presented in Appendix A of ref.[41].

A second problem in evaluating (31) arises from the possibility that one of the poles approaches an integration limit. Because :

$$\int_a^b \frac{f(x)}{p + i\varepsilon - x} dx = \int_a^b dx \frac{f(x) - f(p)}{p + i\varepsilon - x} - f(p) \ln \frac{b - p - i\varepsilon}{a - p - i\varepsilon} \qquad (35)$$

a logarithmic singularity might occur for $|\nu| = |\frac{k^2}{2} \pm k|$ if $T(+1,k)$ differs from zero. From (34) it is obvious that $H(\mp 1,k) = 0$, as in also clear from figure 1.

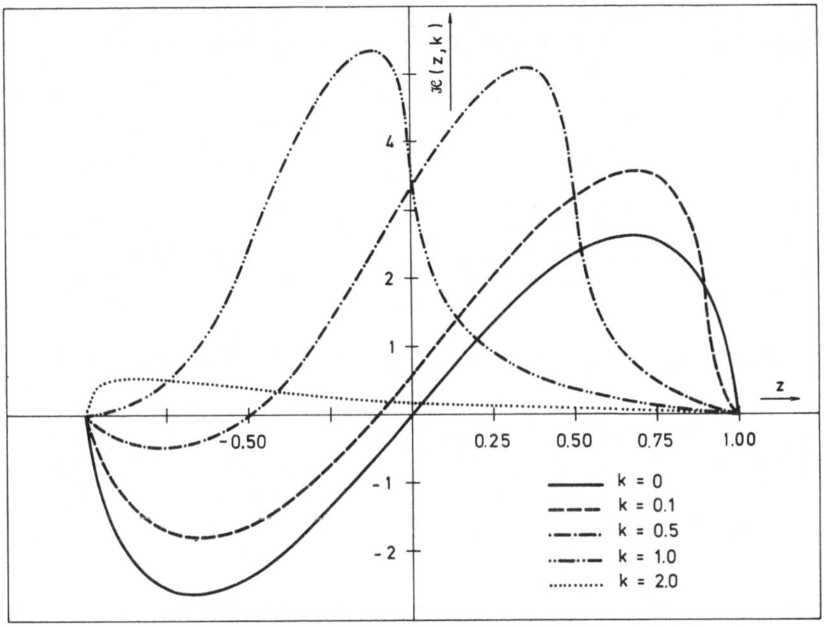

Fig. 1. The function $H(z,k)$ defined in (34) for various values of k.

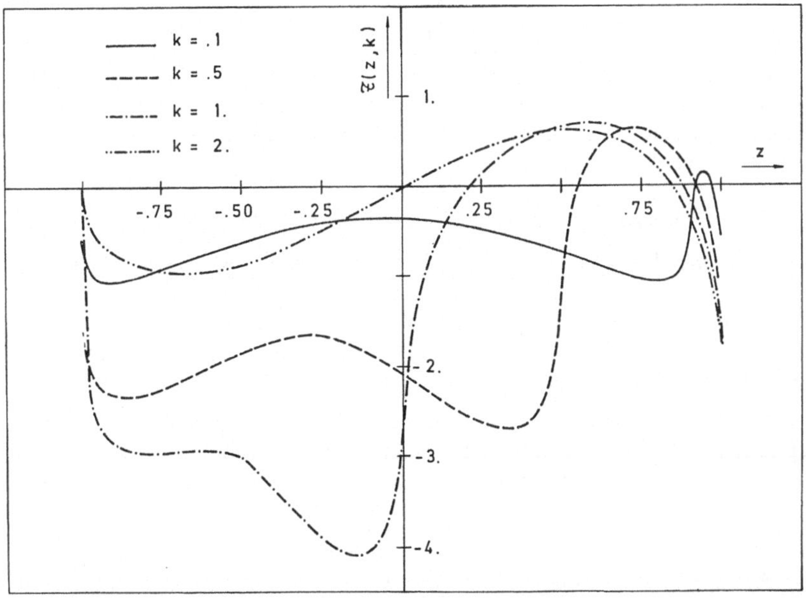

Fig. 2. The function $T(z,k)$ defined in (32) for various values of k.

Furthermore it follows from (33) that $F(\pm 1,o) = 0$. Therefore (32) yields that $T(\pm 1,k) = F(\pm 1,k)$, which results in :

$$T(\pm 1,k) = k^2 \frac{k \pm 2}{|k \pm 1|} \ln \left| \frac{1 + |k \pm 1|}{1 - |k \pm 1|} \right| + (k \pm 1) \frac{1 - (k \pm 1)^2}{|k \pm 1|} \ln \left| \frac{1 + |k \pm 1|}{1 - |k \pm 1|} \right| \tag{36}$$

Consequently, $G(kk_F, 2\nu E_F/\hbar)$ shows a logarithmic singularity in the region $|\nu| \simeq |k^2/2 \pm k|$, because $T(\pm 1,k) \neq 0$, provided that $f(k,\nu)$ differs from zero :

$$G(kk_F, 2\nu E_F/\hbar)\Big|_{|\nu| \simeq |k^2/2 \pm k|} \simeq \pm f(k,\nu) \frac{2\pi^2}{k^2} k \frac{2 \pm k}{1 \pm k} \ln \left| \frac{2 \pm k}{k} \right| \ln \left| \nu - \left| \frac{k^2}{2} \pm k \right| \right| \tag{37}$$

From the expression (30) for $f(k,\nu)$ in terms of the Lindhard polarizability Q_o, it follows that $f(k,\nu) \sim k^2$ for $k \to 0$ in the static limit $\nu = 0$. The logarithmic singularity (37) therefore smooths out near the origin. Also in the static limit at $k = 2$, the singularity is cancelled by the factor $(2-k)$ multiplying the logarithmic functions.

Two immediate consequences result from the singularity (37). At the parabolas $|\nu| = |k^2/2 \pm k|$, the real part of the dielectric function equals 1, whereas the imaginary part of the inverse dielectric function becomes zero. This is readily seen from (26). Thus the dynamical structure factor, including exchange, has a zero as a function of frequency at the parabola $\nu = |k - k^2/2|$, whereas in the RPA approximation only a discontinuity in the derivative occurs. This phenomenon was already discussed in a previous paper [31] and is again illustrated in a few plots of $-\mathrm{Im}\ \varepsilon^{-1}(q,\omega)$ in figure 3.

As a second consequence of this logarithmic singularity, the plasmon branch does not penetrate into the particle-hole continuum, but only approaches it asymptotically. Because at

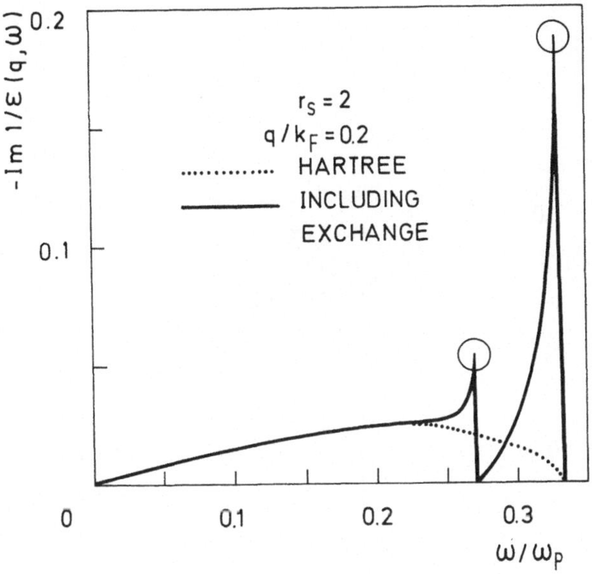
-Im 1/ε (q,ω)
0.2
0.1
$r_s = 2$
$q/k_F = 0.2$
HARTREE
INCLUDING EXCHANGE
0
0.1
0.2
0.3
ω/ω_p

Fig. 3a

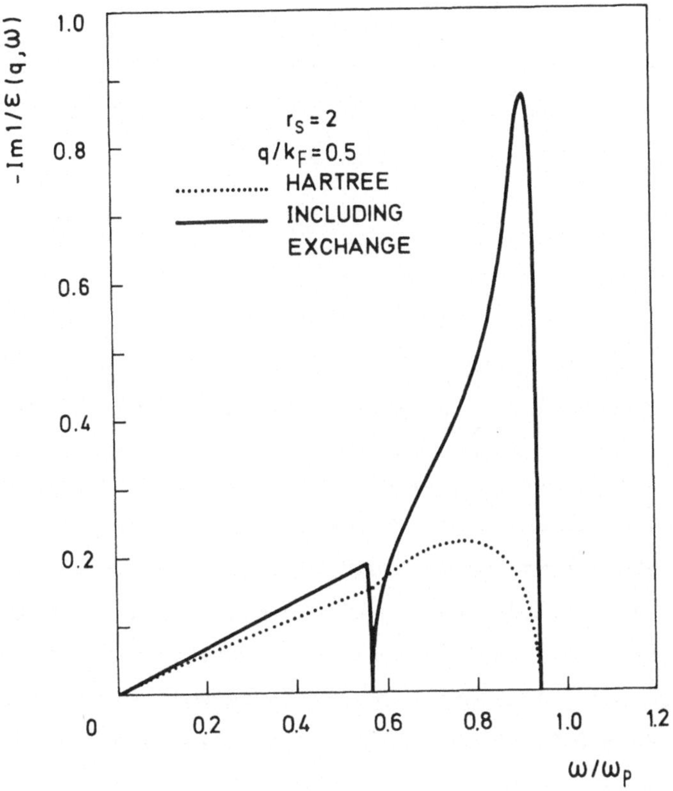
-Im 1/ε (q,ω)
1.0
0.8
0.6
0.4
0.2
$r_s = 2$
$q/k_F = 0.5$
HARTREE
INCLUDING EXCHANGE
0
0.2
0.4
0.6
0.8
1.0
1.2
ω/ω_p

Fig. 3b

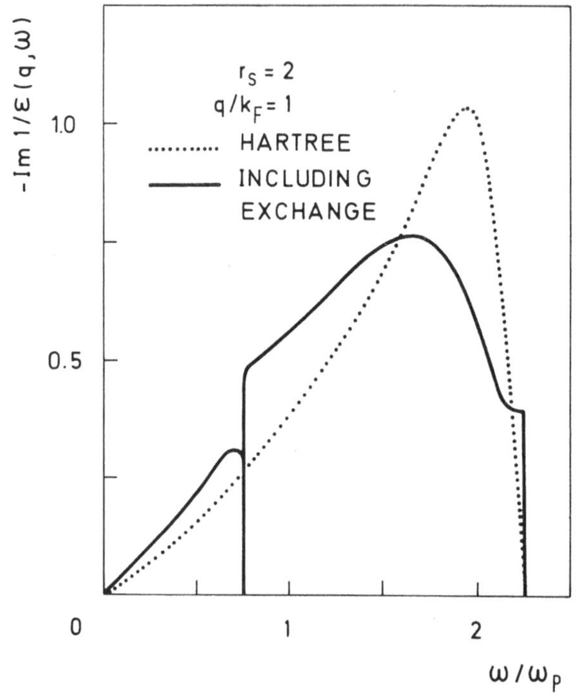

Fig. 3c

very high frequencies $Q_o(q,\omega)$ is negative, and $G(q,\omega)$ is positive as we have shown in Appendix A of ref.[41], far above the continuum $1 - G(q,\omega)\, Q_o(q,\omega)$ is positive. At the upper boundary of the continuum, $G(q,\omega) = -\infty$ from (37), and because $Q_o(q,\omega)$ is negative, it follows that $1 - G(q,\omega)\, Q_o(q,\omega) = -\infty$ at this upper boundary. Thus with decreasing frequency, $1 - G(q,\omega)\, Q_o(q,\omega)$ decreases from some positive value at very high frequency to $-\infty$ at $\omega = \hbar(qk_F + q^2/2)/m$, passing through zero at some critical value ω_c. From (26) it thus follows that $\varepsilon(q,\omega)$ has a pole at

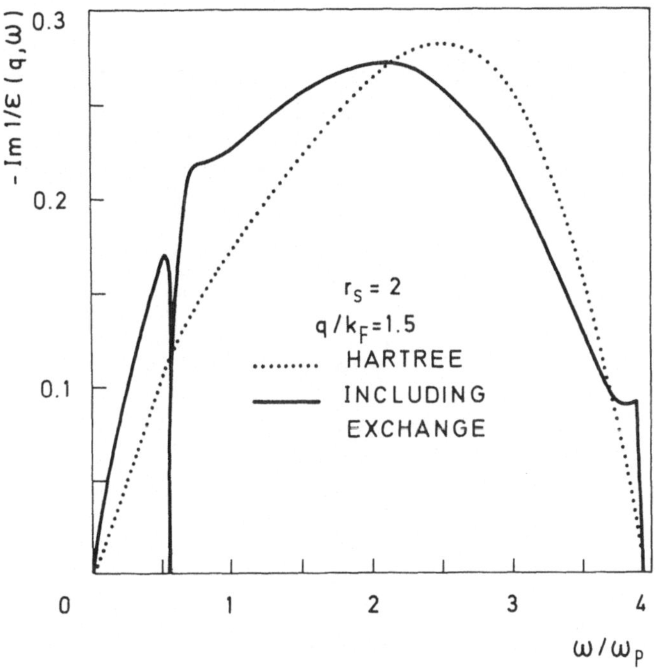

Fig. 3d

$\omega = \omega_c$, and because $Q_o(q,\omega)$ is negative, $\varepsilon(q,\omega)$ tends to $-\infty$ if ω decreases towards ω_c. But because $\varepsilon(q,\omega)$ is positive at very high frequency, a zero in $\varepsilon(q,\omega)$ has to be found above ω_c. Therefore the plasmon frequency cannot penetrate into the continuum. It should be noted that a similar singular behaviour of Re $\varepsilon(q,\omega)$ also shows up above $\omega = \hbar(qk_F - q^2/2)/m$, but because $G(q,\omega)$ and $Q_o(q,\omega)$ have an imaginary part, the pole in Re $\varepsilon(q,\omega)$ is replaced there by a strongly peaked structure. In figure 4, Re $\varepsilon(q,\omega)$ is shown as a function of frequency for several values of the wave vector.

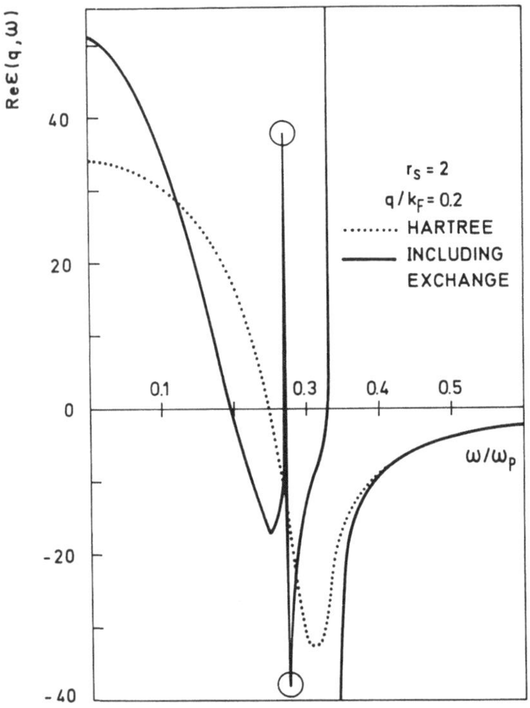
Re ε(q, ω)
40
20
0
-20
-40
0.1
0.3
0.4
0.5
ω/ωp
rs = 2
q/kF = 0.2
.... HARTREE
—— INCLUDING EXCHANGE

Fig. 4a

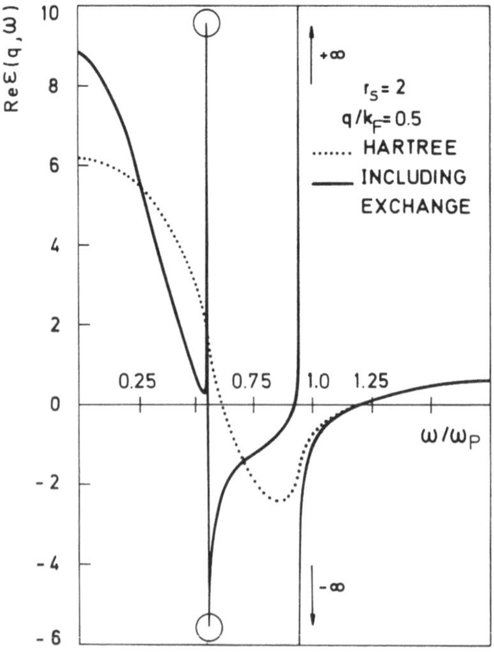
Re ε(q, ω)
10
8
6
4
2
0
-2
-4
-6
0.25
0.75
1.0
1.25
ω/ωP
+∞
-∞
rs = 2
q/kF = 0.5
...... HARTREE
—— INCLUDING EXCHANGE

Fig. 4b

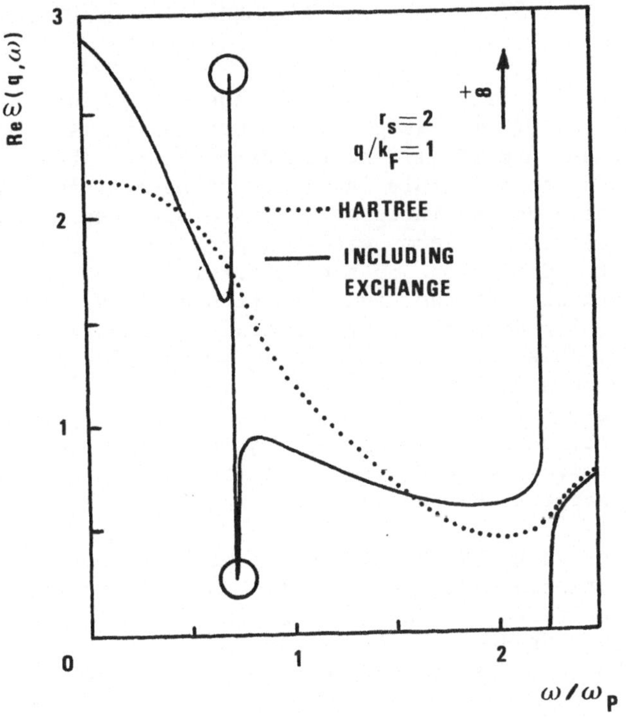
3
2
1
0
1
2
Re ε(q,ω)
ω/ωp
+∞
rs= 2
q/kF= 1
HARTREE
INCLUDING
EXCHANGE

Fig. 4c

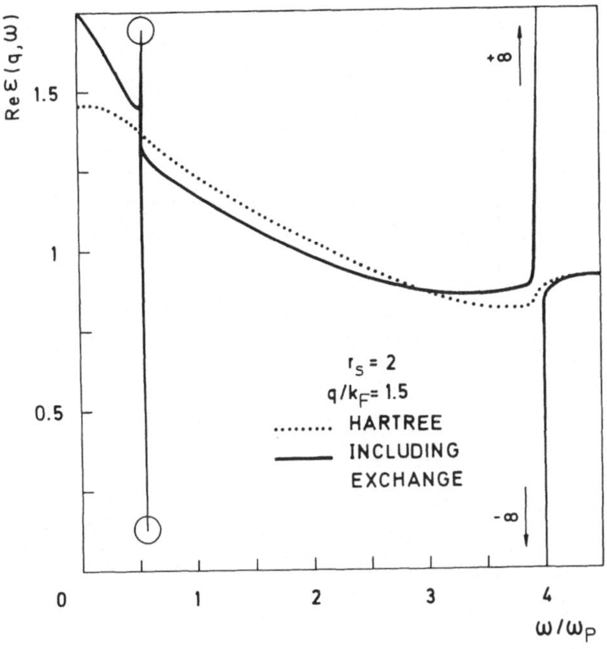
1.5
1
0.5
0
1
2
3
4
Re ε(q,ω)
ω/ωP
+∞
-∞
rs = 2
q/kF= 1.5
HARTREE
INCLUDING
EXCHANGE

Fig. 4d

APPENDIX A - EVALUATION OF G(q,ω) BY ELEMENTARY METHODS

In section III, the result of the evaluation of the integral (28) for G(q,ω) was given as obtained in [41] by taking the Fourier transform of all the functions in the integrand, which originally leads to a 17-fold integral. By several subsequent transformations a double integral was then obtained, and even a single integral for the terms with a second order pole. This reduction of the dimensions of the integral is also possible with elementary methods, but the final result has the disadvantage that the remaining double integral contains a double principal value problem. This leads to increasing computation time and numerical inaccuracy. Nevertheless, the method has its own merits, not only because the treatment of the second order pole in (28) in more elegant, but merely because a quite different expression is obtained, which therefore yields an independent check on the analytical treatment and the numerical results.

To evaluate (28) by elementary methods, one first performs the translations, transforming the integration regions in (28) into unit spheres, centered at the origin. Denoting then the terms with a product of first order poles by G_{11}, and collecting the poles of second order in G_{20}, one obtains :

$$G(kk_F, 2\nu E_F/\hbar) = f(k,\nu)\,[G_{11}(k,\nu) - G_{20}(k,\nu)] \tag{A.1}$$

where f(k,ν) is defined in (30), and where G_{11} and G_{20} are given by :

$$G_{11}(k,\nu) = \tag{A.2}$$

$$\int d^3r \int d^3r'\, N(r)N(r') \left\{ \frac{1}{|\vec{r}-\vec{r}\,'|^2} \frac{1}{\nu + i\varepsilon + k^2/2 - \vec{r}.\vec{k}} \frac{1}{\nu + i\varepsilon + k^2/2 - \vec{r}\,'.\vec{k}} + \frac{1}{|\vec{r}-\vec{r}\,'|^2} \frac{1}{\nu + i\varepsilon - k^2/2 - \vec{r}.\vec{k}} \frac{1}{\nu + i\varepsilon - k^2/2 - \vec{r}\,'.\vec{k}} - \frac{1}{|\vec{r}-\vec{r}\,'+\vec{k}|^2} \frac{1}{\nu + i\varepsilon - k^2/2 - \vec{r}.\vec{k}} \frac{1}{\nu + i\varepsilon + k^2/2 - \vec{r}\,'.\vec{k}} - \frac{1}{|\vec{r}-\vec{r}\,'-\vec{k}|^2} \frac{1}{\nu + i\varepsilon + k^2/2 - \vec{r}.\vec{k}} \frac{1}{\nu + i\varepsilon - k^2/2 - \vec{r}\,'.\vec{k}} \right\}$$

$$G_{20}(k,\nu) = \int d^3r \int d^3r' \, N(r)\, N(r') \left\{ \begin{array}{l} \dfrac{1}{|\vec{r}-\vec{r}'|^2} \dfrac{1}{(\nu + i\varepsilon + k^2/2 - \vec{r}.\vec{k})^2} \\[2ex] + \dfrac{1}{|\vec{r}-\vec{r}'|^2} \dfrac{1}{(\nu + i\varepsilon - k^2/2 - \vec{r}.\vec{k})^2} \\[2ex] - \dfrac{1}{|\vec{r}-\vec{r}'-\vec{k}|^2} \dfrac{1}{(\nu + i\varepsilon + k^2/2 - \vec{r}.\vec{k})^2} \\[2ex] - \dfrac{1}{|\vec{r}-\vec{r}'+\vec{k}|^2} \dfrac{1}{(\nu + i\varepsilon - k^2/2 - \vec{r}.\vec{k})^2} \end{array} \right\} \quad \text{(A.3)}$$

The function G_{11} can be reduced into a double integral. Using the identity

$$\frac{1}{a-\vec{r}.\vec{k}} \frac{1}{b-\vec{r}.\vec{k}} = \frac{1}{b-a+(\vec{r}-\vec{r}').\vec{k}} \left[\frac{1}{a-\vec{r}.\vec{k}} - \frac{1}{b-\vec{r}'.\vec{k}} \right]$$

and interchanging $\vec{r}$ and $\vec{r}'$ in the terms containing $(\nu + i\varepsilon \pm k^2/2 - \vec{r}'.\vec{k})^{-1}$, (A.2) becomes an expression, where all terms have a denominator of the form $(\nu + i\varepsilon + k^2/2 - \vec{r}.\vec{k})^{-1}$. Expressing then the integrals by cylindrical coordinates, with the z-axis along $\vec{k}$, and recombining terms, one finds for (A.2) :

$$G_{11}(k,\nu) = \frac{2\pi^2}{k} \int_{-1}^{1} dz \left[\frac{H(z,o) - H(z,k)}{\nu + i\varepsilon - k^2/2 - zk} + \frac{H(z,o) - H(z,-k)}{\nu + i\varepsilon + k^2/2 - zk} \right] \quad \text{(A.4)}$$

where

$$H(z,k) = \frac{1}{\pi^2} \int_{-1}^{1} dz' \int_0^{\sqrt{1-z^2}} \rho\, d\rho \int_0^{\sqrt{1-z'^2}} \rho' d\rho' \int_0^{2\pi} d\varphi \int_0^{2\pi} d\varphi' \; \frac{1}{z-z'+k} \frac{1}{\rho^2 + \rho'^2 + (z-z'+k)^2 - 2\rho\rho'\cos\varphi'} \quad \text{(A.5)}$$

We note that (A.4) is of the same form as (59), but the expression (A.5) for $H(z,k)$ is formally different from the expression (60) obtained in section C. However, both expressions will turn out to be equal, and thus the use of the same symbol is not confusing.

In appendix D of ref.[41], it is shown that :

$$\frac{1}{\pi^2}\int_0^{\sqrt{u}} \rho\, d\rho \int_0^{\sqrt{v}} \rho' d\rho' \int_0^{2\pi} d\varphi \int_0^{2\pi} d\varphi' \frac{1}{\rho^2+\rho'^2+s_k^2-2\rho\rho' \cos\varphi'}$$

$$= A(s_k^2,u,v) - u\, L(s_k^2,u,v) - v\, L(s_k^2,v,u) \tag{A.6a}$$

where :

$$A(s_k^2,u,v) = \frac{1}{2}\{W(s_k^2,u,v) - u - v - s_k^2\} \tag{A.6b}$$

$$L(s_k^2,u,v) = -\ln \frac{v-u+s_k^2+W(s_k^2,u,v)}{2s_k^2} \tag{A.6c}$$

with :

$$W(s_k^2,u,v) = \sqrt{(u-v)^2 + 2s_k^2(u+v) + s_k^4} \tag{A.6d}$$

By introducing the short hand notations :

$$s_k = z - z' + k \tag{A.7a}$$

$$u = 1 - z^2 \tag{A.7b}$$

$$v = 1 - z'^2 \tag{A.7c}$$

$$A_k = A(s_k^2,u,v) \tag{A.7d}$$

$$L_k = L(s_k^2,u,v) \tag{A.7e}$$

$$L'_k = L(s_k^2,v,u) \tag{A.7f}$$

$$W_k = W(s_k^2,u,v) \tag{A.7g}$$

the 5-fold integral (A.5) for $H(z,k)$ becomes a single integral :

$$H(z,k) = \int_{-1}^{1} dz' \frac{A_k - uL_k - vL'_k}{s_k} \tag{A.8}$$

where one should remember that from (A.7) all the symbols used depend on the integration variables z and z'.

As is readily seen from (A.6.c), the integrand of (A.8) contains a singularity of the form $s_k^{-1} \ln s_k^2$, which is hard to handle numerically. Therefore, further analytical simplifications are needed.

Considering first the special case k = 0, which explicitly occurs in (A.4), the short hand notations (A.7) for the functions defined in (A.6), can then be written as :

$$s_o = z - z' \tag{A.9a}$$

$$v = u + 2zs_o - s_o^2 \tag{A.9b}$$

$$W_o = 2\sqrt{s_o^2} \tag{A.9c}$$

$$L_o = -\ln \frac{zs_o + s_o^2 + \sqrt{s_o^2}}{s_o^2} \tag{A.9d}$$

$$L'_o = -\ln \frac{-zs_o + s_o^2 + \sqrt{s_o^2}}{s_o^2} \tag{A.9e}$$

$$A_o = \sqrt{s_o^2} - u - zs_o \tag{A.9f}$$

Introducing then s_o as the new integration variable, the integral (A.8) becomes, in the limit k = 0 :

$$H(z,o) = -2z + \int_{z-1}^{z+1} ds_o \left\{\frac{\sqrt{s_o^2}}{s_o} - \frac{u}{s_o} - u\,\frac{L_o + L'_o}{s_o} - 2zL'_o + s_o L'_o\right\} \tag{A.10}$$

In this integral, all symbols except for $u = 1 - z^2$, depend on s_o, as given in (A.9).

By noting from (A.4) that the only values of interest for z satisfy $|z| \leqslant 1$, some elementary integrals can be performed :

$$\int_{z-1}^{z+1} dx_o \frac{\sqrt{s_o^2}}{s_o} = 2z \tag{A.11a}$$

$$\int_{z-1}^{z+1} ds_o \frac{1}{s_o} = \ln \frac{1+z}{1-z} \tag{A.11b}$$

$$\int_{z-1}^{z+1} ds_o L'_o = -4\ln 2 + 2(1-z)\ln(1-z) + 2(1+z)\ln(1+z) \tag{A.11c}$$

$$\int_{z-1}^{z+1} ds_o s_o L'_o = -4z\ln 2 - (1-z)^2\ln(1-z) + (1+z)^2\ln(1+z) \tag{A.11d}$$

yielding for (A.10) :

$$H(z,o) = -2z(1+z)\ln(1+z) - 2z(1-z)\ln(1-z) + 4z\ln 2 - (1-z^2)K(z) \tag{A.12}$$

where

$$K(z) = \int_{z-1}^{z+1} ds_o \frac{L_o + L'_o}{s_o} \tag{A.13}$$

The remaining integral (A.13) still contains the singularity $s_o^{-1} \ln s_o^2$ in the integrand, but by a simple transformation a numerically tractable integral is obtained. Using (A.9), it follows that $K(z=o) = 0$, because for $z = 0$ the integrand in (A.13) changes sign if s_o is replaced by $-s_o$. By differentiating (A.13) with respect to z, one finds for $z \neq 0$:

$$\frac{dK(z)}{dz} = \frac{\ln(1+z) - \ln 2}{1+z} + \frac{\ln(1-z) - \ln 2}{1-z} + \int_{z-1}^{z+1} ds_o \left\{- \frac{1}{zs_o + \sqrt{s_o^2}} + \frac{1}{-zs_o + s_o^2 + \sqrt{s_o^2}}\right.$$

$$= [\ln(1+z) + \ln(1-z) - 2\ln 2]\left[\frac{1}{1+z} + \frac{1}{1-z}\right]$$

Thus, integrating again, and using the boundary condition $K(o) = 0$, one finds

$$K(z) = \frac{1}{2}\ln^2(1+z) - \frac{1}{2}\ln^2(1-z) - 2\ln 2[\ln(1+z) - \ln(1-z)]$$

$$+ \int_{-z}^{z} dx \frac{\ln(1+x)}{1-x} \tag{A.14}$$

Because $|z| \leqslant 1$, the remaining integral in (A.14) can be evaluated by standard numerical methods. A possible difficulty from $|z| \simeq 1$ is removed by the fact that from (A.12) $K(z)$ has to be multiplied by $(1 - z^2)$.

The more general integral $H(z,k)$ for $k \neq 0$, also contains the singularity $s_k^{-1} \ln s_k^2$. This problem too can be reduced to the standard principal value problem with a singularity of the form s_k^{-1}.

Noting that $W_k \to |u-v|$ if $s_k \to 0$, one obtains from (A.6) and (A.7) :

$$W_k = |u - v| + s_k^2 \frac{2u + 2v + s_k^2}{W_k + |u-v|}$$

$$L_k = -\ln \left[\frac{v - u + |v - u|}{2s_k^2} + \frac{1}{2} \frac{W_k + |u - v| + 2u + 2v + s_k^2}{W_k + |u - v|} \right]$$

The logarithmic singularity for $s_k \to 0$ thus only occurs in L_k if $u < v$. Explicitly separating it from the other terms, one obtains:

$$L_k = \ln s_k^2 - \ln \left[v - u + \frac{1}{2} s_k^2 \frac{W_k + 3v + u + s_k^2}{W_k + v - u} \right] \qquad u < v$$

$$= -\ln \frac{W_k + 3u + v + s_k^2}{2(W_k + u - v)} \qquad u > v$$

The corresponding expressions for L_k' are obtained by interchanging u and v. Because $u < v$ implies $-|z| < z' < |z|$ from (A.7), the logarithmic singularity in the integral (A.8) for $H(z,k)$ can then be integrated out analytically, and one is left with :

$$H(z,k) = -k(2z+k)\left[\ln^2|z + |z| + k| - \ln^2|z - |z| + k|\right]$$

$$-[1 - (z+k)^2]\left[\ln^2|z + 1 + k| - \ln^2|z - 1 + k|\right]$$

$$+ (3z + 3k - |z|)(z + |z| + k)\ln|z + |z| + k|$$

$$- (3z + 3k + |z|)(z - |z| + k)\ln|z - |z| + k|$$

$$- (3z + 3k - 1)(z + 1 + k)\ln|z + 1 + k|$$

$$+ (3z + 3k + 1)(z - 1 + k)\ln|z - 1 + k|$$

$$- 6(z + k)(1 - |z|) + \int_{-1}^{1} dz' \frac{A_k + \mathcal{L}_k}{z - z' + k} \tag{A.15a}$$

where :

$$\mathcal{L}_k = \begin{cases} u \ln(v - u + s_k^2 D_k) + v \ln D_k & u < v \\ v \ln(u - v + s_k^2 D_k) + u \ln D_k & u > v \end{cases} \tag{A.15b}$$

with :

$$D_k = \begin{cases} \dfrac{W_k + 3v + u + s_k^2}{2(W_k + v - u)} & u < v \\ \dfrac{W_k + 3u + v + s_k^2}{2(W_k + u - v)} & u > v \end{cases} \tag{A.15c}$$

The remaining integral in (A.15a) now only contains the standard principal value problem, due to $(z - z' + k)^{-1}$.

Combining then (A.4), (A.12), (A.14) and (A.15), the product of factors in G_{11} containing poles of first order, leads to a double integral. But in the numerical evaluation one has to calculate a principal value twice, whereas in the method with Fourier transforms only one principal value problem remained.

For completeness, it should be emphasized that the expression (34) in section C. is identical to (A.15). The derivation of this equivalence, using integral transforms, is rather lengthy and not very relevant. The main aim of (A.15) is to give an independent check of the numerical results, and the fact that the values from (34) and (A.15) agreed to within 4 digits, gives confidence in the numerical methods used.

Going over now to the evaluation of G_{20}, defined in (A.3), we note that (A.3) can be written as :

$$G_{20}(k,\nu) = X_{\vec{k}}(\nu + k^2/2,o) + X_{\vec{k}}(\nu - k^2/2,o)$$
$$- X_{\vec{k}}(\nu + k^2/2,\vec{k}) - X_{\vec{k}}(\nu - k^2/2,-\vec{k}) \qquad (A.16)$$

where :

$$X_{\vec{k}}(\omega,\vec{\gamma}) = \int d^3r \, \mathcal{N}(\vec{r}) \, \mathcal{J}(\vec{r} - \vec{\gamma}) \frac{1}{(\omega + i\varepsilon - \vec{r}.\vec{k})^2} \qquad (A.17)$$

with :

$$\mathcal{J}(t) \equiv \int d^3r \, \mathcal{N}(\vec{r}) \frac{1}{|\vec{r} - \vec{t}|^2} = 2\pi \left\{1 + \frac{1-t^2}{2t} \ln \left|\frac{1+t}{1-t}\right|\right\} \qquad (A.18)$$

From (A.16), it is readily seen that the integral (A.17) has only to be considered for $\vec{\gamma}$ along the $\vec{k}$-axis. Putting then :

$$\gamma_k = \begin{cases} k & ; \quad \vec{\gamma} = \vec{k} \\ 0 & ; \quad \vec{\gamma} = 0 \\ -k & ; \quad \vec{\gamma} = -\vec{k} \end{cases} \qquad (A.19a)$$

and expressing (A.17) in cylindrical coordinates with the z-axis along $\vec{k}$, one obtains :

$$X_{\vec{k}}(\omega,\gamma_k \frac{\vec{k}}{k}) =$$

$$= \int_{-1}^{1} dz \frac{1}{(\omega + i\varepsilon - kz)^2} \int_0^{\sqrt{1-z^2}} \rho \, d\rho \int_0^{2\pi} d\varphi \, \mathcal{J}(\sqrt{z^2 + \rho^2 + \gamma_k^2 - 2z\gamma_k})$$

$$= \pi \int_{-1}^{1} dz \frac{1}{(\omega + i\varepsilon - zk)^2} \int_{z^2 + \gamma_k^2 - 2z\gamma_k}^{1 + \gamma_k^2 - 2z\gamma_k} du \, \mathcal{J}(\sqrt{u}) \qquad (A.19b)$$

Using :

$$\frac{1}{(\omega + i\varepsilon - zk)^2} = \frac{1}{k} \frac{d}{dz} \frac{1}{\omega + i\varepsilon - zk}$$

and performing an integration by parts, one obtains :

$$\chi_{\vec{k}}(\omega, \gamma_k \frac{\vec{k}}{k}) =$$

$$= \frac{4\pi^2}{k} \int_{-1}^{1} \frac{dz}{\omega + i\varepsilon - zk} \left\{ z + \gamma_k \frac{2z\gamma_k - \gamma_k^2}{2\sqrt{1 - 2\gamma_k z + \gamma_k^2}} \ln \frac{1 + \sqrt{1 - 2z\gamma_k + \gamma_k^2}}{1 - \sqrt{1 - 2z\gamma_k + \gamma_k^2}} + (z - \gamma_k) \frac{1 - (z - \gamma_k)^2}{2\sqrt{(z - \gamma_k)^2}} \ln \frac{1 + \sqrt{(z - \gamma_k)^2}}{1 - \sqrt{(z - \gamma_k)^2}} \right\} \quad (A.20)$$

Combining then (A.16), (A.19) and (A.20) and collecting terms, one is left with the expression :

$$G_{20}(k,\nu) = \frac{2\pi^2}{k} \int_{-1}^{1} dz \left\{ \frac{F(z,o) - F(z,k)}{\nu + i\varepsilon - k^2/2 - kz} + \frac{F(z,o) - F(z,-k)}{\nu + i\varepsilon + k^2/2 - zk} \right\} \quad (A.21)$$

where $F(z,k)$ is precisely the function given in (33), obtained by the method of Fourier transforms.

REFERENCES

[1] D. Bohm and D. Pines, Phys. Rev. 92, 609 (1958).

[2] J. Lindhard, Kgl. Danske Videnskab. Selskab, Mat.-Fys. Medd. 28 (8) (1954).

[3] A.J. Glick and R.A. Ferrell, Ann. Phys. (USA) 11, 359 (1960).

[4] J. Hubbard, Proc. Roc. Soc. A243, 336 (1957); Phys. Letters 25A, 709 (1967).

[5] D.J.W. Geldart and S.H. Vosko, Canad. J. Phys. 44, 2137 (1966).

[6] L. Kleinman, Phys. Rev. 160, 585 (1967); 172, 383 (1968).

[7] K.S. Singwi, M.P. Tosi, R.H. Land and A. Sjölander, Phys. Rev. 176, 589 (1968).

[8] K.S. Singwi, A. Sjölander, M.P. Tosi and R.H. Land, Sol. Stat. Comm. 7, 1503 (1969); Phys. Rev. B1, 1044 (1970).

[9] M.P. Tosi, Nuovo Cimento 1, 160 (1969).

[10] D.C. Langreth, Phys. Rev. 181, 753 (1969).

[11] L. Hedin and S. Lundqvist, in "Solid State Physics", F. Seitz and D. Turnbull, eds., Vol.23, Academic Press, New York (1969), p.1.

[12] D.J.W. Geldart and R. Taylor, Canad. J. Phys. 48, 155 (1970).

[13] R.W. Shaw, Jr., J. Phys. C3, 1140 (1970).

[14] A.W. Overhauser, Phys. Rev. B3, 1888 (1971).

[15] T. Schneider, Physica 52, 481 (1971).

[16] P. Vashishta and K.S. Singwi, Phys. Rev. B6, 875 (1972).

[17] L.J. Sham, Phys. Rev. B7, 4357 (1973).

[18] K.N. Pathak and P. Vashishta, Phys. Rev. B7, 3649 (1973).

[19] A.K. Gupta, P.K. Aravind and K.S. Singwi, Sol. Stat. Comm. 26, 49 (1978).

[20] R. Dekeyser, Physica (Utrecht) 38, 189 (1968).

[21] F. Toigo and T.O. Woodruff, Phys. Rev. B2, 3958 (1970); 4, 371 (1971).

[22] A.K. Rajagopal and K.P. Jain, Phys. Rev. A5, 1475 (1972).

[23] A.K. Rajagopal, Phys. Rev. A6, 1239 (1972).

[24] A.R.P. Rau and A.K. Rajagopal, Phys. Rev. B11, 3604 (1975); 14, 3052 (1976).

[25] G. Niklasson, Phys. Rev. B10, 3052 (1974).

[26] A.A. Kugler, J. Stat. Phys. 12, 35 (1975).

[27] F. Brosens, L.F. Lemmens and J.T. Devreese, Phys. Stat. Sol. (b), 74, 45 (1976).

[28] D.N. Tripathy and S.S. Mandal, Phys. Rev. B16, 231 (1977).

[29] J.C. Kimball, Phys. Rev. A7, 1648 (1973); B14, 2371 (1976).

[30] F. Brosens, L.F. Lemmens and J.T. Devreese, Phys. Stat. Sol. (b) 81, 551 (1977).

[31] F. Brosens, J.T. Devreese and L.F. Lemmens, Phys. Stat. Sol. (b) 80, 99 (1977).

[32] F. Brosens, L.F. Lemmens and J.T. Devreese, Phys. Stat. Sol. (b) 82, 117 (1977).

[33] J.T. Devreese, F. Brosens and L.F. Lemmens, Phys. Stat. Sol. (b) 91, 349 (1979).

[34] P.E. Batson, C.H. Chen and J. Silcox, Phys. Rev. Letters 37, 937 (1976).

[35] W.E. Brittin and W.R. Chappell, Rev. Mod. Phys. 34, 620 (1962).

[36] D.R. Hamann and A.W. Overhauser, Phys. Rev. B143, 183 (1966).

[37] N.E. Brener and J.L. Fry, Phys. Rev. B19, 1720 (1979); 22, 2737 (1980).

[38] M. Gell-Mann and K.A. Brueckner, Phys. Rev. 106, 364 (1957).

[39] P. Nozières and D. Pines, Nuovo Cimento 9, 470 (1958).

[40] A. Holas, P.K. Aravind and K.S. Singwi, Phys. Rev. B20, 4912 (1979).

[41] J.T. Devreese, F. Brosens and L.F. Lemmens, Phys. Rev. B21, 1349 (1980).

[42] F. Brosens, J.T. Devreese and L.F. Lemmens, Phys. Rev. B21, 1363 (1980).

PART III - DISCUSSION AND COMPARISON WITH FIRST-ORDER PERTURBATION THEORY

A. INTRODUCTION

In part II, dynamical exchange effects were included in the dielectric function $\varepsilon(q,\omega)$ of the homogeneous electron gas, by performing the dynamical exchange decoupling in the equation of motion for the Wigner distribution function. The dielectric function can then be written in the form :

$$\varepsilon(q,\omega) = 1 + \frac{Q_o(q,\omega)}{1 - G(q,\omega)Q_o(q,\omega)} \tag{1}$$

where $Q_o(q,\omega)$ is the Lindhard polarizability. The local field correction $G(q,\omega)$ depends not only on the wave vector, but also on the frequency in contrast to the local field correction, introduced by Hubbard [1] . In equation (21) of part II, an explicit expression for $G(q,\omega)$ as a sixfold integral was obtained. This integral could hardly be evaluated numerically for reasons discussed in part II, and was analytically reduced to a tractable double integral. Expressing the wave vector in units of the Fermi wave vector k_F, and the frequency in units of twice the Fermi energy E_F :

$$\begin{aligned} q &= kk_F \\ \hbar\omega &= 2\nu E_F \end{aligned} \tag{2}$$

it was shown that $G(kk_F, 2\nu E_F/\hbar)$ is a universal function of k and ν for all densities, as given in equation (28) of part II.

This sixfold integral was reduced to a double integral by two different methods, and two formally different, but equivalent expressions were obtained. The most convenient form for numerical integration is found in equation (31-34) of part II. Some results for Re $\varepsilon(q,\omega)$ and Im $1/\varepsilon(q,\omega)$ were displayed in part II, where it was shown that $G(q,\omega)$ has a logarithmic singularity at the parabolas $|\omega| = \hbar|q^2/2 \pm q\, k_F|/m$, which define distinct regions of the particle-hole continuum.

In this third part, we give a preliminary evaluation of this method for including dynamical exchange effects. In section B, the internal consistency of the dynamical exchange decoupling and its relation to several sum rules is discussed. In section C, we summarize some numerical results. In the static limit it is shown that an instability of the paramagnetic state at low electron densities arises quite naturally. A similar instability was previously discussed by Hamann and Overhauser [2] in studying the spin susceptibility. Furthermore, at finite frequency, some results for $G(q,\omega)$ and Im $\varepsilon^{-1}(q,\omega)$ are presented, which show that dynamical exchange effects shift the frequencies of the maxima in the structure factor to lower frequencies than in RPA. This shift substantially improves the agreement with experimental data on the plasmon dispersion in aluminum. Finally, in section E, the variational method which we proposed is compared with the calculation of the diagrams for the proper polarizability to first order in the electron-electron interaction [3].

B. CONSISTENCY REQUIREMENTS

In comparing theoretical predictions from the dielectric response function of the homogeneous electron gas to experimental results in real metals, two fundamentally different sources for possible discrepancies can arise. The description of a metal with the jellium model, i.e. replacing the discrete lattice of atoms by a uniform neutralizing background, is an oversimplification. But even this simplified theoretical model is not exactly soluble, and several approximations have to be made within the framework of the model to derive the dielectric properties. The nature of the jellium model and the mathematical approximations made in the study of the model both influence the results, and comparison to experiment does not provide a unique way to judge the validity of the theoretical assumptions. Therefore, exact interrelations among various quantities in the electron gas model, which also can be calculated from the dielectric function, form a powerful tool to study the consistency of the approximations made in deriving $\varepsilon(q,\omega)$. Several sum rules were already discussed in section D of part I, and will now be applied to the dielectric function which we obtained.

From the explicit expression (28) for $G(q,\omega)$ in part II, one easily derives [4-6] that in the static limit :

$$G(q,o) \underset{q\to o}{\to} \frac{1}{4}\left(\frac{q}{k_F}\right)^2 \tag{3}$$

and therefore :

$$\varepsilon(q,o) \underset{q\to o}{\to} \frac{k_{FT}^2}{q^2} \frac{1}{1-\frac{r_s}{\pi}\left(\frac{4}{9\pi}\right)^{1/3}} \tag{4}$$

As is well known, the long wavelength result (4) is also obtained by using the Hartree-Fock compressibility in the compressibility sum rule (given in (55) of Part I). This means that the limit (3) for G(q,o) is precisely expected from the Hartree-Fock approximation for the ground state energy [see eq. (55-57) of Part I].

Also in the static limit, but at sufficiently large wave vector, one easily proves that with dynamical exchange decoupling :

$$G(q,o) \underset{q\to\infty}{\to} \frac{1}{3} \tag{5}$$

which is in agreement with the large q limit of the first order HF correction [7] which is dominant at large q. Combined with the Niklasson relation (I.58), the limit (6) thus implies that g(o) = 1/2. The same result for g(o) is obtained from the static structure factor in the HF approximation.

It should be noted that the frequency dependent approach to $G(q,\omega)$, worked out by Toigo and Woodruff [8] does not reproduce the expected limit (3), as discussed in [9], due to an implicit averaging over all frequencies in their derivation.

The agreement between the HF approximation and the static limit of the dielectric function with dynamical exchange effects in deriving the compressibility and the pair correlation function, is not surprising, because in the static limit the dynamical exchange decoupling reduces into the HF decoupling. This agreement simply indicates that the internal consistency of the decoupling method in the static limit is not disturbed by the variational procedure used to solve the equation of motion for the induced electron density.

More severe consistency relations however involve the frequency dependence of $G(q,\omega)$. In eq.(64) of part I, the pair correlation function g(o) at the origin was obtained from the frequency dependent dielectric function via the Kimball-Niklasson relation [10-11]. This calculation of g(o) requires the evaluation of the static structure S(q) in the limit $q \to \infty$.

For our dielectric function with dynamical exchange decoupling, this evaluation can be performed analytically. The derivation is given in the appendix of ref.[13], and results in

$$\lim_{q\to\infty} q^4 [S(q) - 1] = - \frac{4\pi n}{a_o} \tag{6}$$

We already presented this result in [14] without derivation.

Using eq.(64) of part I, one thus readily obtains that $g(o) = 1/2$, in agreement with the pair correlation function as obtained from the static limit (5), and with the HF pair correlation function. As already mentioned in section D of part I, this internal consistency between the frequency dependent dielectric function and its static limit can not be reached by any static approximation for $G(q,\omega)$.

It should be noted that the frequency dependent approximation by TW [8], also leads to the limiting behaviour (6) and therefore $g(o) = 1/2$ is also obtained from the high frequency limit of their dielectric function by combining (I.64) and (6). But at $\omega = 0$, the TW local field correction becomes 2/3 [9] in the large q limit, and the static relation (I.58) then implies $g(o) = 0$. With the TW approximation, one thus obtains a different value for g(o) if derived from the static limit or from the high frequency limit of $G(q,\omega)$.

An other interesting consistency requirement can be obtained from the third frequency moment sum rule (I.70) and the general analytical properties of $G(q,\omega)$, from which the exact relations (I.71-72) in part I are derived.

In Appendix A of ref.[12], the high frequency limit of our expression for $G(q,\omega)$ was derived. For large and small wave vectors respectively, the following expressions were obtained :

$$G(q,\infty) \underset{q\to\infty}{\to} \frac{1}{3}\left[1 - \frac{4}{7}\left(\frac{k_F}{q}\right)^2 + \ldots\right] \tag{7}$$

$$G(q,\infty) \underset{q\to o}{\to} \frac{3}{20}\left(\frac{q}{k_F}\right)^2 + \ldots \tag{8}$$

Comparing (7-8) with the exact relations (I.71-72) of Part I, it follows that the dynamical exchange decoupling implies that $\langle KE\rangle = \langle KE\rangle_o$, and that

$$\langle V\rangle = -\frac{9}{16}\,\frac{m\omega_p^2}{k_F^2} = -\frac{3}{4\pi}\,e^2 k_F \tag{9}$$

which is precisely the exchange energy in the HF approximation. Furthermore, this identification again leads to $g(o) = 1/2$, as obtained previously from the static limit and the frequency dependent behaviour of $G(q,\omega)$ with dynamical exchange decoupling.

Finally, we mention that the high-frequency limit of $G(q,\omega)$, as calculated in Appendix A of ref.[12], exactly gives the expression for the third frequency moment (I.70), with the HF static structure factor inserted in (I.70b).

In summary, the static limit of the dielectric function obtained with the dynamical exchange decoupling method, presented in part II, is consistent with the HF results for both the compressibility and the pair correlation function at the origin. This is quite reasonable, because the HF approximation is the static limit of the dynamical exchange decoupling. Furthermore, as shown above, this dynamical exchange treatment in the high frequency limit is internally consistent with the static results for the energy and the pair correlation function, which gives confidence in the procedure used to solve the equation of motion for the Wigner distribution function. Because the approximate solution of this equation, derived in part II, rigorously satisfies the equation of motion for the charge density and the current density, it is not surprising that the dielectric function with dynamical exchange decoupling conserves frequency moments to infinite order in the HF ground state as shown by an alternative derivation of the same expression for $G(q,\omega)$ in [6].

C. DISCUSSION OF NUMERICAL RESULTS

§ 1. Static limit

From the two-dimensional integral for $G(q,\omega)$, given in eq.(31-34) of part II, numerical values can be obtained by elementary numerical methods. Some preliminary results in the static limit $\omega = 0$ have been reported in [15]. In [13], a rather broad overview is presented, including a table of values of $G(kk_F, 2\nu E_F/\hbar)$ for several values of k and ν. Because, as a function of k and ν, $G(kk_F, 2\nu E_F/\hbar)$ is independent of the density, these tabulated values can easily be used in practical computations for arbitrary r_s.

For an overview and a detailed discussion of these results we refer to ref.[13]. In the present paper, we only repeat the main points of interest, and some representative figures.

In figure 1, the static limit $G(q,o)$ is plotted, and compared to the local field correction from other theories. In the static limit, an alternative procedure has been proposed

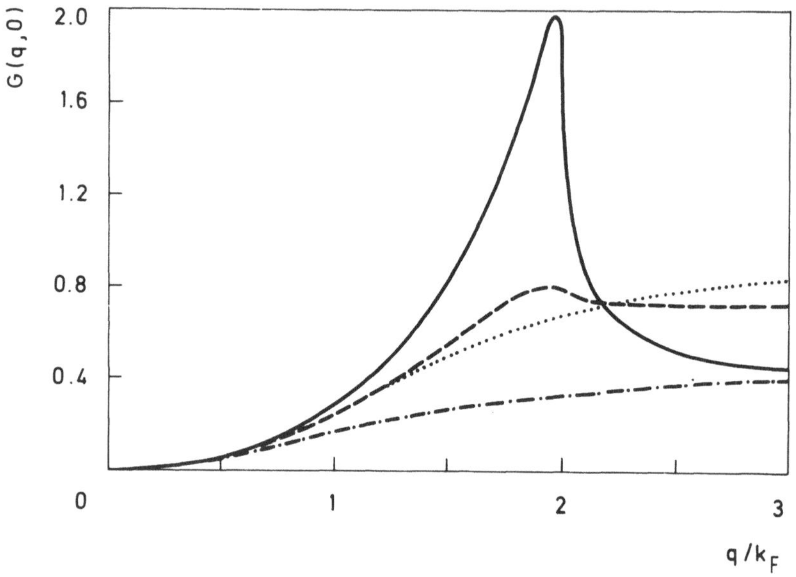

Figure 1. Local field correction G(q,o) in the static limit as calculated from dynamical exchange decoupling (full line), compared to the approximations of Hubbard [2] (-·-·-·), Toigo and Woodruff [8] (---) and Vashishta and Singwi [26] at $r_s = 3$ (.....).

to evaluate G(q,o) from a threefold numerical integral, and the values obtained do not significantly differ from ours. G(q,o) can also be obtained formally by considering the susceptibility from a diagrammatic expansion [7] as the first two terms in a geometric series in powers of e^2. A theoretical justification of this procedure is given in [15].

The most striking feature of figure 1 is that G(q,o) as derived with the dynamical exchange decoupling, exhibits a pronounced structure about $q \simeq 2k_F$. This peak is a trace left from the logarithmic singularities in $G(kk_F, 2\nu E_F/\hbar)$ for $\nu_{\pm} \simeq |k^2/2 \pm k|$ (as discussed in part II), and which cancel each other at the intersection of both parabolas for k = 2. The rather pronounced peak might be related to singularities in the off-diagonal matrix elements of the non-local exchange operator [16].

Due to the peak in G(q,o), the static dielectric function not only diverges in the long wavelength limit, but also at finite wave vector for sufficiently low electron densities, because the denominator $[1 - G(q,o)\ Q_o(q,o)]$ of G(q,o) in (1) has a zero for some critical r_s-value. Since $G(kk_F,o)$ is a universal function of k, and $Q_o(kk_F,o)$, considered as a function of k, is proportional to r_s, the critical r_s value for this singularity is found to be :

$$\frac{1}{r_s^{critical}(k)} = G(kk_F,o)\ [Q_o(kk_F,o)]_{r_s=1} \tag{10}$$

In figure 2, these critical values for r_s are plotted and compared to the critical r_s values, as obtained from the Hubbard approximation [2] :

$$G_{Hubbard}(q) = \frac{1}{2}\frac{q^2}{q^2 + k_s^2} \tag{11}$$

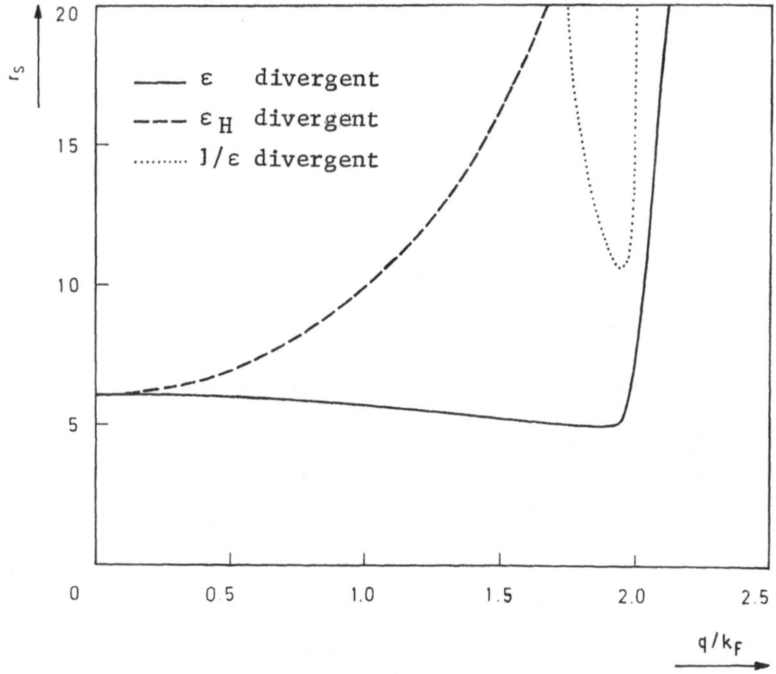

Figure 2. Critical r_s values as a function of q/k_f for which ε(q,o) with dynamical exchange diverges (full line) compared to these critical values from the Hubbard approximation (---). The dotted curve indicates the r_s value for which ε(q,o) has a zero, and thus 1/ε(q,o) shows a pole.

where the screening constant k_s^2 is put equal to $2k_F^2$ in order to satisfy the compressibility sum rule.

Because of this choice of k_s^2, the singularities in $\varepsilon(q,o)$ resulting from the Hubbard approximation and from the dynamical exchange decoupling coincide at $r_s = 6.0292$ in the long wavelength limit. This is exactly at the density for which the paramagnetic HF state becomes unstable, as reflected in the sign inversion of the compressibility.

For increasing wavevector, the critical r_s value increases monotonically in the Hubbard approximation. This is in contrast to the dynamical exchange decoupling method, where the rapid increase of $G(q,o)$ to a peak near $q = 2k_F$, results in a decrease of the critical r_s values to a minimum at $r_s = 5$ near $q \simeq 1.85\ k_F$. As discussed in detail in ref. [13], these critical r_s-values are related to the instability in the susceptibility as derived by Hamann and Overhauser [2], hereafter referred to as HO.

Another consequence of the peak in $G(q,o)$ is the occurence of a zero in the static dielectric function for $r_s \simeq 10.6$. The r_s-values for which $\varepsilon(q,o) = 0$ are also plotted in figure 2. A minimum in these critical values of r_s as a function of q is found for $r_s \simeq 10.6$ near $q \simeq 1.94\ k_F$. These critical r_s-values determine a (q,r_s) region with $1/\varepsilon(q,o)$ negative, and thus where the electron gas in unstable relative to the occurence of charge density waves, because $1/\varepsilon(q,o)$ is the ratio between the total effective potential and the external applied potential. If $\varepsilon(q,o)$ is negative, the induced charge density overcompensates the external charge density.

In figure 3, $1/\varepsilon(q,o)$ is plotted as a function of q for $r_s = 3, 4.92, 5.64, 7.67$ and 12.79, which are r_s-values also used by HO. At $r_s = 3$ (figure 3a), no special structure in $1/\varepsilon(q,o)$ arises from exchange effects. But at $r_s = 4.92$, i.e. just below the minimal critical value of r_s, defined by (20), pronounced structure compared to RPA starts showing up for $q \simeq 2k_F$ (fig.3b). For $r_s \gtrsim 5$, a region appears where $1/\varepsilon(q,o)$ is negative, as shown in figure 3c for $r_s = 5.64$. With still increasing r_s (> 6.0292), the region of negative $1/\varepsilon(q,o)$ spreads out and includes $q = 0$, according to the compressibility sum rule, as shown in figure 3d for $r_s = 7.67$.

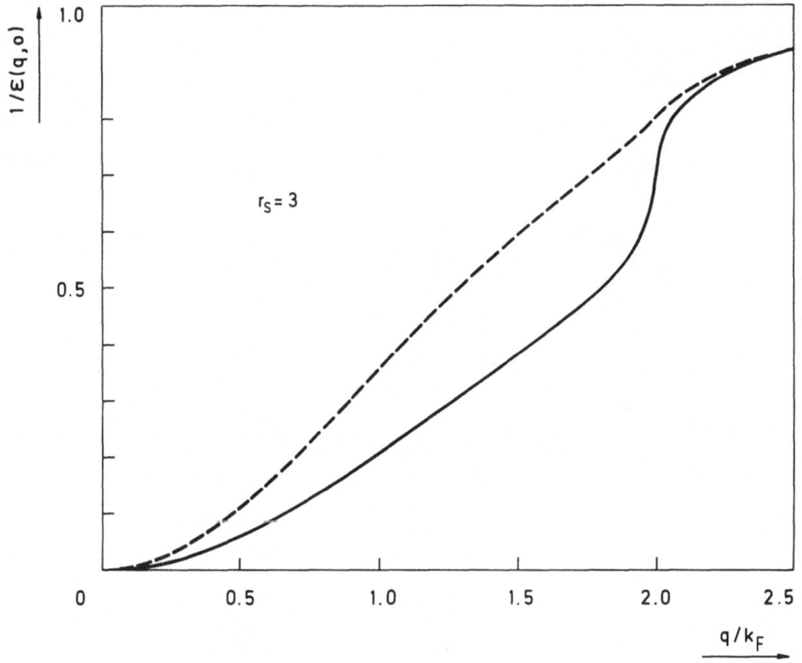

1/ε(q,o)
1.0
0.5
0
$r_s = 3$
0.5
1.0
1.5
2.0
2.5
q/k_F

Fig. 3a

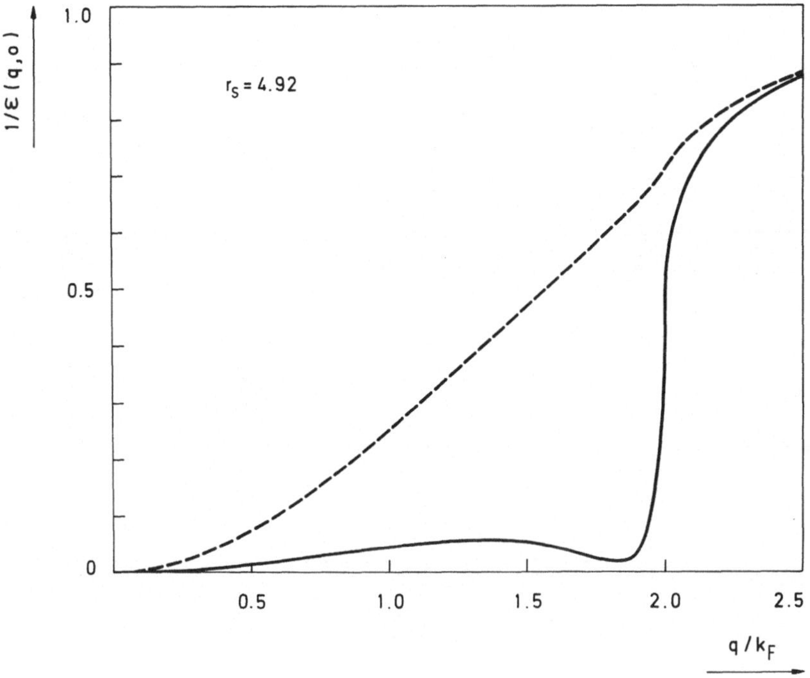

1/ε(q,o)
1.0
0.5
0
$r_s = 4.92$
0.5
1.0
1.5
2.0
2.5
q/k_F

Fig. 3b

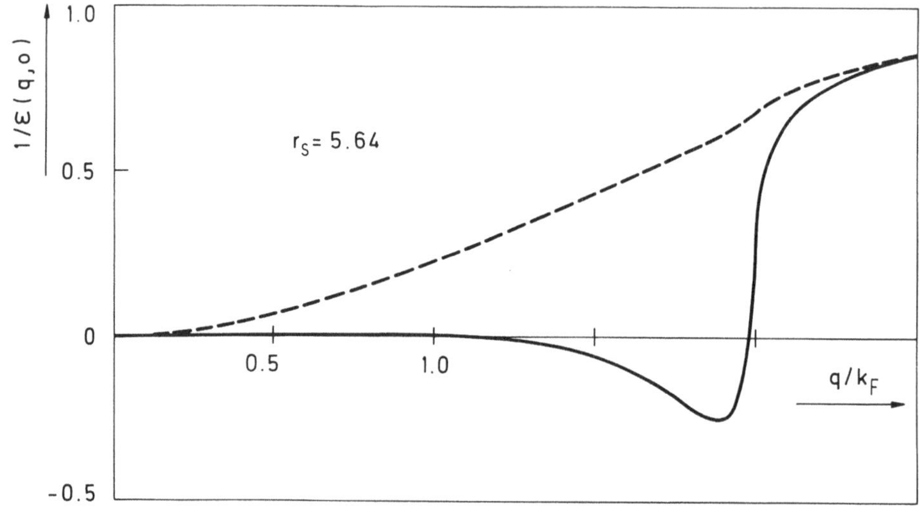

Fig. 3c

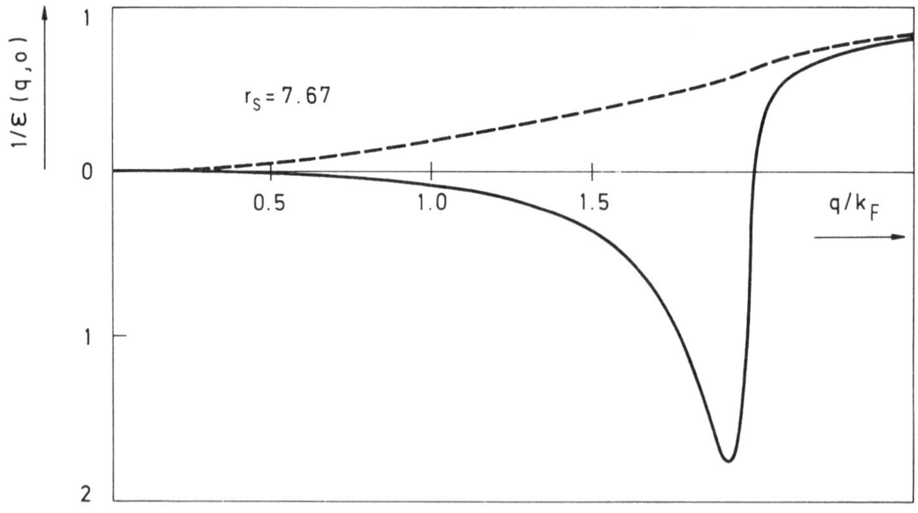

Fig. 3d

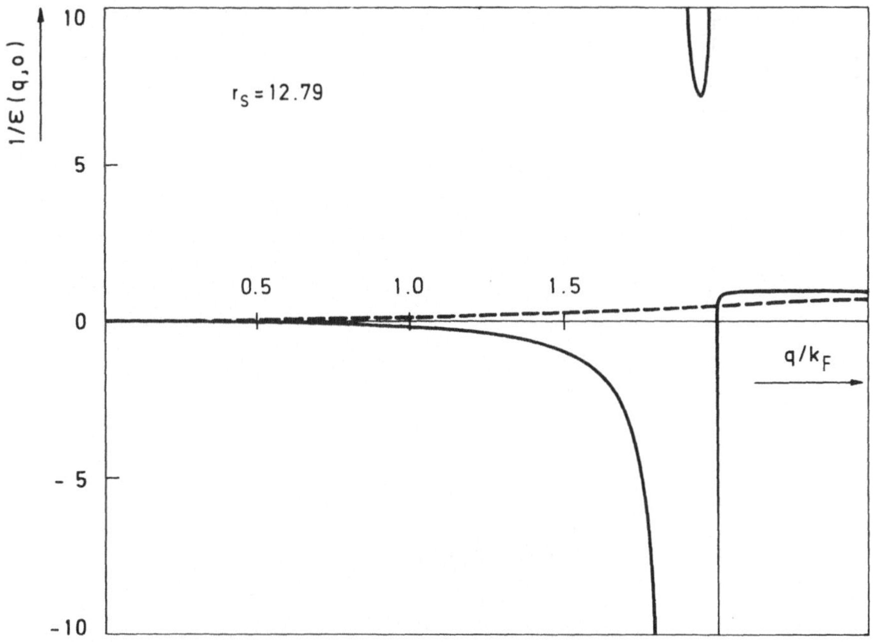

Figure 3(a-e). Inverse static dielectric function with dynamical exchange decoupling as a function of wavevector (full line), and compared to RPA (---) for r_s = 3 (a), 4.92 (b), 5.64 (c), 7.67 (d) and 12.79 (e).

Furthermore, if $1/\varepsilon(q,o)$ is negative, it shows a steep structure near $q = 2k_F$. The negative peak increases in magnitude with increasing r_s, and tends to $-\infty$ at $r_s = 10.6$, where $\varepsilon(q,o)$ has zeros. For $r_s > 10.6$, a region of positive $1/\varepsilon(q,o)$ appears in between the two q-values satisfying $\varepsilon(q,o) = 0$, as is shown in figure 3e for $r_s = 12.79$.

In figure 4, the corresponding plots for $\varepsilon(q,o)$ are given for r_s = 3,4.92, 5.64, 7.67 and 12.79.

An instability, similar to the one obtained here from the dynamical exchange decoupling, has been derived and discussed by HO [2], who studied the stability of the electron gas in the paramagnetic state, relative to small deformations of the spin magnetization. In the HO approach the spin susceptibility was to be found from the solution of an integral equation for the self-

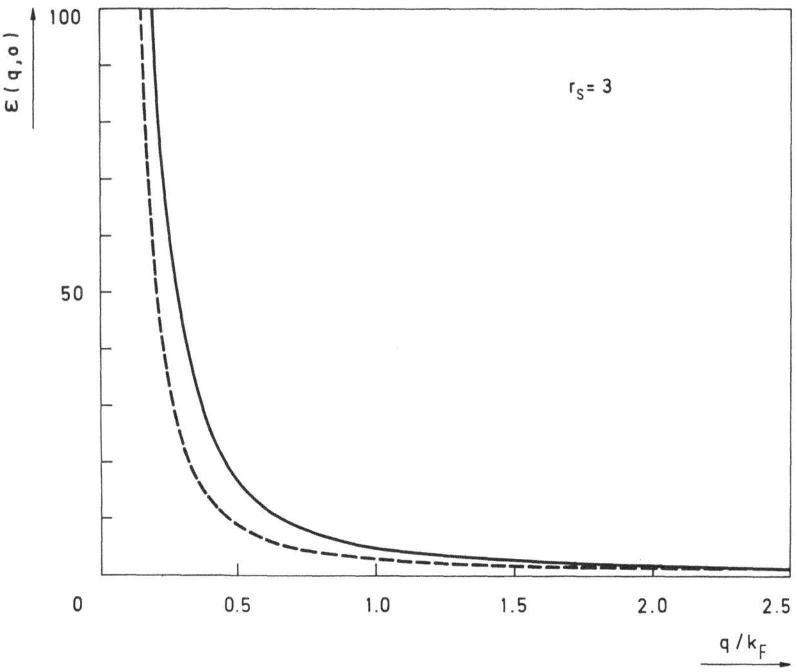
100
ε (q,o)
50
0
0.5
1.0
1.5
2.0
2.5
q / k_F
r_S = 3

Fig. 4a

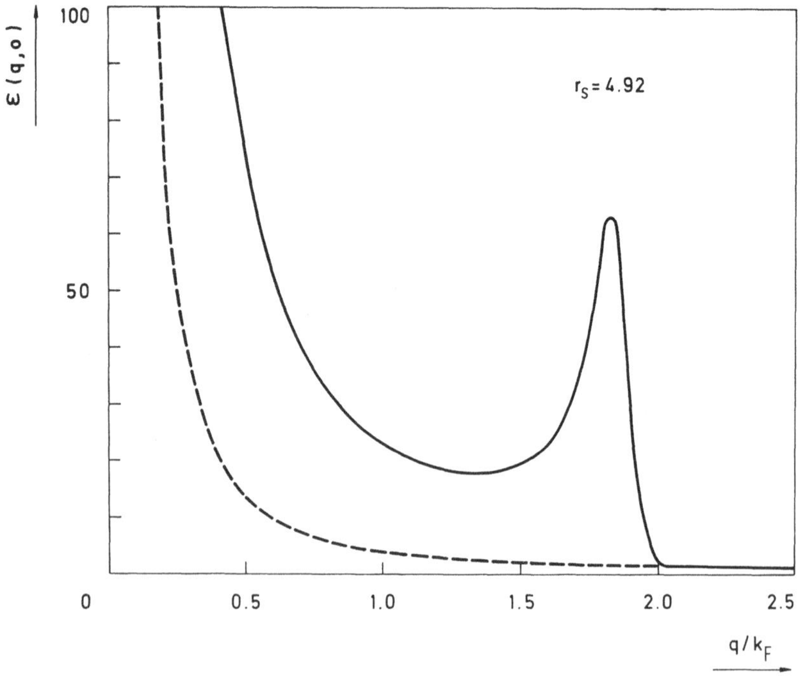
100
ε (q,o)
50
0
0.5
1.0
1.5
2.0
2.5
q / k_F
r_S = 4.92

Fig. 4b

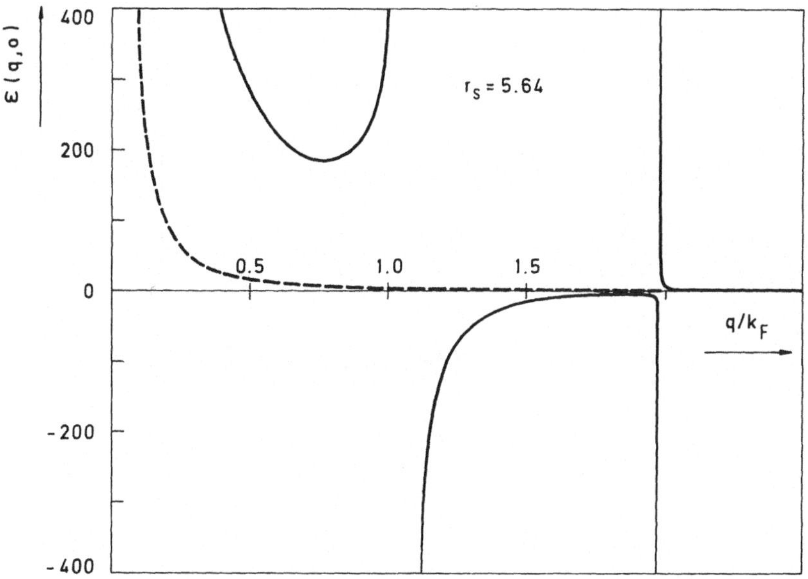

ε(q,o)
400
200
0
-200
-400
r_s = 5.64
0.5
1.0
1.5
q/k_F

Fig. 4c

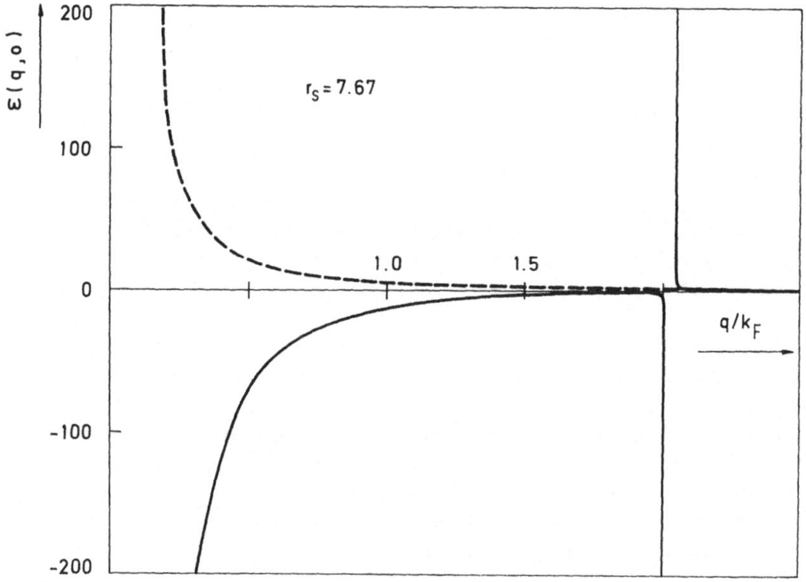

ε(q,o)
200
100
0
-100
-200
r_s = 7.67
1.0
1.5
q/k_F

Fig. 4d

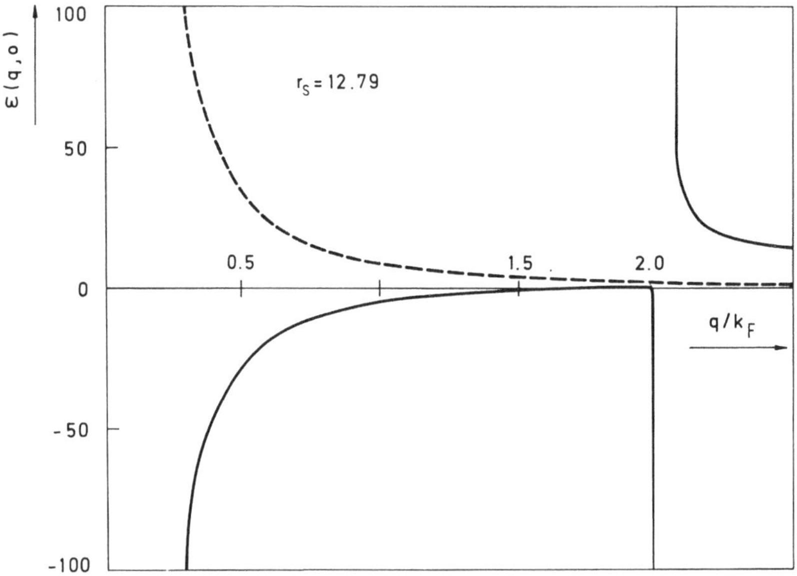

Fig. 4(a-e). Static dielectric function with dynamical exchange decoupling as a function of wavevector (full line), and compared to the RPA (---) for r_s = 3 (a), 4.92 (b), 5.64 (c), 7.67 (d) and 12.79 (e).

consistent effective field, which is of the same nature as the integrated equation of motion the present authors derived in equation (25a) of part II. For a more detailed discussed of the relation between their and our results, we refer to [13] .

It should be noted that the singular behaviour of the response function in the critical region, does not provide a reliable description of the instability. It only indicates that in the critical region the possibility of spin or charge density waves should have been included in the equilibrium conditions. A discussion of these instabilities and their consequences on the equilibrium properties of the electron gas is given in [17] , and in Overhauser's paper in the present volume.

§ 2. Dynamical behaviour of $G(q,\omega)$

The frequency dependent local field correction $G(q,\omega)$ has been evaluated numerically in two independent ways, as described in part II.

In order to save computation time, it is worthwhile to remember the theorem that in the units $q = kk_F$ and $\omega = 2\nu E_F/\hbar$, $G(kk_F, 2\nu E_F/\hbar)$ is a universal function of k and ν for all densities. In part II it was shown that

$$G(kk_F, 2\nu E_F/\hbar) = \frac{2\pi^2}{k} f(k,\nu) \int_{-1}^{1} dz\, T(z,k) \left[\frac{1}{\nu + i\varepsilon - k^2/2 - kz} - \frac{1}{\nu + i\varepsilon + k^2/2 + kz} \right] \quad (12)$$

where $f(k,\nu)$ and $T(z,k)$ are given in eq.(20) and eq.(31-34) of part II and where $T(z,k)$ can be obtained by performing a single integral numerically. It thus follows that tabulating $T(z,k)$ at given k, as a function of z for $-1 \leqslant z \leqslant 1$, allows to compute $G(kk_F, 2\nu E_F/\hbar)$ at arbitrary ν by performing a single numerical integral for each ν. Numerical results for ReG and ImG are plotted in figure 5 for several values of k. Numerical values are listed in table I of ref.[13].

For $\nu = 0$, $G(kk_F, 2\nu E_F/\hbar)$ tends to the static limit, displayed in figure 1 as a function of k. At $|\nu| = |k^2/2 \pm k|$, the real part of the integral in (12) diverges logarithmically, as discussed in part II. Because the factor $f(k,\nu)$ contains $1/Q_o^2(kk_F, 2\nu E_F/\hbar)$ this divergent real part contributes to $\mathrm{Im}G\,(kk_F, 2\nu E_F/\hbar)$ at $\nu = |k^2/2-k|$, where $\mathrm{Im}Q_o\,(kk_F, 2\nu E_F/\hbar)$ differs from zero. Consequently, with dynamical exchange decoupling, both $\mathrm{Re}G\,(kk_F, 2\nu E_F/\hbar)$ and $\mathrm{Im}G\,(kk_F, 2\nu E_F/\hbar)$ show a logarithmic singularity at $\nu = |k^2/2-k|$.

On the other hand, at $\nu = k^2/2+k$, $\mathrm{Im}Q_o\,(kk_F, 2\nu E_F/\hbar)$ equals zero, and thus only the imaginary part of the integral in (12) contributes to $\mathrm{Im}G\,(kk_F, 2\nu E_F/\hbar)$. At the upper boundary of the particle-hole continuum, the divergence only appears in $\mathrm{Re}G\,(kk_F, 2\nu E_F/\hbar)$, while $\mathrm{Im}G\,(kk_F, 2\nu E_F/\hbar)$ remains finite.

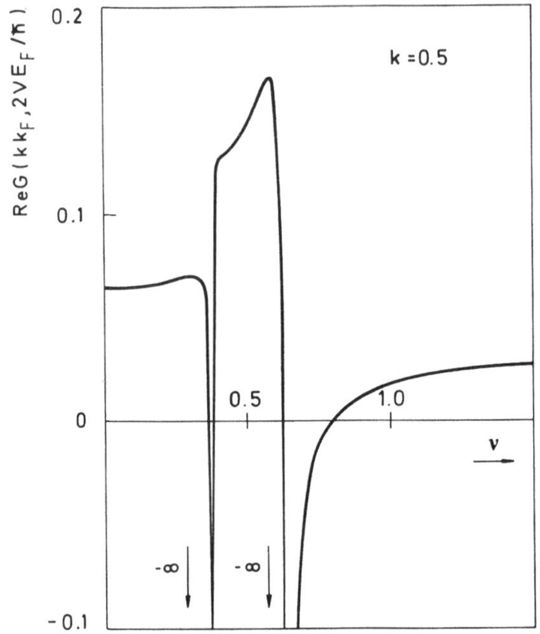
ReG(kk$_F$,2VE$_F$/ħ)
0.2
0.1
0
-0.1
k =0.5
0.5
1.0
v
-∞
-∞

Fig. 5a

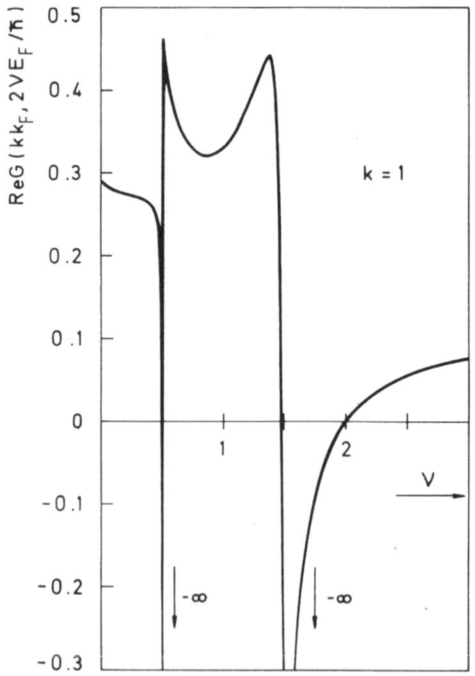
ReG(kk$_F$,2VE$_F$/ħ)
0.5
0.4
0.3
0.2
0.1
0
-0.1
-0.2
-0.3
k = 1
1
2
V
-∞
-∞

Fig. 5b

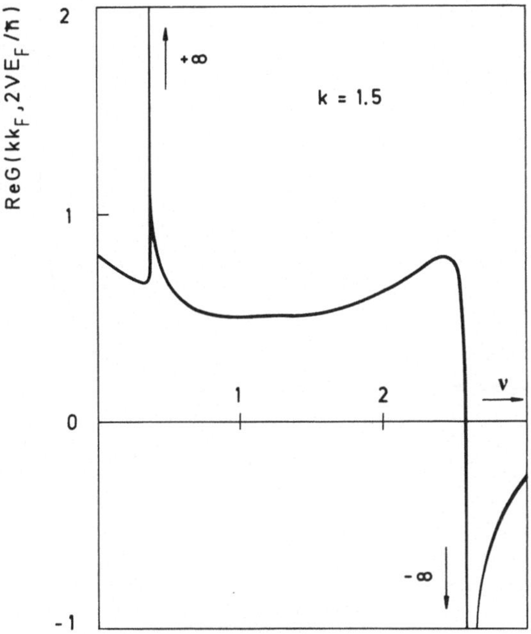
ReG(kk_F, 2VE_F/ħ)
2
1
0
-1
+∞
k = 1.5
1
2
ν
-∞

Fig. 5c

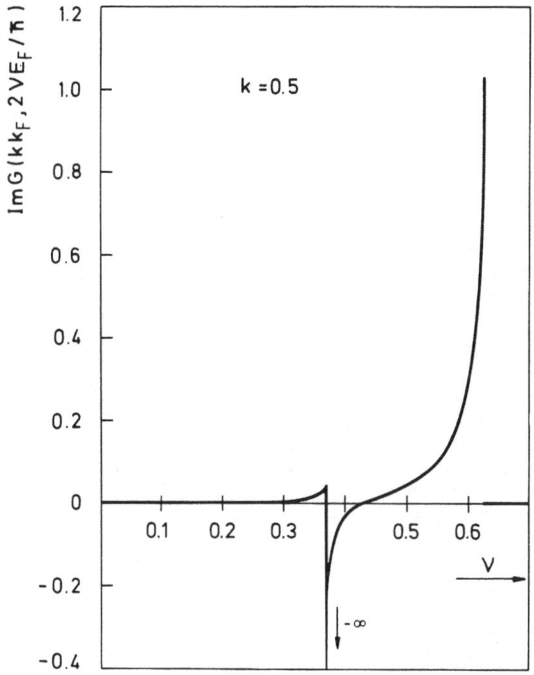
ImG(kk_F, 2VE_F/ħ)
1.2
1.0
0.8
0.6
0.4
0.2
0
-0.2
-0.4
k =0.5
0.1
0.2
0.3
0.5
0.6
ν
-∞

Fig. 5d

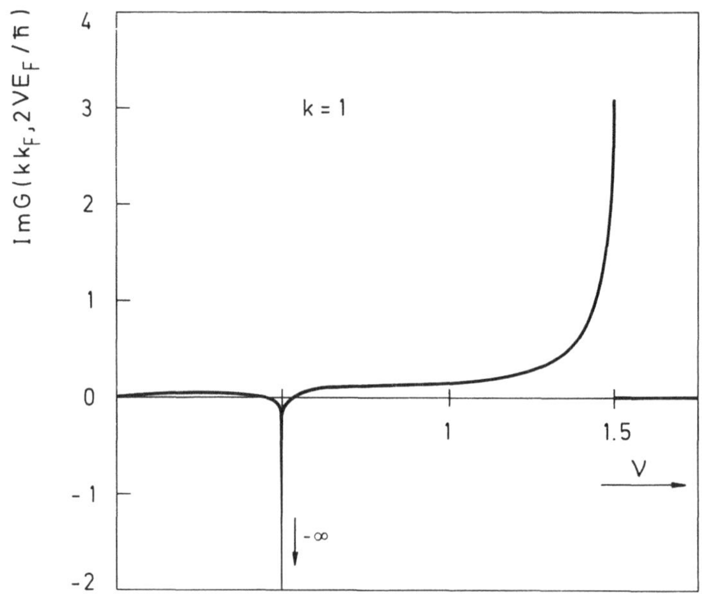
ImG(kk_F,2VE_F/ħ)
4
3
2
1
0
-1
-2
k = 1
1
1.5
V
-∞

Fig. 5e

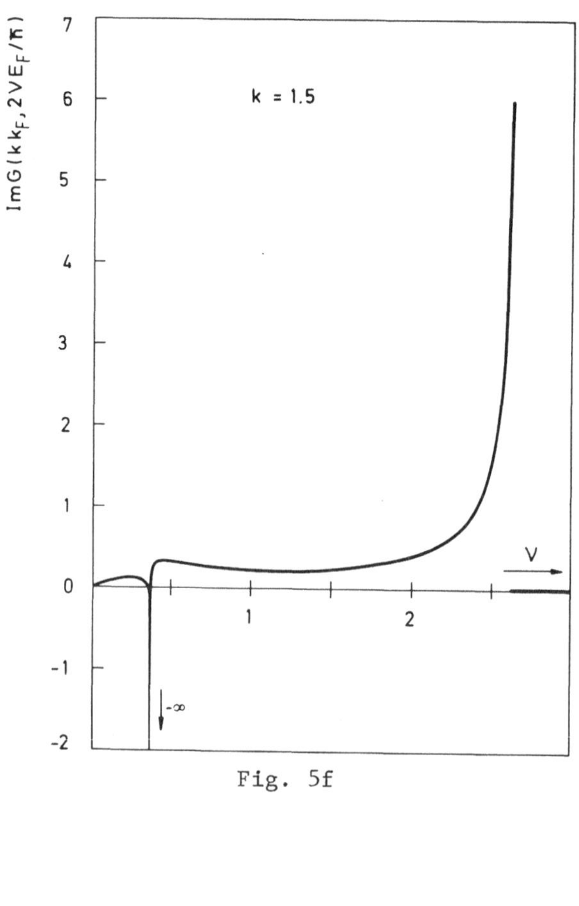
ImG(kk_F,2VE_F/ħ)
7
6
5
4
3
2
1
0
-1
-2
k = 1.5
1
2
V
-∞

Fig. 5f

For frequencies above the continuum, both $\mathrm{Im}Q_o(kk_F, 2\nu E_F/\hbar)$ and the imaginary part of the integral in (12) are zero, and thus $\mathrm{Im}G(kk_F, 2\nu E_F/\hbar)$ is zero above the continuum, resulting in a discontinuity at $\nu = |k^2/2+k|$.

With further increasing frequency, $G(kk_F, 2\nu E_F/\hbar)$ approaches the high frequency limit, calculated in Appendix A of ref.[12].

§ 3. Dynamical behaviour of $\varepsilon(q,\omega)$ and plasmon dispersion

Because $G(kk_F, 2\nu E_F/\hbar)$ is a universal function of k and ν for all densities, the density dependence in $\varepsilon(kk_F, 2\nu E_F/\hbar)$ as a function of k and ν only enters via the Lindhard polarizability $Q_o(kk_F, 2\nu E_F/\hbar)$, which as a function of k and ν only depends on the density by the proportionality factor r_s. Therefore, the numerical evaluation of $\varepsilon(kk_F, 2\nu E_F/\hbar)$ at an arbitrary density is rather straightforward, given the values of $G(kk_F, 2\nu E_F/\hbar)$ in table I of ref.[13]. Because of the logarithmic singularity, it is rather important that this table contains many points near $\nu = |k^2/2 \pm k|$.

For relatively small r_s, the dielectric function with dynamical exchange included is not much different from RPA, except for the exchange effects near $\nu = |k^2/2 \pm k|$. But with increasing r_s, overall exchange effects appear in the numerical values, as shown in figure 6 where Re $\varepsilon(q,\omega)$ is plotted as a function of ω for $r_s = 1$, 3 and 5 and for $q = k_F$. For $r_s = 2$, we refer to figure 4 in part II.

The singularity in Re $\varepsilon(q,\omega)$ near $\omega = \hbar(q^2/2 + qk_F)/m$ is related to the logarithmic singularity in $G(q,\omega)$. Although a singularity also occurs in $G(q,\omega)$ at $\omega = \hbar|q^2/2 - qk_F|/m$, it does not produce a pole in Re $\varepsilon(q,\omega)$, but only a strongly peaked structure, because $Q_o(q,\omega)$ has an imaginary part in this region.

In figure 7, Im $1/\varepsilon(q,\omega)$ is shown for $r_s = 1$, 3 and 4 and for $q = k_F$. For $r_s = 2$ we refer to figure 3 of part II. The logarithmic singularity in $G(q,\omega)$ at $\omega = \hbar|q^2/2 - qk_F|/m$

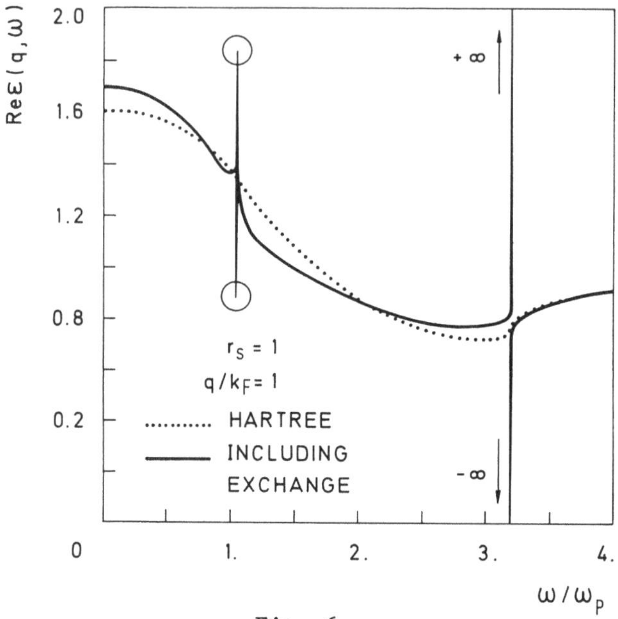
Re ε(q,ω)
2.0
1.6
1.2
0.8
0.2
0
1.
2.
3.
4.
ω/ω_p
+∞
-∞
r_S = 1
q/k_F = 1
HARTREE
INCLUDING EXCHANGE

Fig. 6a

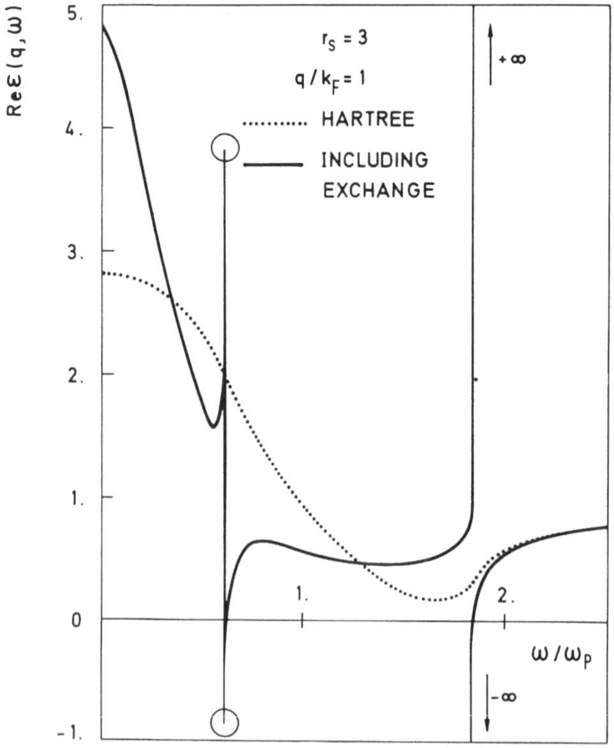
Re ε(q,ω)
5.
4.
3.
2.
1.
0
-1.
1.
2.
ω/ω_p
+∞
-∞
r_S = 3
q/k_F = 1
HARTREE
INCLUDING EXCHANGE

Fig. 6b

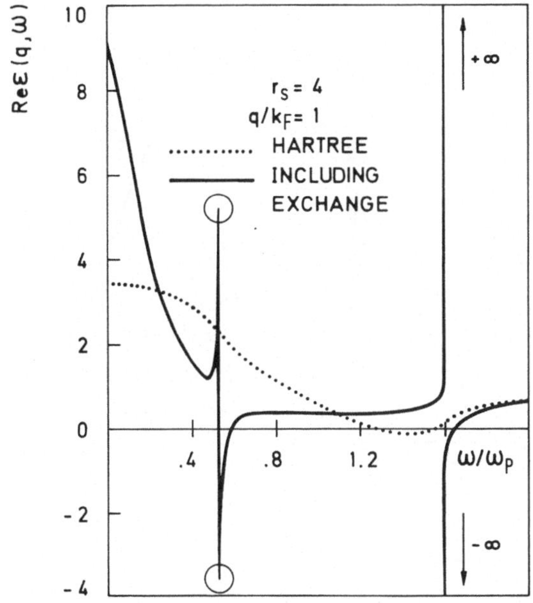

Fig. 6(a-c). Real part of the dielectric function $\varepsilon(q,\omega)$ as a function of frequency for $q = k_F$ at the electron densities, characterized by $r_s = 1$ (fig.6a), 3 (fig.6b) and 4 (fig.6c). The RPA dielectric function is represented by the dotted line. The open circles indicate that the magnitude at the peaks is not known due to numerical inaccuracy.

induces a zero in Im $1/\varepsilon(q,\omega)$. However, the width of this dip is so small that one hardly expects it to be important for comparison with experiment.

Another consequence of the logarithmic singularity is found in the behaviour of the plasmon dispersion. As already mentioned in [18], the plasmon, defined by Re $\varepsilon(q,\omega) = 0$ outside the continuum, does not penetrate into the particle-hole continuum, but approaches it asymptotically. However, the physically relevant property is the frequency at the maximum in the structure factor. The oscillator strength of the plasmon is proportional to $[d\varepsilon(q,\omega)/d\omega]^{-1}$ at the plasmon frequency. From figure 6, it is clear that Re $\varepsilon(q,\omega)$ is a very steep function of ω near its zero when this zero approaches the continuum. Consequently, the

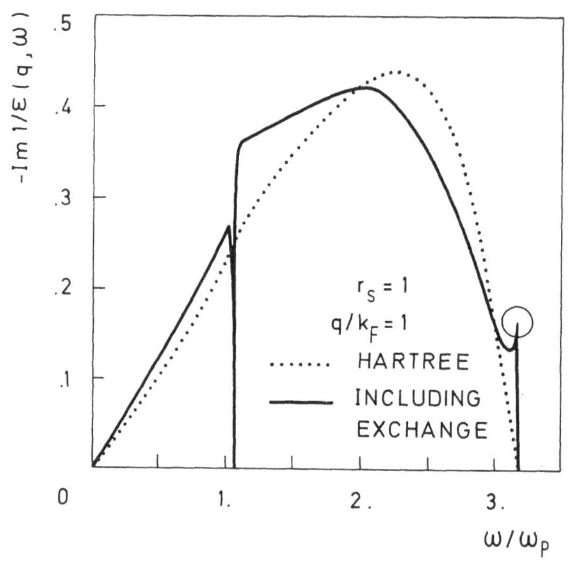
-Im 1/ε(q,ω)
.5
.4
.3
.2
.1
0
1.
2.
3.
ω/ωp
rs = 1
q/kF = 1
HARTREE
INCLUDING
EXCHANGE

Fig. 7a

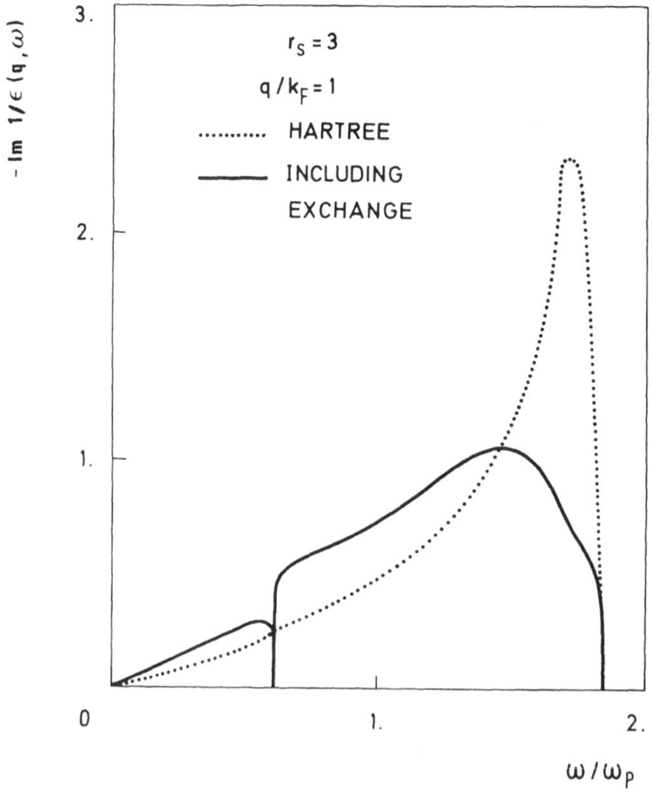
-Im 1/ε(q,ω)
3.
2.
1.
0
1.
2.
ω/ωp
rs = 3
q/kF = 1
HARTREE
INCLUDING
EXCHANGE

Fig. 7b

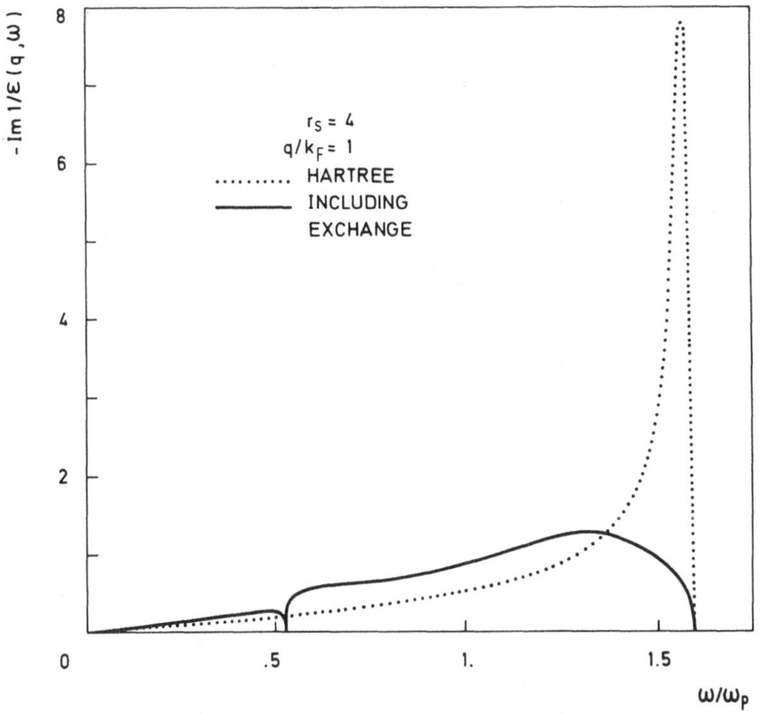

Fig. 7(a-c). The imaginary part of the inverse dielectric function versus frequency (full line) compared to RPA (....) for $q = k_F$ for $r_s = 1$ (fig.7a), 3 (fig. 7b) and 4 (fig. 7c). The open curcles indicate that the magnitude of the peaks is not precisely known due to numerical inaccuracy.

derivative with respect to frequency at the plasmon frequency becomes very large, and thus the plasmon oscillator strength decreases rapidly with increasing wavevector. The maximum in the structure factor is then no longer defined by the plasmon peak position, but by the maximum of the inverse dielectric function inside the continuum.

At finite wavevector, the frequencies of the maxima in the structure factor with dynamical exchange decoupling are appreciably lowered compared to RPA. This is a trend which is confirmed by recent electron scattering experiments in Al. [19]. In figure 8, the maxima in Im $1/\varepsilon(q,\omega)$ with dynamical exchange decoupling are shown for $r_s = 2$, and compared to RPA and to the experimental data for Al ($r_s = 2.07$). The position of the plasmon peak in both the exchange and the RPA treatments, are also indicated. Keeping in mind that the oscillator strenght of the plasmon peak with dynamical exchange decoupling disappears

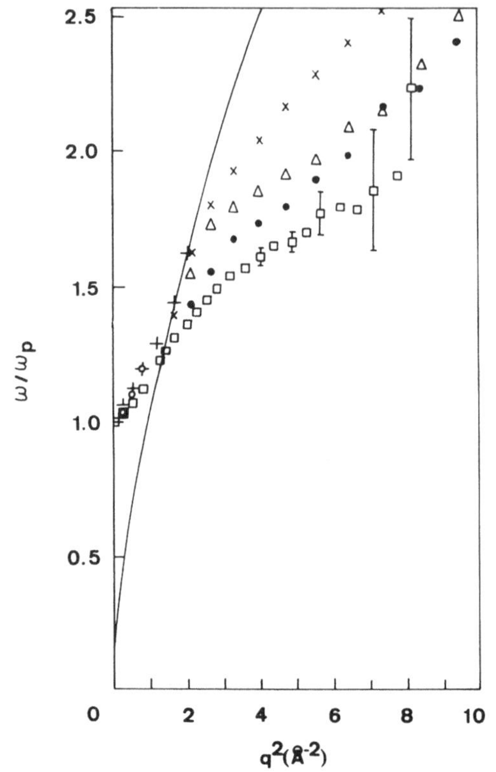

Figure 8. Plasmon frequency as defined by Re $\varepsilon(q,\omega) = 0$ and peak position of the dynamical structure factor $S(q,\omega)$, versus the square of the wavevector at $r_s = 2$. The RPA plasmon is indicated by (0), and the plasmon including dynamical exchange effects by (+). When the plasmon including exchange, approaches the particle-hole continuum (thin line), its oscillator strength is taken over by the maxima of $S(q,\omega)$ (o), which are at substantially lower frequencies than the RPA maxima (X). The experimental data (□) for Al ($r_s = 2.07$) are taken from P.E. Batson, C.H. Chen and J. Silcox [19]. The triangles represent the result of ref.[26].

when it approaches the continuum and is taken over by the maximum in Im $1/\varepsilon(q,\omega)$, figure 8 clearly shows that at finite q, the exchange effects considerably improve the agreement with experiment, compared to RPA.

Finally, we mention that the experiments, as reported in the present volume by Schnatterly, seem to confirm that $G(q,\omega)$ as a function of k and ν is almost independent of r_s. This point however requires further study.

§ 4. Remark on spin density waves

In the derivation of $G(q,\omega)$ via the dynamical exchange decoupling in the equation of motion for the Wigner distribution function, it has been assumed, in line with RPA, that the ground state distribution of the electron gas is homogeneous in space. This does not seem unreasonable because it is generally accepted that the transition to a Wigner lattice occurs at densities far below the metallic range [20] .

Furthermore, it was assumed that the ground state is paramagnetic. This assumption is inspired by the fact that the HF ground state in the ferromagnetic state has only lower energy than the paramagnetic HF ground state for $r_s \gtrsim 5.45$. Moreover, the inclusion of correlation presumably shifts the transition into the ferromagnetic state to $r_s \gtrsim 6.03$ [21] . However, spin density waves lower the ground state energy of the electron gas in the HF approximation [22] . Therefore, it is indicated to examine whether the dielectric function including dynamical exchange effects, reflects an instability relative to deformations of the paramagnetic state.

Following the derivation of the dielectric function with dynamical exchange decoupling, starting from the equation of motion (11-14) in part II, the variational procedure can easily be applied for each spin state separately, by introducing a spin dependent function $\gamma^{\sigma}_{\vec{q}\omega}(p)$ in eq.(17) of part II :

$$\tilde{f}_{\sigma}(\vec{p},\vec{q},\omega) = \tilde{f}^{L}_{\sigma}(\vec{p},\vec{q},\omega)\ \gamma^{\sigma}_{\vec{q}\omega}(\vec{p}) \tag{13}$$

Again determining a momentum independent function $\gamma^{\sigma}_{\vec{q}\omega}$ by the variational principle, one then obtains :

$$\gamma^{\sigma}_{\vec{q}\omega} = - \frac{\frac{1}{2}\,(e\varphi_{\vec{q}\omega} + U_{\vec{q}\omega})\ Q_o(q,\omega) - \frac{4\pi e^2}{q^2}\, n^{\sigma}_{q\omega}}{\frac{1}{2}\, e\varphi_{\vec{q}\omega}\ \frac{Q_o^2(q,\omega)\ G(q,\omega)}{1 + Q_o(q,\omega)}} \tag{14}$$

which replaces eq.(20) of part II. Subsequently calculating the density for both spins from (13), one obtains two coupled equations :

$$n^{\uparrow}_{\vec{q}\omega} \left[Q_o(q,\omega)\, G(q,\omega) - 1 - \frac{1}{2} Q_o(q,\omega)\right] - \frac{1}{2} n^{\downarrow}_{\vec{q}\omega} Q_o(q,\omega)$$

$$= \frac{1}{2} e\varphi_{\vec{q}\omega} \frac{q^2}{4\pi e^2} Q_o(q,\omega) \tag{15}$$

and the same equation with $n^{\uparrow}_{q\omega}$ and $n^{\downarrow}_{\vec{q}\omega}$ interchanged. From these equations, one readily obtains $n_{\vec{q}\omega} \equiv n^{\uparrow}_{\vec{q}\omega} + n^{\downarrow}_{\vec{q}\omega}$ as given by (25b) in part II. However, both equations are only linearly independent, with $n^{\uparrow}_{\vec{q}\omega} = n^{\downarrow}_{\vec{q}\omega}$, if $1 - G(q,\omega)\, Q_o(q,\omega)$ differs from zero. If however the condition

$$1 - G(q,\omega)\, Q_o(q,\omega) = 0 \tag{16}$$

is fulfilled, the relative contribution of spins up and spins down to the density remains undetermined. The condition (16) implies a singularity in the dielectric function.

For the static limit, the instability (16) has already been discussed in § 1. In the dynamical case $\omega \neq 0$ however, $G(q,\omega)$ has a logarithmic singularity at $\omega = \hbar(q^2/2+qk_F)/m$ for all densities, which is responsible for the fact that $\varepsilon(q,\omega)$ with dynamical exchange decoupling has a pole slightly above the particle-hole continuum for all densities. The occurence of this pole is expressed by the condition (16), which is the dynamical extension of the static instability, discussed by HO [2].

The occurence of this pole in $\varepsilon(q,\omega)$ does not necessarily imply the existence of spin density waves from dynamical exchange effects in the homogeneous electron gas. It only means that the stability of the paramagnetic state has to be re-examined with respect to small density deformations. This possibility of an instability already should be included in the initial state. Up to now, the consequences on the dynamical exchange decoupling remain to be studied.

Furthermore, the question arises whether this possibility for the occurence of spin density waves is a property of the electron gas model, or is a consequence of the neglect of correlations.

Although in the static limit, these correlation effects have to some extent been studied in [2], the existence of a pole in the <u>dynamical</u> dielectric function, and its relation to the possibility of spin density waves, has not been pointed out before.

This pole is due to dynamical exchange effects starting from a paramagnetic ground state, but the influence of dynamical correlation effects, and of deviations from the paramagnetic initial state, remains to be studied.

D. COMPARISON WITH FIRST-ORDER PERTURBATION THEORY

As already mentioned in part I, Holas, Aravind and Singwi [3] derived a frequency dependent exchange correction to the dielectric function, which is strongly related to the variational result we obtained, and which we henceforth denote by $G^{var}(q,\omega)$ in order to avoid confusion :

$$\varepsilon^{var}(q,\omega) = 1 + \frac{Q_o(q,\omega)}{1 - G^{var}(q,\omega)\, Q_o(q,\omega)} \tag{17}$$

The derivation in [3] (hereafter referred to as HAS) is an exact calculation of the proper polarization diagrams, to first order in the electron-electron interaction :

$$\varepsilon^{p}(q,\omega) = 1 + Q_o(q,\omega) + Q_1(q,\omega) \tag{18a}$$

Explicitly writing down the first-order correction $Q_1(q,\omega)$, one obtains :

$$\varepsilon^{p}(q,\omega) = 1 + Q_o(q,\omega) + G^{var}(q,\omega)\, Q_o^2(q,\omega) \tag{18b}$$

Note that $G^{var}(q,\omega)$ in (18b) is exactly the function $G^{var}(q,\omega)$ which we obtained from the variational solution for the Wigner distribution function, and which is explicitly given in eq.(21) of part II. Apparently, the perturbative dielectric function (18) is readily obtained by expanding (17) in the electron-electron interaction. This relation between εp and ε^{var} is not fortuitous. Indeed, the perturbation dielectric function (18b) can be obtained from the equation of motion for the Wigner distribution function (II.11-14), if one solves this equation by iteration to first order in the exchange contribution $X_\sigma(q,\omega)$, and then calculates $\varepsilon(q,\omega)$ to first order in these exchange effects. Details can be found in ref.[15] . The variational solution thus approximately sums all subsequent terms in the iteration.

This is also suggested if one forces $\varepsilon^{p}(q,\omega)$ in the commonly used form :

$$\varepsilon^p(q,\omega) = 1 + \frac{Q_o(q,\omega)}{1 - G^p(q,\omega)\, Q_o(q,\omega)} \tag{19}$$

By comparing with (18b) and (17) one obtains :

$$G^{var}(q,\omega) = \frac{G^p(q,\omega)}{1 - Q_o(q,\omega)\, G^p(q,\omega)} \tag{20}$$

[For the sake of clarity in the notations, we mention that $G^{var}(q,\omega)$ is denoted as G^1A in [3], whereas these authors use $G^{pr}A$ for G^p.]

It should be emphasized that the calculation of $G^p(q,\omega)$ by HAS [3] completely confirms our calculations [14,18,26,27]. This forms a third independent check of the evaluation which is summarized in part II.

Although both functions G^{var} and G^p obviously only equal each other in the limit $r_s \to 0$, it is interesting to note that the explicit calculation of $G^p(q,\omega)$ from $G^{var}(q,\omega)$ only requires a few algebraic manipulations. Since $G^{var}(q,\omega)$, as a function of q/k_F and $\hbar\omega/E_F$, is independent of r_s, a comparison of both functions in any calculation of electronic properties can easily be performed, starting from the tables we gave in [13].

The first frequency dependent calculation of $G(q,\omega)$ was already reported in ref. [18], where we noticed that $G^{var}(q,\omega)$ shows logarithmic singularities at the boundaries of the particle-hole continuum. The existence of these singularities in $G^{var}(q,\omega)$ and $G^p(q,\omega)$ is not an erroreous result as stated in [23], but can easily be demonstrated analytically [24]. The calculation by HAS [3] confirms this existence in the mathematical expressions. However, they conclude that these singularities are physically irrelevant, because in the region of the singularities perturbation theory becomes invalid.

It should be emphasized that this conclusion does not necessarily hold for the variational solution. Although we do not claim that the singularities describe a physical phenomenon in the electron gas, there are some indications that they play an important role in a more accurate description of the electron-electron interaction. As pointed out in the previous section, their existence indicates that spin- or charge-density waves have to be included in the equation of motion at the level of introducing the equilibrium distribution. In this context, some further study might be useful in connection with the singularities in the off-diagonal matrix elements of the exchange operator, as shown by Overhauser [16] . Some further insight might be gained by including the properties of the correlation operators [25] . But the highly non-local behaviour of these operators substantially complicates the investigation along these lines.

The simple interrelation (20) between $G^{var}(q,\omega)$ and $G^{p}(q,\omega)$ considerably simplifies the application of the consistency checks on $G^{p}(q,\omega)$, as discussed in section A. In fact, it turns out that $G^{p}(q,\omega)$ satisfies most of the sum rules we applied on $G^{var}(q,\omega)$. The only exception concerns the compressibility sum rule. From (3) and (20) one obtains that :

$$\varepsilon^{p}(q,o) \underset{q\to o}{\to} \frac{k_{FT}^{2}}{q^{2}} \{1 + \frac{r_s}{\pi} (\frac{4}{9\pi})^{1/3}\} \tag{21}$$

If one applies the compressibility sum rule (see eq.(55) of part I), it thus follows that the perturbative dielectric function does not involve the HF compressibility, in contrast to $G^{var}(q,o)$, but only its expansion to first order in r_s. The compressibility sum rule thus seems to favor the variational solution. However, a decisive test would require the calculation of the ground state energy, in which until now we did not succeed within reasonable accuracy.

In the long wavelength limit, but at sufficiently high frequency, it follows from (20) that $G^{var}(q,\omega)$ and $G^{p}(q,\omega)$ are equal to order q^2 [5] :

$$G^{var}(q,\omega) \underset{q\to o}{\to} \frac{4}{15} \frac{e^4 k_F^4}{m^2 \pi^2} \frac{1}{Q_o^2(q,\omega)} \tag{22}$$

From the long wavelength limit of $Q_o(q,\omega)$:

$$Q_o(q,\omega) \underset{q\to o}{\to} - \frac{\omega_p^2}{\omega^2} (1 + \frac{q^2}{\omega^2} \frac{3}{5} \frac{\hbar^2 k_F^2}{m^2}) \tag{23}$$

one thus finds for the zeros of the dielectric function :

$$\omega_p^{var}(q) \underset{q \to o}{\to} \omega_p \{1 + \frac{q^2}{\omega_p^2} \frac{3}{5} \frac{\hbar^2 k_F^2}{m^2} (1 - \frac{\alpha r_s}{3\pi})\}^{1/2} \tag{24}$$

where $\alpha = (4/9\pi)^{1/3}$. The same long wavelength limit for the plasmon frequency is found from the perturbative result (up to order q^2). Compared to the experimental data, as e.g. presented by Schnatterly in the present volume, both approximations improve upon the RPA plasmon dispersion for $q \to 0$, since they shift the $\omega_p(q)$ to lower frequencies than in RPA. At larger wave vector, i.e. in the pair creation region, the maxima in the structure factor with the perturbative method are shown [3] to occur at slightly lower frequencies than with our variational method.

Concerning the plasmon dispersion, it should be emphasized that the physical quantity to study is the structure factor, which shows a δ-function like peak outside the continuum for $\mathrm{Re}\ \varepsilon(q,\omega) = 0$, and a broader peak in the particle hole continuum. Due to the singularities in $G^{var}(q,\omega)$, the plasmon peak from $\mathrm{Re}\ \varepsilon(q,\omega) = 0$ in the perturbative and the variational method, does not penetrate into the continuum. But with increasing wave vector, its oscillator strength is taken over by the maximum of $S(q,\omega)$ in the continuum.

Since the dynamical structure factor is proportional to the scattering cross section, the imaginary part of the dielectric function should not be negative. In this respect, $\mathrm{Im}\ \varepsilon^P(q,\omega)$ shows a rather serious disadvantage compared to $\mathrm{Im}\ \varepsilon^{var}(q,\omega)$. As we will show in a forthcoming paper, $\mathrm{Im}\ \varepsilon^P(q,\omega)$ becomes negative in a finite frequency region for $q > 2k_F$ and $\omega > \hbar\ (q^2/2 - qk_F)/m$ at arbitrary density. Moreover, even for $q < 2k_F$, a finite frequency region can be found for $r_s \gtrsim 2$, for which $\mathrm{Im}\ \varepsilon^P(q,\omega) < 0$. With the variational method, we find that $\mathrm{Im}\ \varepsilon^{var}(q,\omega) > 0$ in general, except possibly for an infinitesimally small frequency region $\omega \simeq |q^2/2 - qk_F|/m$. The latter possibility is currently being investigated, but decisive results are not yet obtained.

In conclusion, both the perturbative and the variational method to include dynamical exchange effects in the dielectric function, substantially improve the RPA and subsequent static local field corrections. The perturbative dielectric function $\varepsilon^P(q,\omega)$ has to be considered as the limit for $r_s \to 0$ of the variational dielectric function $\varepsilon^{var}(q,\omega)$. The latter is easy to use in practice, given the tables of ref.[13], because $G^{var}(q,\omega)$ as a function of q/k_F and ω/E_F is independent of r_s. Finally, consistency requirements and the behaviour of Im $\varepsilon(q,\omega)$ seem to favor the variational inclusion of dynamical exchange effects.

REFERENCES

[1] J. Hubbard, Proc. Roy. Soc. A 243, 336 (1957).

[2] D.R. Hamann and A.W. Overhauser, Phys. Rev. B 143, 183 (1966).

[3] A. Holas, P.K. Aravind and K.S. Singwi, Phys. Rev. B 20, 4912 (1979).

[4] A.K. Rajagopal, Phys. Rev. A 6, 1239 (1972).

[5] F. Brosens, L.F. Lemmens and J.T. Devreese, Phys. Stat. Sol. (b), 74, 45 (1976).

[6] D.N. Tripathy and S.S. Mandal, Phys. Rev. B 16, 231 (1977).

[7] D.J.W. Geldart and R. Taylor, Canad. J. Phys. 48, 155 (1970).

[8] F. Toigo and T.O. Woodruff, Phys. Rev. B 2, 3958 (1970); 4, 371 (1971).

[9] D.J.W. Geldart, R. Taylor and M. Rasolt, Phys. Rev. B 5, 2740 (1972).

[10] J.C. Kimball, Phys. Rev. A 7, 1648 (1973); B 14, 2371 (1976).

[11] G. Niklasson, Phys. Rev. B 10, 3052 (1974).

[12] J.T. Devreese, F. Brosens and L.F. Lemmens, Phys. Rev. B 21, 1349 (1980).

[13] F. Brosens, J.T. Devreese, L.F. Lemmens, Phys. Rev. B 21, 1363 (1980).

[14] F. Brosens, L.F. Lemmens and J.T. Devreese, Phys. Stat. Sol. (b) 82, 117 (1977).

[15] F. Brosens, L.F. Lemmens and J.T. Devreese, Phys. Stat. Sol. (b), 81, 551 (1977).

[16] A.W. Overhauser, Phys. Rev. B 2, 874 (1970).

[17] A.W. Overhauser, Adv. Phys. 27, 343 (1978).

[18] F. Brosens, J.T. Devreese and L.F. Lemmens, Phys. Stat. Sol. (b) 80, 99 (1977).

[19] P.E. Batson, C.H. Chen and J. Silcox, Phys. Rev. Letters 37, 937 (1976).

[20] E.P. Wigner, Trans. Faraday Soc. 34, 678 (1938).

[21] N. Wiser and M.M. Cohen, J. Phys. C (Solid State Physics) 2, 193 (1969).

[22] A.W. Overhauser, Phys. Rev. 128, 1437 (1962).

[23] B.K. Rao, S.S. Mandal and D.N. Tripathy, J. Phys. F : Metal Phys. 9, L51 (1979); 10, L32 (1980).

[24] F. Brosens, J.T. Devreese, L.F. Lemmens, J. Phys. F : Metal Phys. 10, L27 (1980).

[25] K.J. Duff and A.W.Overhauser, Phys. Rev. B 5, 2799 (1972).

[26] P. Vashishta and K.S. Singwi, Phys. Rev. B 6, 875 (1972).

LIQUID ALKALI METALS AND ALKALI-BASED ALLOYS AS ELECTRON-ION PLASMAS

M.P. Tosi

GNSM-CNR, Instituto di Fisica Teorica dell'Università
Trieste, Italy, and International Centre for
Theoretical Physics, Trieste, Italy

ABSTRACT

The article reviews the theory of thermodynamic and structural properties of liquid alkali metals and alkali-based alloys, within the framework of linear screening theory for the electron-ion interactions.

1. INTRODUCTION

A liquid alkali metal can be viewed as a dense plasma of classical ions coupled to a nearly degenerate electron gas. In such a system near freezing, the strength of the bare ion-ion coupling is caracterized by a value of the order of 200 for the ratio between the mean potential energy and the mean kinetic energy per ion. The corresponding ratio for the bare electron-electron coupling lies between 3 and 6, depending on the particular metal. A primitive model of the liquid metal can be constructed from the physical properties of its ionic and electronic components if they are assumed to be only weakly coupled.

In this viewpoint Bohm and Staver[1] calculated long ago the speed of sound in metals by screening the ionic plasma frequency $\Omega_p = (4\pi\rho_i Z^2 e^2/M)^{1/2}$ with the dielectric function of the electron gas in the long-wavelength limit, $\varepsilon(k) \to 1 + k_e^2/k^2$.

The argument leads to long-wavelength phonons with an acoustic-type dispersion relation $\omega(k) = [\Omega_p^2/\varepsilon(k)]^{1/2} \to ck$, with a value $c = (\frac{1}{3} Zm/M)^{1/2} v_F$ for the speed of sound when the inverse screening length k_e is estimated from the equation of state of the ideal Fermi gas, $k_e = (6\pi\rho_e e^2/\varepsilon_F)^{1/2}$. This simple estimate of c is in reasonable agreement with experiment.

Many authors have developed this model in dealing with the phonon dispersion curves[2] and the cohesive properties[3] of alkali metals in the solid phase, and this work will be very briefly recalled in section 2. The model leads to the introduction of effective ion-ion pair potentials, that have subsequently been used with success[4] in computer simulation studies of metals in the liquid phase. In this scheme the liquid metal is effectively treated as a monatomic liquid, except that one should pay attention to the density dependence of the effective pair potential, arising from the underlying system of conduction electrons.

We are instead aiming here at presenting a viewpoint of liquid alkali metals which, in parallel with the solid-state work mentioned above, preserves explicitly their constitution as liquids of ions and electrons. The essential stimulus to the recent developments along this line has come from the progress made over the last decade in the understanding of classical ionic plasmas at high density[5]. The main emphasis will be on the thermodynamic and structural properties of the liquid alkali metals and their alloys, but a brief mention of other interesting alkali-based liquids will be made in the last section.

2. SOME RESULTS OF ELECTRON SCREENING THEORY FOR CRYSTALLINE METALS

At the simplest level of discussion the electron-ion interaction is described by a model potential $V_{ei}(r)$ that is supposedly sufficiently weak to be treated by lowest order perturbation theory. An ionic density fluctuation $\rho_i(\underline{k},\omega)$ can be viewed as an external disturbance on the homogeneous electron gas, which induces a polarization

$$\rho_e(\underset{\sim}{k},\omega) = \chi_e(k,\omega)V_{ei}(k)\rho_i(\underset{\sim}{k},\omega) \quad , \tag{2.1}$$

where $\chi_e(k,\omega)$ is the linear density response function of the electron gas, related to its dielectric function by

$$\frac{1}{\varepsilon(k,\omega)} = 1 + \frac{4\pi e^2}{k^2}\chi_e(k,\omega) . \tag{2.2}$$

The adiabatic approximation for the electron-ion coupling is introduced by replacing $\chi_e(k,\omega)$ in (2.1) by its static value $\chi_e(k)$. The total potential with which the ionic density fluctuation acts on an ion can thus be written

$$V_{ii}(k)\rho_i(\underset{\sim}{k},\omega) + V_{ei}(k)\rho_e(\underset{\sim}{k},\omega) \equiv V_{eff}(k)\rho_i(\underset{\sim}{k},\omega), \tag{2.3}$$

which defines the effective ion-ion potential

$$V_{eff}(k) = V_{ii}(k) + \tilde{v}(k), \quad \tilde{v}(k) = \frac{V_{ei}^2(k)}{4\pi e^2/k^2}\left[\frac{1}{\varepsilon(k)} - 1\right] \tag{2.4}$$

as the sum of the bare ion-ion potential $V_{ii}(k)$ (for which we shall take the Coulomb value $4\pi Z^2e^2/k^2$) and of the indirect term $\tilde{v}(k)$ mediated by the conduction electron gas. The latter term is clearly dependent on the electron density ρ_e.

2.1. Phonon Dispersion Curves

The dynamical matrix of the metal can correspondingly be written, following Toya[2], as the sum of the dynamical matrix of a Coulomb lattice on an inert neutralizing background and of the contribution due to the indirect term $\tilde{v}(k)$. Many calculations of phonon dispersion curves, with a variety of inputs for the bare electron-ion interaction and for the electronic dielectric function, have been carried out within this scheme, but we focus for the present purposes on the calculations of Price et al.[6] for the alkali metals, noting for later reference the following main points :

(i) the phonon dispersion curves of each alkali metal can be rather accurately described with the simple one-parameter form proposed by Ashcroft[7] for $V_{ei}(r)$,

$$V_{ei}(r) = \begin{cases} -Ze^2/r & \text{for } r > r_c \\ 0 & \text{for } r < r_c \end{cases} \tag{2.5}$$

or

$$V_{ei}(k) = -\frac{4\pi Ze^2}{k^2}\cos(kr_c) \tag{2.6}$$

The parameter in this expression is the 'core radius' r_c, at which a cut-off in the bare electron-ion Coulomb potential is introduced to mock up the effect of orthogonalization of the wave functions of the conduction electrons to the occupied core states. This simple scheme seems particularly appropriate for sodium and potassium, since the value of r_c obtained from the phonon dispersion curves agrees closely with values obtained from Fermi surface properties and from the liquid metal resistivity.

(ii) the role of the indirect interaction $\tilde{v}(k)$ in bringing the dispersion curves from those of a bare Coulomb lattice to those of the metal is crucial for the longitudinal phonons (as shown by the Bohm-Staver argument : the LA branch emerges from a cancellation of the Coulomb singularities in the two terms of $V_{eff}(k)$) but only minor for the transverse phonons, especially in the region of $\underline{k}$-space around the Brillouin zone centre.

For a recent discussion of effective interionic potentials in metals, reference can be made to the review of Faber[8].

2.2. Cohesive Energy

The corresponding expression for the cohesive energy of a static ionic lattice screened by the electron gas consists of four terms[3] :

$$E = E_g + E_H + E_{es} + E_{bs} \; . \tag{2.7}$$

These are, respectively, the ground state energy E_g of the homogeneous electron gas on an inert background and its shift E_H (the 'Hartree' energy) due to the average non-Coulomb part of $V_{ei}(r)$,

$$E_H = \rho_e \lim_{k\to 0} [V_{ei}(k) + \frac{4\pi Ze^2}{k^2}] \tag{2.8}$$

$(= 2\pi\rho_e/Ze^2r_c^2$ for the Ashcroft potential);

the Madelung energy of the bare Coulomb lattice on an inert background,

$$E_{es} = -0.896\, Z^2e^2(4\pi\rho_i/3)^{1/3} \quad \text{(bcc lattice)}; \tag{2.9}$$

and the 'band structure' energy due to the indirect ion-ion interactions,

$$E_{bs} = \frac{1}{2}\rho_i \sum_{\underset{\sim}{G}(\neq 0)} \tilde{v}(G) \quad (\underset{\sim}{G} = \text{reciprocal lattice vectors}) \tag{2.10}$$

The equation of state, determining the equilibrium lattice parameter, and the bulk modulus of the metal follow from (2.7) by successive differentiations with respect to the density.

An important question is the 'thermodynamic consistency' of the theory, i.e. the equality between the value B of the bulk modulus derived from the cohesive energy and the value $\tilde{B}$ obtained from the long-wavelength limit of the phonon dispersion curves. Even if the electronic dielectric function used in the calculations is thermodynamically consistent[9], agreement between B and $\tilde{B}$ requires the inclusion of non-linear effects of the electron-ion interactions[10]. The two values of the bulk modulus differ precisely in that the expression for B contains an additional term arising from the density dependence of the effective ion-ion potential. The same term appears in the expression for $\tilde{B}$ only when one keeps account also of the shift in average potential felt by the electrons, through a fourth-order perturbation calculation of the phonon dispersion curves.

At the level of the present discussion, therefore, the two approaches to the bulk modulus cannot be consistent. Good agreement with experiment for the cohesive properties of the crystalline alkalis, including the bulk modulus B, was nevertheless obtained in ref. 6 with the value of the core radius r_c derived from the phonon dispersion curves through an arbitrary increase of the value (2.8) for the Hartree energy by a factor 1.218.

3. ELECTRON SCREENING THEORY FOR LIQUID METALS

We return to the discussion given at the beginning of section 2 and think of the ionic density fluctuation $\rho_i(\underline{k},\omega)$ as a polarization induced by a weak external potential $V_i^{eff}(\underline{k},\omega)$ fictitiously applied to the ionic system. In terms of linear response functions for the homogeneous fluid of ions and electrons we can write

$$\begin{aligned} \rho_i(\underset{\sim}{k},\omega) &= \chi_i(k,\omega)\, V_i^{ext}(\underset{\sim}{k},\omega) \\ \rho_e(\underset{\sim}{k},\omega) &= \chi_{ei}(k,\omega)\, V_i^{ext}(\underset{\sim}{k},\omega) \end{aligned} \qquad (3.1)$$

Comparison with eqn.(2.1) for weak electron-ion coupling yields in the adiabatic approximation

$$\chi_{ei}(k,\omega) = \chi_e(k)\, V_{ei}(k)\, \chi_i(k,\omega) \qquad (3.2)$$

Under the same assumption of weak coupling between ions and electrons, we can also describe the ion fluid polarization $\rho_i(\underline{k},\omega)$ as that of a bare plasma responding to an internal potential which is the sum of the external potential and of the potential due to the electron fluid polarization :

$$\rho_i(\underset{\sim}{k},\omega) = \chi_i^o(k,\omega)\,[V_i^{ext}(\underset{\sim}{k},\omega) + V_{ei}(k)\,\rho_e(\underset{\sim}{k},\omega)] \quad (3.3)$$

Comparison with (3.1) and (3.2) yields[11] :

$$\chi_i(k,\omega) = \chi_i^o(k,\omega)/[1 - \tilde{v}(k)\chi_i^o(k,\omega)] \ . \quad (3.4)$$

In these expressions $\chi_i^o(k,\omega)$ is the density response function of a bare ionic plasma on an inert background.

The zeroes of $\chi_i^{-1}(k,\omega)$ give the longitudinal collective modes of the liquid metal. Their discussion on the basis of eqn. (3.4) clearly parallels the theory of the phonon dispersion curves for the crystalline phase recalled in section 2.1. A connection with structural properties of the liquid can also be established from this equation in the static limit $\omega \to 0$, by virtue of the fluctuation-dissipation theorem[12] which for a classical fluid relates directly the static response to the structure factor describing the diffraction pattern of the liquid. One obtains for the structure factor S(k) of the liquid metal the expression[13]

$$S(k) = S_o(k)/[1 + \beta\rho_i\tilde{v}(k)S_o(k)] \qquad (\beta = 1/k_BT) \ , \quad (3.5)$$

where $S_o(k)$ is the structure factor of the one-component classical plasma (OCP) on an inert background. Similarly, eqn. (3.2) yields for the ion-electron structure factor $S_{ei}(k)$, which describes the linearly distorted electron density around an ion in the liquid, the expression

$$S_{ei}(k) = (\rho_i/\rho_e)^{1/2}\chi_e(k)\;V_{ei}(k)\;S(k) \ , \quad (3.6)$$

on the assumption that quantum effects can be neglected in this quantity. We shall here continue the discussion for alkali metals in parallel with section 2, and only later take up the calculation of their structure in the present scheme.

3.1. Compressibility by the Method of Long Waves

The analysis of (3.4) in the limit of long wavelength and low frequency yields[11] longitudinal collective excitations with a dispersion relation appropriate to sound waves attenuated by viscosity. The value for the isothermal speed of sound agrees with that obtained from the liquid-metal compressibility K_T as derived directly from (3.5) through the use of the Ornstein-Zernike relation[12],

$$\lim_{k\to 0} S(k) = \rho_i k_B T K_T \tag{3.7}$$

The latter approach involves the use of the long-wavelength form for the structure factor of the OCP,

$$\lim_{k\to 0} S_o(k) = \frac{k^2}{k_D^2}\left(1 + \frac{k^2}{k_i^2}\right)^{-1} \quad (k_D^2 = 4\pi\beta\rho_i Z^2 e^2) \tag{3.8}$$

and leads to the result

$$\rho_i k_B T K_T = \left(\frac{k_D^2}{k_e^2} + \frac{k_D^2}{k_i^2} + k_D^2 r_c^2\right)^{-1} . \tag{3.9}$$

The first term on the right-hand side is the Bohm-Staver result, but k_e now is the inverse screening length of the interacting electron gas. The other terms arise from the dispersion of the ionic plasma excitation and from the non-Coulombic term in the bare electron-ion interaction.The numerical results obtained[14] with the values of the core radius from the phonon analysis discussed in section 2.1 and the values of the 'inverse screening length' k_i of the OCP derived from its free energy[15] are reported in table 1. The good agreement with experiment shows that the theory is giving a consistent account of sound waves in the liquid and in the solid.

We also note that (3.6) in the long wavelength limit leads to[16]

Table 1. Isothermal compressibility and temperature derivative of the speed of sound for liquid alkali metals near freezing (from D.K. Chaturvedi et al., ref. 14).

		Na	K	Rb	Cs
$\rho_i k_B T K_T$	theory	0.0215	0.0231	0.0236	0.0235
	expt	0.1233[a] 0.0240[b]	0.0225[a] 0.0247[b]	0.0220[a]	0.0237[a]
$-(\partial c/\partial T)_P$ (m/secK)	theory	0.69	0.56	0.40	0.38
	expt[(a)]	0.52	0.53	≈0.4	≈0.3

(a) From G.M.B. Webber and R.W.B. Stephens, in 'Physical Acoustics' vol.IVB, p.53 (edited by W.P. Mason; Academic Press, New York 1968).

(b) From A.J. Greenfield et al., ref.26.

We also note that (3.6) in the long wavelength limit leads to[16].

$$\lim_{k\to 0} [S_{ei}(k) = Z(\rho_i/\rho_e)^{1/2} S(k)]. \qquad (3.10)$$

In terms of the radial distribution functions g(r) and $g_{ie}(r)$ (see (3.12) and (3.13) below) this can be rewritten as

$$Z\rho_i \int d\underset{\sim}{r}[g(r) - 1] - \rho_e \int d\underset{\sim}{r} [g_{ie}(r) - 1] = -Z \qquad (3.10')$$

which expresses the condition that the total (ionic plus electronic) charge surrounding an ion must be equal to the negative of the charge of the ion (electroneutrality condition). A similar relation exists between the electron-electron structure factor in the liquid metal and S(k), but is not included in the present treatment.

3.2. Cohesive Properties

The internal energy per ion in the liquid metal[12] can be written

$$E = T_i + T_e + \frac{1}{2\rho_i} \int d\underset{\sim}{r}[\rho_e^2 g_e(r)V_{ee}(r) + \rho_i^2 g(r)V_{ii}(r) + 2\rho_i\rho_e g_{ie}(r)V_{ie}(r)] \qquad (3.11)$$

where T_i and T_e are the kinetic energies of ions and electrons in the metal, the V's are the bare interaction potentials, and the g's are the radial distribution functions, related to the structure factors by

$$S(k) = 1 + \rho_i \int d\underset{\sim}{r} [g(r) - 1] \exp(i\underset{\sim}{k}.\underset{\sim}{r}) \qquad (3.12)$$

(with a similar expression for $g_e(r)$) and

$$S_{ie}(k) = (\rho_i\rho_e)^{1/2} \int d\underset{\sim}{r} [g_{ie}(r) - 1] \exp(i\underset{\sim}{k}.\underset{\sim}{r}) . \qquad (3.13)$$

For weak electron-ion interactions we can consider the switching on of $V_{ie}(r)$ as a process of polarization of the homogeneous electron gas by the ions in their fixed configuration as described by g(r). By a standard result of linear response theory, one-half of the energy gained by the interaction of the polarization with the polarizing field cancels against the 'quasi-elastic' energy spent in creating the polarization, i.e.

$$\frac{1}{2} \rho_e \int d\underset{\sim}{r} [g_{ie}(r) - 1]V_{ie}(r) =$$

$$E_g - T_e - \frac{\rho_e^2}{2\rho_i} \int d\underset{\sim}{r} \, [g_e(r) - 1] V_{ee}(r) \tag{3.14}$$

Using this equation and eqn.(3.6) to eliminate $g_e(r)$ and $g_{ei}(r)$, the internal energy can be written

$$E = T_i + E_1 + E_2 + E_3 \tag{3.15}$$

where

$$E_1 = \frac{1}{2} \rho_i \int d\underset{\sim}{r} \, g(r) \, V_{eff}(r) \; , \tag{3.16}$$

$$E_2 = \frac{1}{2} [V_{eff}(r) - V_{ii}(r)]_{r=0} \; , \tag{3.17}$$

$$E_3 = E_g - \frac{1}{2} \rho_i \int d\underset{\sim}{r} \, V_{eff}(r) + \rho_e \int d\underset{\sim}{r} [V_{ie}(r) + \frac{Ze^2}{r}] \tag{3.18}$$

The same expression was derived by Price[17] by direct analogy with eqns.(2.7)-(2.10) for the solid. Notice that E_1 is the conventional expression for the mean potential energy of a monatonic liquid with pair-wise interactions, while E_2 and E_3 are purely density-dependent

Table 2. Equilibrium properties of sodium (in Rydberg/ion; from D.L. Price, ref.17).[a]

	E_{solid}(OK)	E_{liquid}	VP_{liquid}	VB_{liquid}	$\tilde{VB}_{liquid}$
T_i	0.0010	0.0038	0.0025	0.0025	0.0025
E_1	-0.0160	-0.0114	0.0203	0.1030	0.0950
E_2	-0.2902	-0.2882	-0.0194	-0.0350	-
E_3	-0.1880	-0.1857	-0.0284	-0.0212	-
ΔE_H	0.0308	0.0280	0.0280	0.0560	-
Total	-0.4624	-0.4535	0.0030	0.1053	0.0975
	$E_\ell - E_s$ =	0.0089			
Expt	-0.460 $\pm$0.002	0.00895 $\pm$0.00002	0.0000	0.0973	

(a) The liquid is at 393 K, i.e. close to freezing.

terms. On the basis of the numerical results for the solid, Price added to (3.15) a shift ΔE_H of the Hartree energy by an amount $0.218\ E_H$ (E_H is the last term in (3.18)). His numerical results for the solid and liquid phases of sodium are reported in table 2: g(r) for the liquid was taken from computer simulation results.

Price also discussed the relation between the 'thermodynamic bulk modulus' B obtained from E (column 4 of table 2) and the 'dynamic bulk modulus' $\tilde{B}$ obtained from (3.7) (column 5). The latter is equivalent to the bulk modulus of a conventional monatomic liquid in which the density dependence of the interactions is absent - so that, in particular, only the term E_1 contributes, through the density dependence of g(r) and the r dependence of $V_{eff}(r)$ - while B involves also the density dependence of the interactions.

4. STRUCTURE FACTOR OF LIQUID ALKALI METALS

The use of eqn (3.5) at arbitrary wave number preliminary requires a qualitative understanding and a quantitative theory of the structure of the OCP. The strength of the interactions in this model fluid, relative to the kinetic energy, is measured by the so-called plasma parameter $\Gamma = Z^2e^2/(ak_BT)$ with $a = (4\pi\rho_i/3)^{-1/3}$. For liquid alkali metals near freezing, this parameter takes values in the range 150-200. We are thus interested in the structure of the strongly coupled OCP.

4.1. Structure of the OCP at Strong Coupling

Computer simulation work on the OCP shows[18] that at strong coupling the Coulomb correlation hole surrounding each point charge in the plasma is sharply defined, the radial distribution function $g_o(r)$ being zero over a range of r comparable with a and rising henceforth very sharply to reach its main peak in correspondence with the first-neighbour shell. This qualitative feature suggested to Gillan[19] the possibility of a theory in which the short-range correlations would be mocked up by endowing each point particle of the plasma with a hard-core diameter σ, such that $g_o(r) = 0$ for $r \leqslant \sigma$. The imposition of this condition for $r = \sigma$, and not just for $r < \sigma$, serves to determine the value of σ as a function of Γ and is crucial from the thermodynamic point of view : it ensures that the hard cores remain fictitious and do not contribute to the virial pressure through hard-core collisions. This assumption on short-range correlations was combined by Gillan with an assumption on the long-range correlations due to the Coulomb interactions, for which he took, in essence, the Debye-Hückel approximation for $r \geqslant \sigma$[20]. These assumptions allow one to obtain an analytic solution for the structural problem, in the form of an analytic expression for $S_o(k)$ in terms of parameters determined by the value of Γ.

The results of Gillan for the structure of the OCP are very promising, but the approach is not hermodynamically consistent, in the sense already mentioned in ref.15 above : it violates the relation between k_i^2 in (3.8) and the compressibility obtained from the free energy of the OCP. This approach has recently been modified[21], however, by allowing for deviations from the Debye-Hückel form of the correlations in the range $r \geqslant \sigma$, in such a manner as to embody thermodynamic consistency and the known equation of state, while retaining the advantage of an analytic solution for $S_o(k)$. An example of the accuracy attained by the theory is reported in figure 1 by comparison with computer simulation data at $\Gamma = 160$. This approach works with similar accuracy over a wide range of Γ, but seems to fail for $\Gamma \lesssim 20$: effects of penetration in the Coulomb hole are clearly indicated by computer simulation results at lower Γ's ($\lesssim 4$).

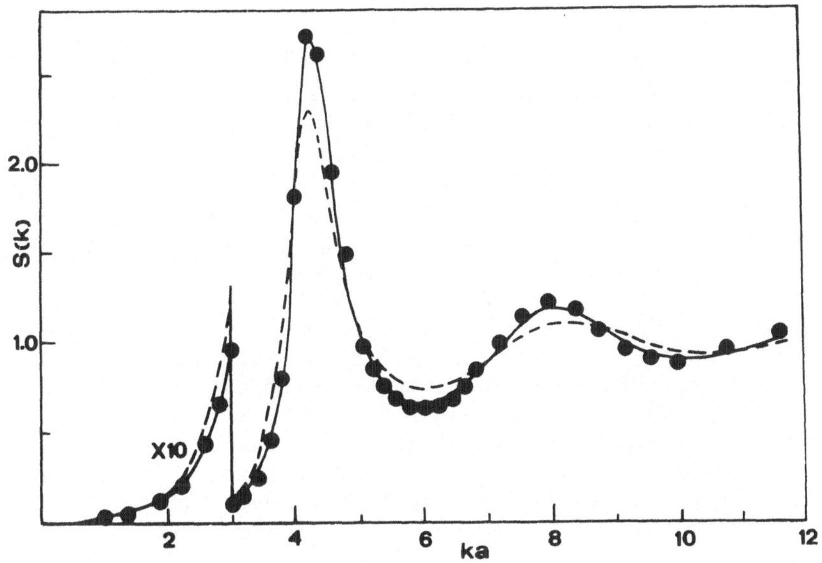

Figure 1. Structure factor of the OCP at $\Gamma = 160$ (notice that the vertical scale has been enlarged by a factor 10 at low wavenumber). The dots are computer simulation results (ref.18), while the curves are theoretical results based on the charged-hard-spheres model, without (dashed curve) and with (full curve) thermodynamic consistency. From D.K. Chaturvedi et al., ref.21.

The qualitative point that we wish to stress again before proceeding is that the pair structure of the strongly coupled OCP is very closely described by that of a fluid of charged hard spheres, provided that the thermodynamic properties (the virial pressure, the free energy, and the 'compressibility sum rule') are preserved.

4.2. Structure of Liquid Alkali Metals

Equation (3.5) as it stands is not a useful formula to apply in the wave number region of the main peak of S(k). This is most simply seen by noticing that the second term in the denominator is negative and attains here a magnitude of a few to several tenths of unity, with an uncertainty by a factor of order 4 as gauged from different forms for the bare electron-ion interaction that are available in the literature. The applicability of linear theory in this region of wave number, which corresponds to the first star of reciprocal lattice vectors in the crystal, is also open to serious question.

The simplest empirical modification[14] that can be introduced in (3.5) is to omit completely the screening correction for wave number k above the first node of $V_{ie}(k)$, corresponding to $k_n = \pi/(2r_o) \approx 2k_F$ for the Ashcroft potential in the alkali metals. This amounts to assuming that the structure factor of the liquid alkali metal coincides with that of the bare ionic plasma[22] for $k \gtrsim 2k_F$ and thus, according to the discussion given just above for the structure of the OCP, with that of a suitably chosen fluid of charged hard spheres. At lower wave numbers electron screening is instead allowed to intervene, leading for $k \to 0$ to the behaviour discussed in section 3.1.

This simple empirical scheme yields very good agreement with the available experimental data for the structure of the alkali metals near freezing. There is some indication (outside the likely uncertainty in the measured structure factors, which is not negligible in the region of the main peak of S(k)) that some lowering of the value of Γ for the underlying bare plasma, below the value that would be appropriate to the temperature and density of the metal, may be needed for high quantitative accuracy. This is not a major matter, however, and indeed to a first approximation one may assert that the values of Γ for the alkali metals at freezing are all the same and equal to that value at which the OCP is believed to freeze[23]. This implies an approximate scaling law for the structure factors of the alkali metals at their respective freezing points, for $k \gtrsim 2k_F$, since the structure factor of the OCP is a function only of ka at given Γ. The measured structure factors of liquid Na, K and Cs at freezing satisfy this scaling[24].

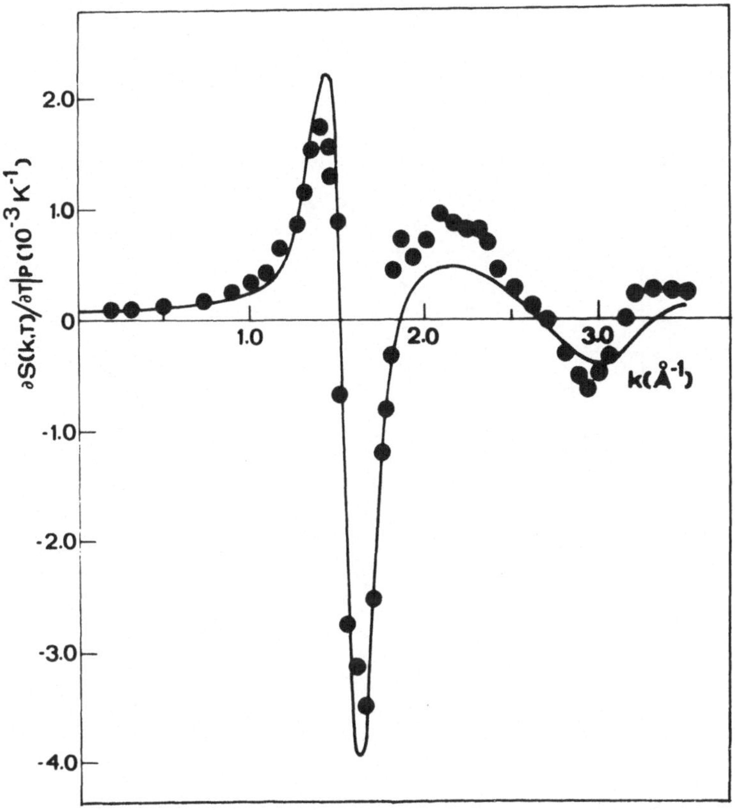

Figure 2. Temperature derivative of S(k) at constant pressure for liquid potassium near freezing, compared with X-ray diffraction data of Greenfield et al. (ref.26). From D.K. Chaturvedi et al., ref.14.

We illustrate the merits of this approach by reporting in figure 2 the calculated temperature derivative of the structure factor of potassium near freezing against the X-ray diffraction results of Greenfield et al.[25]. The calculated temperature dependence arises from that of the structure factor of the OCP and (at low wave numbers) from thermal expansion effects in the electronic screening. A similar comparison with neutron diffraction data[26] for the variation in position and height of the main peak with density in expanded rubidium is reported in figure 3. We stress that no adjustment of parameters is involved in these calculations and that in figure 3 and in the main part of figure 2 we are in fact comparing properties of a bare OCP with experimental data for a liquid metal.

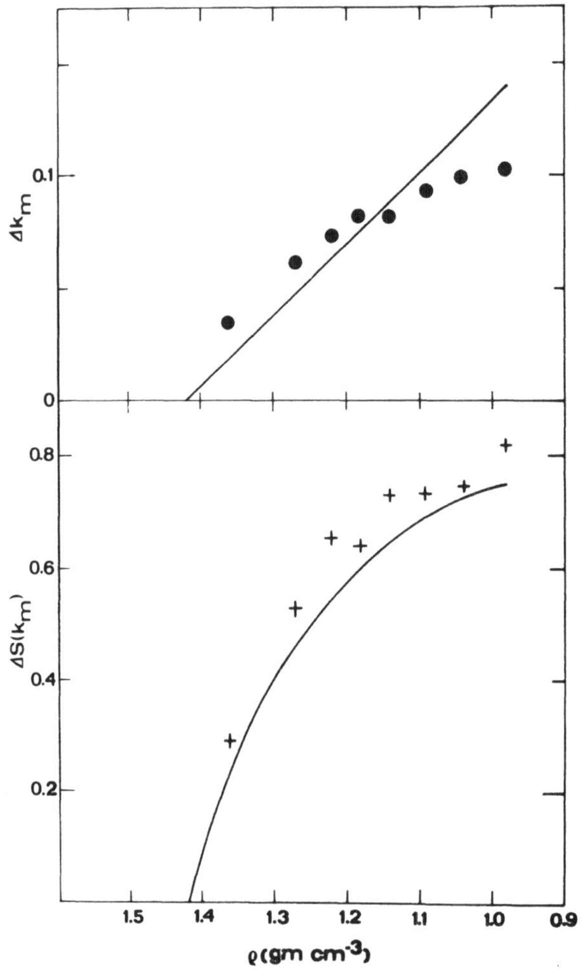

Figure 3. Variation with density of the position (top) and height (bottom) of the main peak of S(k) for liquid rubidium, compared with neutron diffraction data of Block et al. (ref.27). From D.K. Chaturvedi et al., ref.14.

As a corollary to the above illustrations, one may also note that the details of the ionic effective interactions in these liquid metals are mostly reflected in small-angle scattering data. A direct comparison of the results of eqn.(3.5), based on computer simulation data[13] for the structure factor of the OCP at $\Gamma = 160$ and the interionic potential of Price et al.[6] with X-ray diffraction data[25] in the small-angle scattering region is presented in figure 4 for sodium at 200°C (where $e^2/(ak_BT) = 163$). A high degree of consistency between the phonon dispersion curves in the solid and the small-angle scattering from the liquid is thus indicated.

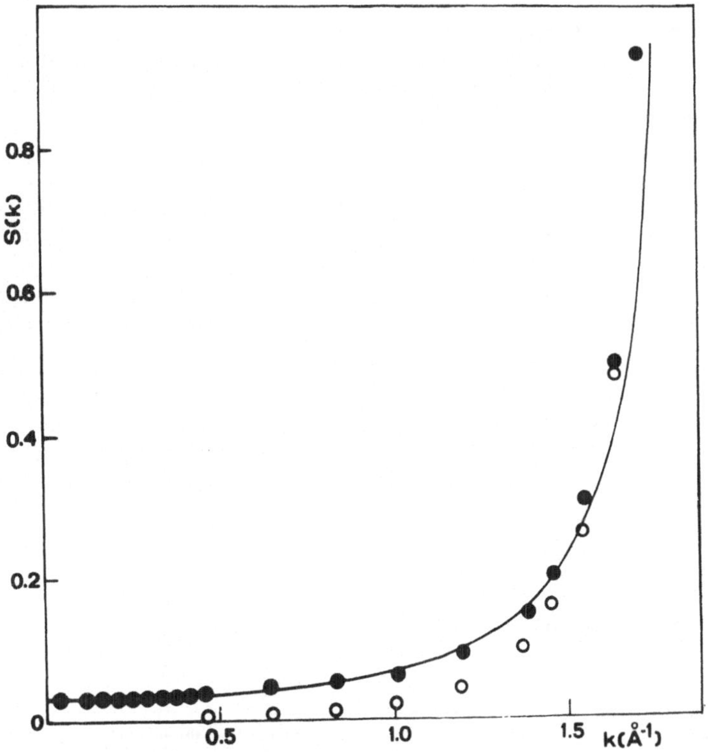

Figure 4. Small-angle scattering from liquid sodium at 200°C: the dots are calculated from eqn.(3.5) and the curve are X-ray diffraction results of Greenfield et al. (ref.26). The circles show the structure factor of the bare OCP at $\Gamma = 160$, from simulation data of Galam and Hansen (ref.13).

A basic justification for the empirical scheme that we have been discussing may now be suggested. Equation (3.5) can be viewed as combining the structural effects of a very strong repulsive interaction (the ion-ion Coulomb repulsion, leading to the charged-hard-spheres-like structure of the strongly coupled OCP) with those of a smooth attractive interaction (the screening correction $\tilde{v}(r)$) by means of an RPA treatment of the latter. Such a treatment can be improved by an optimization procedure that replaces $\tilde{v}(r)$ by an effective interaction having the properties that (a) it differs from $\tilde{v}(r)$ only inside the region where the repulsion is dominant, and (b) it leads to a vanishing radial distribution function g(r) inside the same region. This procedure introduces the effect of the attractive interaction $\tilde{v}(r)$ in the region of space where it is felt, without modifying the excluded-volume effect due to the Coulomb repulsion.

It has indeed been shown[27] that the empirical scheme presented in this section gives already a close realization of such an optimized random-phase approximation for the liquid alkali metals near freezing.

5. THERMODYNAMIC PROPERTIES OF LIQUID ALKALI ALLOYS BY THE METHOD OF LONG WAVES

The diffraction patterns of a binary mixture are described by three partial structure factors $S_{\alpha\beta}(k)$, related to the three radial distribution functions for the various types of atomic pairs. The extension of the Ornstein-Zernike relation (3.7) to multi-component liquids was given a long time ago by Kirkwood and Buff[28]. In terms of the direct correlation functions $c_{\alpha\beta}(k)$, defined through the inverse of the matrix of structure factors by

$$c_{\alpha\beta}(k) = \delta_{\alpha\beta} - S^{-1}_{\alpha\beta}(k) \; , \tag{5.1}$$

the Kirkwood-Buff relations read

$$\lim_{k\to 0} c_{\alpha\beta}(k) = \delta_{\alpha\beta} - \frac{(\rho_\alpha \rho_\beta)^{1/2}}{k_B T} \left(\frac{\partial \mu_\alpha}{\partial \rho_\beta}\right)_{T,V,\rho_{\bar{\beta}}} \tag{5.2}$$

where μ_α is the chemical potential of the α-th component and $\bar{\alpha}$ denotes all the other components. For a binary alloy, in particular, the thermodynamic quantities in (5.2) can be related to its isothermal compressibility, to the coefficient for the dependence of the density on concentration, and to the 'concentration-concentration structure factor' S_{cc}, given by the second derivative of the Gibbs free energy with respect to concentration c ($c = c_2 = 1 - c_1$) :

$$S_{cc} = Nk_B T / \left(\frac{\partial^2 G}{\partial c^2}\right)_{T,P,N} \tag{5.3}$$

This quantity describes the correlations between thermodynamic fluctuations of concentration in the binary mixture, just as the Ornstein-Zernike relation relates the correlations between thermodynamic density fluctuations to the compressibility. Deviations of S_{cc} from its value for an ideal solution ($S^{ideal}_{cc} = c_1 c_2$) thus measure the degree of relative ordering of the components of the mixture : an enhancement over S^{ideal}_{cc} reflects a tendency to segregation, while a depression reflects a tendency to 'alternation' of the components (sublattice formation).

We focus for the purposes of the present discussion on the Na-K alloy, for which all the necessary thermodynamic data have been available. Two main points have emerged from the analysis of these data :

(i) the ionic partial structure factors as functions of concentration[29] can be successfully fitted[30] by the model of conformal solutions, in which, in particular, S_{cc} has the expression

$$S_{cc} = c_1 c_2 / [1 - 2c_1 c_2 w/k_B T] \tag{5.4}$$

where w is the 'interchange free energy', a function of P and T but not of concentration;

(ii) the use of electroneutrality of the alloy allows one to relate[31] (as in eqn.(3.10) for the pure metal) the partial structure factors involving the electronic component of the alloy to the ionic partial structure factors,

$$\lim_{k\to 0} [S_{e\alpha}(k) = \left(\frac{\rho_\alpha}{\rho_e}\right)^{1/2} Z_\alpha S_{\alpha\alpha}(k) + \left(\frac{\rho_{\bar\alpha}}{\rho_e}\right)^{1/2} Z_{\bar\alpha} S_{\alpha\bar\alpha}(k)] \tag{5.5}$$

S_{eNa} and S_{eK} are found to be strong functions of concentration, and specifically S_{eNa} increases rapidly with concentration, starting from either pure Na or pure K, while S_{eK} correspondingly decreases. In terms of the corresponding electron-ion radial distribution functions, this behaviour implies an electronic charge transfer from K ions to Na ions in the alloying process, relative to a situation in which the electron distribution in the alloy is kept uniform.

Let us examine this problem from the point of view of electron screening theory. The extension of eqn.(3.4) to the binary alloy[11] yields the following results :

(i) the dynamical modes of the screened two-component plasma at long wavelength and low frequency are[11], in agreement with the prediction of linearized hydrodynamics, a sound wave mode attenuated by viscosity and by interdiffusion, as well as a relaxation mode of interdiffusion of the components;

(ii) the ionic partial structure factors are obtained[14] by inversion of eqn.(5.1), where

$$c_{\alpha\beta}(k) = c^o_{\alpha\beta}(k) + (\rho_\alpha \rho_\beta)^{1/2} \frac{k^2 V_{\alpha e}(k)\, V_{\beta e}(k)}{4\pi e^2 k_B T} [1 - 1/\varepsilon(k)] \tag{5.6}$$

$c^o_{\alpha\beta}(k)$ being the direct correlation functions of the bare ionic plasma;

(iii) in particular for $k \to 0$,

$$\left(\frac{\partial\mu_\alpha}{\partial\rho_\beta}\right)_{T,V,\rho_{\bar\beta}} = \left(\frac{\partial\mu^o_{\alpha}}{\partial\rho_\beta}\right)_{T,V,\rho_{\bar\beta}} + \frac{4\pi Z_\alpha Z_\beta e^2}{k_e^2} + 2\pi Z_\alpha Z_\beta e^2 (r^2_{c\alpha} + r^2_{c\beta}) \tag{5.7}$$

(iv) the electron-ion structure factors are given by[32]

$$S_{\alpha e}(k) = \chi_e(k)\left[\left(\frac{\rho_\alpha}{\rho_e}\right)^{1/2} V_{\alpha e}(k)\, S_{\alpha\alpha}(k) + \left(\frac{\rho_{\bar\alpha}}{\rho_e}\right)^{1/2} V_{\bar\alpha e}(k) S_{\alpha\bar\alpha}(k)\right], \tag{5.8}$$

reducing to (5.5) for $k \to 0$ where $\chi_e(k) \to -k^2/(4\pi e^2)$ and $V_{\alpha e}(k) \to -4\pi Z_\alpha e^2/k^2$.

From the quantitative viewpoint, whereas the predictions of the theory are in reasonable agreement[14] with experiment for the dependence of the compressibility and of the density of the Na-K alloy on concentration, we focus here on S_{cc} since it is very sensitive to the details of the theory. Equation (5.7) leads to an expression for S_{cc} of the form (5.4), with w given for homovalent ions by

$$w/k_B T = \frac{1}{2}\rho \frac{(v_2 - v_1)^2}{k_T} \tag{5.9}$$

This quantity, where $v_2 - v_1$ is the difference in partial molar volumes of the components in the alloy, clearly represents the elastic work in the alloying process. Its value can be obtained directly from experimental data, since

$$\rho(v_2 - v_1) = -\frac{1}{\rho}\left(\frac{\partial\rho}{\partial c}\right)_{T,P,N} \tag{5.10}$$

and is larger than the value of w obtained[30] from the conformal-solution analysis of the data ($w/k_BT \simeq 1.1$) by an order of magnitude. Thus linear screening is completely missing the effect of electronic charge transfer from K to Na in alloying. Similar conclusions have been illustrated recently[32] in relation to the formation of chemical bonding in semiconductors.

Some light on this problem is shed by empirical analyses of the heats of formation of metallic alloys[33] and by their discussion in the framework of a pseudopotential formalism[34]. Contributions of opposite signs to the heat of formation, which cancel each other to a large extent in the Na-K system, arise from elastic deformation effects and from electron transfer effects. Attempts at estimating w by this route[35], however, do not lead to sufficiently accurate predictions. It should be stressed that the energy scale of interest in this problem is very small (w is a few meV in the Na-K system) and that very high precision is required. The problem remains an important one, also in view of the availability of small-angle scattering data[36].

6. SOME OTHER ALKALI-BASED LIQUIDS

We briefly mention in this concluding section two other types of liquid systems of special topical interest, whose theoretical treatment requires a detailed electron-ion model.

6.1. Solutions of Molten Alkali Metals and Molten Alkali Halides

The alkali metals and the alkali halides form true solutions in the liquid state, with partial miscibility at lower temperatures and a critical point above which the two liquids become miscible in any proportion[37]. The simultaneous presence of a critical point and of a metal-nonmetal transition is a special reason for interest in these systems.

At the metal-rich end of the phase diagram, the halogen X is believed[38] to be present in the solution as an X^- negative ion. Its measured effect on the resistivity of the liquid metal[39] is large. Interesting questions are how such an ion is screened by the conduction electrons, the nature of its coordination shell of alkali ions, and the origin of the large contribution to the resistivity[40].

At low metal concentrations the most interesting question concerns the nature of the states of the valence electrons of the alkali atoms added to the molten alkali halide. An electronic contribution to the conductivity in the concentration range 0.01-0.1 mol of metal has been approximately extracted from the experimental data, and is found to be appreciable at these high temperatures and to increase with temperature. These features are qualitatively consistent with a phenomenon of Anderson localization[41]. At still lower concentrations broad bands of optical absorption have been reported[42], whose peak frequency and width are correlated with those of the F-centre absorption band in the alkali halide crystals. The existence of a bound electronic state analogous to the F centre in a molten alkali halide to which an alkali ion and a free electron have been added, with an excitation spectrum in approximate agreement with the observations, have been demonstrated[43].

Extensive studies by neutron scattering techniques on some of these systems have been very recently reported by Jal[44].

6.2. CESIUM-GOLD ALLOY

The behaviour of the electric conductivity and of the optical absorption spectrum of this system[45] indicate that, at stoichiometric composition, it should be viewed as an 'ionic liquid' (or, better, as a strongly polar semiconducting liquid) composed of Cs^+ and Au^- ions. As one moves away from stoichiometry by adding extra cesium, one thus expects qualitative similarities with the behaviour of alkali metal - alkali halide solutions near the salt-rich end of the phase diagram. The existence of localized electron states in a narrow range of concentration away from stoichiometry is indicated[46] by NMR data.

Extensive neutron diffraction data on this liquid alloy at various concentrations have been reported recently[47]. The behaviour of the diffraction pattern at low scattering angles shows qualitative changes occurring in the concentration range between 55 % and 60 % cesium. A shoulder in the diffraction pattern at $k \approx 1.2$ Å^{-1}, which can be associated with alternation of the components characteristic of an ionic liquid[48], is replaced by an enhancement of the scattering for $k \to 0$, as in metallic alloys where the conduction electrons can screen the concentration fluctuations of the components. Work on the calculation of the structure of this alloy is in progress[49].

REFERENCES

1. D. Bohm and T. Staver, Phys. Rev. 84, 836 (1951); see also D. Pines, 'Elementary Excitations in Solids' (Benjamin, New York 1964).

2. T. Toya, J. Res. Inst. Catalysis, Hokkaido Univ. 6, 161 and 183 (1958).

3. N.W. Ashcroft and D.C. Langreth, Phys. Rev. 155, 682 (1967); D.C. Wallace, Phys. Rev. 182, 778 (1969).

4. See e.g. R.H. Fowler, J. Chem. Phys. 59, 3435 (1973); A. Rahman, Phys. Rev. Lett. 32, 52 (1974) and Phys. Rev. A9, 1667 (1974); M. Parrinello and A. Rahman, Phys. Rev. Lett. 45, 1196 (1980).

5. For a review see M. Baus and J.P. Hansen, Phys. Repts 59, 1 (1980).

6. D.L. Price, K.S. Singwi and M.P. Tosi, Phys. Rev. B2, 2983 (1970).

7. N.W. Ashcroft, J. Phys. C1, 232 (1968).

8. T.E. Faber, in 'Physics of Modern Materials', vol. II, p.645 (IAEA, Vienna 1980).

9. This condition for the electron gas is known as the 'compressibility sum rule' : see D. Pines and P. Nozières, 'The Theory of Quantum Liquids' (Benjamin, New York 1966). It states that the inverse screening length k_e in the electronic dielectric function must agree with the value calculated from the ground state energy E_g through the expression $k_e^2 = 4\pi\rho_e^2 e^2 K_e$ where K_e is the electron gas compressibility, determined by the second derivative of E_g with respect to the electron density.

10. E.G. Brovman and Yu. Kagan, Z. Eksp. Teor. Fiz. 52, 557 (1967) and 57, 1329 (1969) [English translations : Sov. Phys. JETP, 25, 365 (1967) and 30, 721 (1970)] ; C. Pethick, Phys. Rev. B2, 1789 (1970).

11. F. Postogna and M.P. Tosi, N. Cimento 55B, 399 (1980).

12. See e.g. N.H. March and M.P. Tosi, 'Atomic Dynamics in Liquids' (Macmillan, London 1976).

13. S. Galam and J.P. Hansen, Phys. Rev. A14, 816 (1976); F. Postogna and M.P. Tosi, ref.11. Strictly speaking, S(k) in (3.5) describes the diffraction pattern from interference between waves scatterd by pairs of ions and does not fully describe the X-ray diffraction pattern, which also contains interference terms involving the conduction electrons; see P.A. Egelstaff, N.H. March and N.C. McGill, Cand. J. Phys. 52, 1651 (1974).

14. D.K. Chaturvedi, M. Rovere, G. Senatore and M.P. Tosi, Physica (in the press). A preliminary report has been given by D.K. Chaturvedi, G. Senatore and M.P. Tosi, Lett. N. Cimento 30, 47 (1981).

15. P. Vieillefosse and J.P. Hansen, Phys. Rev. A12,1106 (1975); H.E. De Witt, Phys. Rev. A14, 1290 (1976); J.P. Hansen, G.M. Torrie and P. Vieillefosse, Phys. Rev. A16, 2153 (1977). The relation between k_i^2 and the free energy of the OCP is analogous to the 'compressibility sum rule' for the electron gas mentioned in ref. 9 above, namely $k_i^2 = 4\pi\rho_i Z^2 e^2 K_i$ where K_i is the isothermal compressibility of the OCP. Notice that the OCP appropriate to alkali metals near freezing is in a strong coupling regime, where k_i^2 takes strongly negative values.

16. M. Watabe and M. Hasegawa, in 'The Properties of Liquid Metals', p.133 (edited by S. Takeuchi; Francis and Taylor, London 1973); J. Chihara, ibid., p.137.

17. D.L. Price, Phys. Rev. A4, 358 (1971).

18. S.G. Brush, H.L. Sahlin and E. Teller, J. Chem. Phys. 45, 2102 (1966); J.P. Hansen, Phys. Rev. A8, 3096 (1973); S. Galam and J.P. Hansen, ref.13.

19. M.J. Gillan, J. Phys. C7, L1 (1974).

20. The Debye-Hückel approximation for the classical plasma, when applied at all values of r, corresponds to the RPA for the degenerate electron gas. It assumes for $S_o(k)$ the form of eqn.(3.8) at all values of k, with k_i^2 replaced by k_D^2.

21. D.K. Chaturvedi, G. Senatore and M.P. Tosi, N. Cimento 62B, 375 (1981).

22. A close similarity between the structure factor of the OCP and those of liquid alkali metals in the region of the main peak and beyond was first noticed by H. Minoo, C. Deutsch and J.P. Hansen, J. Phys. Lettres, 38, L191 (1977).

23. For the freezing of the OCP see J.P. Hansen, ref.18; W.L. Slattery , G.D. Doolen and H.E. De Witt, Phys. Rev. A21, 2087 (1980). A. Ferraz and N.H. March, Solid State Commun. 36, 977 (1980) have proposed a freezing criterion for the alkali metals in accord with the viewpoint expressed here.

24. M.J. Huijben and W. van der Lugt, in 'Liquid Metals, 1976', p.141 (Conference Series No.30, The Institute of Physics, Bristol 1977).

25. A.J. Greenfield, J. Wellendorf and N. Wiser, Phys. Rev. A4, 1607 (1971).

26. R. Block, J.B. Suck, W. Freyland, F. Hensel and W. Gläser, in 'Liquid Metals, 1976', p.126 (Conference Series No.30, The Institute of Physics, Bristol 1977).

27. G. Senatore and M.P. Tosi, Phys. Chem. Liquids (in the press). The optimized random-phase approximation was introduced for argon-like liquids by H.C. Andersen, D. Chandler and J.D. Weeks, J. Chem. Phys. 56, 3812 (1972).

28. J.G. Kirkwood and F. Buff, J. Chem. Phys. 19, 774 (1951).

29. S.P. McAlister and R. Turner, J. Phys. F2, L51 (1972).

30. A.B. Bhatia, W.H. Hargrove and N.H. March, J. Phys. C6, 621 (1973).

31. N.H. March, M.P. Tosi and A.B. Bhatia, J. Phys. C6, L59 (1973).

32. J. Baur, K. Maschke and A. Baldereschi, to be published.

33. A.R. Miedema, F.R. de Boer and P.F. de Châtel, J. Phys. F3, 1558 (1973); A.R. Miedema, P.F. de Châtel and F.R. de Boer, Physica 100B, 1 (1980).

34. M.P. Iniguez and J.A. Alonso, to be published.

35. J.A. Alonso and N.H. March, Phys. Chem. Liquids (in the press).

36. B.P. Alblas and W. van der Lugt, J. Phys. F10, 531 (1980).

37. For general reviews see M.A. Bredig, in 'Molten Salt Chemistry', p.367 (edited by M. Blander; Interscience, New York 1964); J.D. Corbett, in 'Fused Salts', p.341 (edited by B. Sundheim; McGraw-Hill, New York 1964).

38. G.P. Flynn, Phys. Rev. B9, 1984 (1974).

39. H.R. Bronstein, A.S. Dworkin and M.A. Bredig, J. Chem. Phys. 37, 677 (1962) and other references given therein.

40. G. Senatore, M.P. Tosi and P.V. Giaquinta, Physica (in the press).

41. I. Katz and S.A. Rice, J. Amer. Chem. Soc. 94, 4874 (1972); see also P.J. Durham and D.A. Greenwood, Phil. Mag. 33, 427 (1976).

42. N.H. Nachtrieb, Adv. Chem. Phys. 31, 465 (1975) and references given therein.

43. G. Senatore, M. Parrinello and M.P. Tosi, Phil. Mag. B41, 595 (1980).

44. J.F. Jal, Thèse présentée devant l'Université Claude Bernard - Lyon I (July 1981).

45. F. Hensel, Adv. Phys. 28, 555 (1979) and references given therein.

46. R. Dupree, D.J. Kirby, W. Freyland and W.W. Warren Jr, Phys. Rev. Lett. 45, 130 (1980).

47. W. Martin, W. Freyland, P. Lamparter and S. Steeb, Phys. Chem. Liquids 10, 49, 61 and 77 (1980).

48. R. Evans and M.M. Telo da Gama, Phil. Mag. 41, 351 (1980).

49. R.K. Sharma, G. Senatore and M.P. Tosi, to be published.

RESISTANCE MINIMA IN MAGNETIC ALLOYS

J. Ruvalds

Physics Department, University of Virginia
Charlottesville
Virginia 22901, USA

The existence of a magnetic impurity in a metal is known to yield a resistance minimum in the dilute limit, whereas the anomalous resistivity in concentrated alloys $x \sim 5\ \%$ is the subject of recent research. A historical review of resistivity and magnetic susceptibility experiments is presented culminating with the determination of a well defined impurity magnetic moment. The Kondo effect of a ℓnT divergence in the resistivity of dilute alloys is discussed. A new divergence of the form $\rho \sim x^2/T$ arising from coupled magnetic impurities is obtained at low temperatures, and correlated with recent measurements on layered compounds. The latter term yields a resistance minimum for ferromagnetic alignment of the impurity spins. Changes in the RKKY interaction induced by the Fermi surface topology allow for antiferromagnetic alignment in some cases where the resistivity minimum is absent. The influence of the magnetic impurity pairs on superconductivity is also discussed.

I. HISTORICAL SURVEY

The discovery of a resistance minimum in metals at low temperatures dates back to fifty years ago, with the observations of Voight and Meissner[1] on metals such as Mg, Mo, Te, Co, and Pd. At that time efforts were underway to achieve a precise verification of the expected temperatures variation

$$\rho(T) = \rho_o + BT^5 , \qquad (1)$$

where BT^5 represents the electron scattering by thermally excited phonons at low temperatures. Impurity contributions are embedded in the residual resistivity constant ρ_o, which should be a simple linear function of the impurity concentration according to Matthiessen's rule. By contrast, the resistivity of many metals was found to have an additional term with an evident divergence at low temperature, whose origin was attributed to an interesting many-body effect only in 1964. In this section we review the highlights of the experimental and theoretical discoveries spanning the intervening period, with an emphasis on the formation of impurity magnetic moments.

A primary obstacle to experimental methods was the strong effect of extremely dilute impurities such as Fe with concentrations less than $x \leqslant .000001$, which was believed to be influential[2]. Hence a series of studies[3,4] on transition metal alloys were instituted, with the clearcut demonstration of a resistivity minimum for dilute Mn impurities in Mn_xAg with $x \ll 1$ %. Furthermore, an interesting maximum in the resistivity occurs at higher concentrations $x \sim 0.1$ %, and both anomalies are destroyed at yet higher concentrations ($x > 0.5$ %) as shown in fig.1.

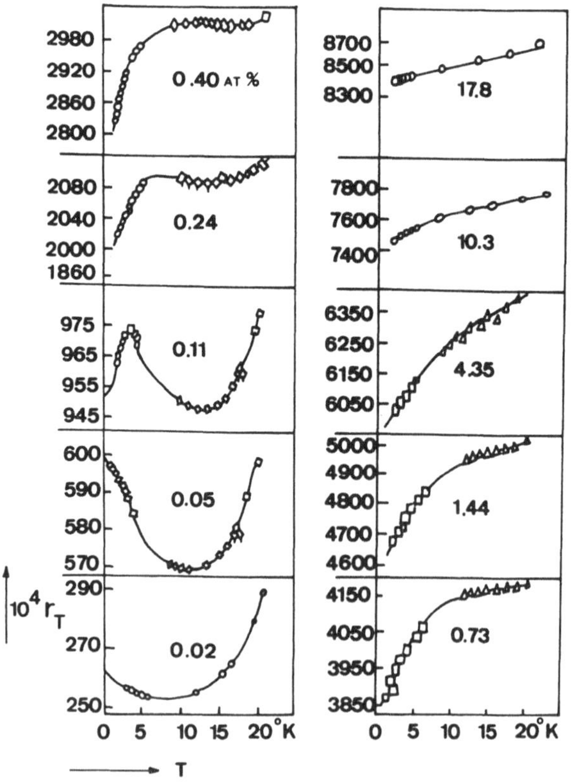

Figure 1. Resistance ratio $r_T = \rho(T)/\rho(273)$ as a function of T for Ag-Mn alloys with atomic concentrations indicated in the graph.

In addition to the resistivity minimum, the Ag-Mn_x alloys exhibit other anomalies which are not yet fully understood. For the present, we mention the quenching of the minimum for $x > 0.5$ %, and the appearance of an anomalous maximum. These are indications of strong impurity-impurity interactions even at such low concentrations of $x \sim 1$ %.

Magnetoresistance measurements[4] produced a coincident series of temperature anomalies in the Ag_x-Mn alloys, which were noted for a negative magneto-resistance at similarly low values of x which show the resistivity minimum in zero field. However, the application of a field H = 20 KOe was sufficient to remove the minimum anomaly, indicating a possible magnetic character for the impurities.

Magnetic susceptibility experiments[5] provided a classic demonstration of the impurity role in the metal host by yielding a distinct Curie-Weiss behavior of the susceptibility

$$\chi = \frac{c}{T - \theta} , \qquad (2)$$

where c is a measure of the effective moment and θ designates the Curie temperature. At low concentrations a very strong T-dependence is observed in many cases, including Ag-Mn_x and Cu-Mn_x which verifies the existence of a magnetic moment associated with a single Mn atom. By contrast, nonmagnetic impurities do not exhibit either the resistivity minimum or a $\chi(T)$ variation.

Interactions among impurities are revealed as changes in the Curie θ in Eq. 2 at rather low concentrations as seen in Table 1.

TABLE I

Host Metal	at % of Mn	Curie Temp θ(°K)
Cu	.03	0 $\pm$ 0.5
Cu	.20	1-2
Cu	1.4	7 $\pm$ 1
Cu	1.40	-3.3
Cu	1.61	-0.3
Cu	2.6	8.3
Cu	4.8	20.-

The change in sign of θ indicates ferromagnetic ordering ($\theta > 0$), at low concentrations $x \geqslant 1.4$ %, which transfers to an antiferromagnetic order at higher concentrations ($\theta < 0$). Physically, the origin of the impurity interaction is the indirect exchange coupling mediated by the conduction electrons and is referred to as the RKKY interaction. As a general rule, the resistivity minimum is highly sensitive to the strength of the RKKY coupling, and is overshadowed by the impurity interaction effects at concentrations exceeding 1 %.

The case of non-magnetic impurities does not give the resistivity anomaly, as demonstrated very precisely for $CuSn_x$[6]. Also a systematic analysis of $\rho(T)$ for Fe_xCu showed convincingly that the temperature at which the minimum occurs varies according to $T_{min} \propto x^{1/5}$. Previously, uncertainties in the data allowed for various types of singularities in $\rho(T)$, whereas this systematic relation imposes a strict constraint on theory.

A comprehensive review of the above, as well as numerous other experiments, is found in the work of Van Den Berg, which also contains an interesting account of theoretical models proposed prior to 1964[7]. It was quite certain by then that the magnetic moment formation was essential to the resistivity anomaly, but a theoretical understanding of the temperature variation of $\rho(T)$ was completely lacking.

Historically, the earliest attempts to explain the unusual resistivity relied on models with sharp structure in the density of states near the Fermi energy. One possible source of such structure is a possible localized electronic state with exceedingly narrow width as proposed by Korringa and Gerritsen[8]. This model yields a qualitative fit to $\rho(T)$ by suitable adjustment of parameters, but does not give the systematic variation of the minimum with impurity content.

An alternate line of approach was the consideration of the magnetic energy levels, whose thermal population may vary on a scale determined by the s-d exchange interaction. A small energy difference between spin aligned and antiparallel states near the Fermi energy will give a temperature variation which is however rather smooth[9-11], and hence incapable of explaining either the resistivity minima or maxima shown in fig.1.

At the lower concentrations, scattering of an electron by a pair of nearest-neighbor impurities may dominate over the molecular field averaged over impurities, and the preferential thermal occupation of aligned spin states may affect the resistivity. Brailsford and Overhauser computed the electron scattering amplitude for a pair of spins which could give a resistivity minimum arising from an additional term with $\rho_{pair} \sim e^{-w/T}$, where w is the spin-spin interaction energy. However, the magnitude of the effect did not scale correctly with the impurity concentration.

Theoretical interest in the problem was then focused on the formation of a magnetic moment at the impurity site. A very elegant demonstration of this effect is provided by the Anderson model[13]. Consider localized d-electrons interacting with the host metal conduction electrons according to the Hamiltonian

$$H = H_s + H_d + H_{sd} , \tag{3}$$

where

$$H_s = \sum_{\underset{\sim}{k}\sigma} \varepsilon_k c^+_{\underset{\sim}{k}\sigma} c_{\underset{\sim}{k}\sigma} \tag{4}$$

$$H_d = \sum_{m\sigma} E_m c^+_{m\sigma} c_{m\sigma} + \frac{1}{2} (U-J) \sum_{m\neq n} c^+_{m\sigma} c_{m\sigma} c^+_{n\sigma} c_{n\sigma} \tag{5}$$

$$+ U \sum_{m,n} c^+_{m\sigma} c_{m\sigma} c^+_{n-\sigma} c_{n-\sigma}$$

$$H_{sd} = \sum_{\underset{\sim}{k}m\sigma} (V_{\underset{\sim}{k}m} c^+_{\underset{\sim}{k}\sigma} c_{m\sigma} + V_{m\underset{\sim}{k}} c^+_{m\sigma} c_{\underset{\sim}{k}\sigma}) \tag{6}$$

The conduction electron Hamiltonian H_s involves the energy $\varepsilon_k = k^2/2m$ and the usual creation and destruction operators $c^+_{\underset{\sim}{k}\sigma}$ and $c_{\underset{\sim}{k}\sigma}$, where k labels the momentum and σ denotes the spin. The localized d-state has energy E_m, with m referring to the d-orbitals and $c^+_{m\sigma}$ ($c_{m\sigma}$) giving its creation (destruction) operators. Finally V_{mk} is the matrix element which results from the hybridization of d and s states. In the Hartree-Fock approximation, the d-state energy is simply

$$E_{m\sigma} = E_d + (U-J) \sum_{n\neq m} \langle n_{n\sigma} \rangle + U \langle n_{m-\sigma} \rangle, \tag{7}$$

and its width is

$$\Gamma = \pi \langle v^2_{mk} \rangle \rho_s (E_{m\sigma}) , \tag{8}$$

where the occupation number averages $\langle n_{n\sigma} \rangle$ are to be determined self-consistently. Physically the Coulomb repulsion U and the exchange interaction J act to split the spin degeneracy of the d-level and thus give two Lorentzian peaks of width Γ in the density of states :

$$\rho_{d\sigma} = \frac{2\ell+1}{\pi} \frac{\Gamma}{(E-E_d^{\sigma})^2 + \Gamma^2} \quad . \tag{9}$$

Hence a primary requirement is that $U >> \Gamma$, and a partial occupation of one of the spin states yields a net magnetic moment $\langle n_{d\uparrow}\rangle - \langle n_{d\downarrow}\rangle$. The critical values of the parameters required for a magnetic state are determined by the magnetic moment stability condition

$$(U + 4J)\ \rho_d^{\ell}\ (E_F) = 1 \ , \tag{10}$$

which exhibits a sharp sensitivity to the position of the Fermi level relative to the d-state energy. A schematic demonstration of the impurity levels in a Cu host is shown in fig.2.

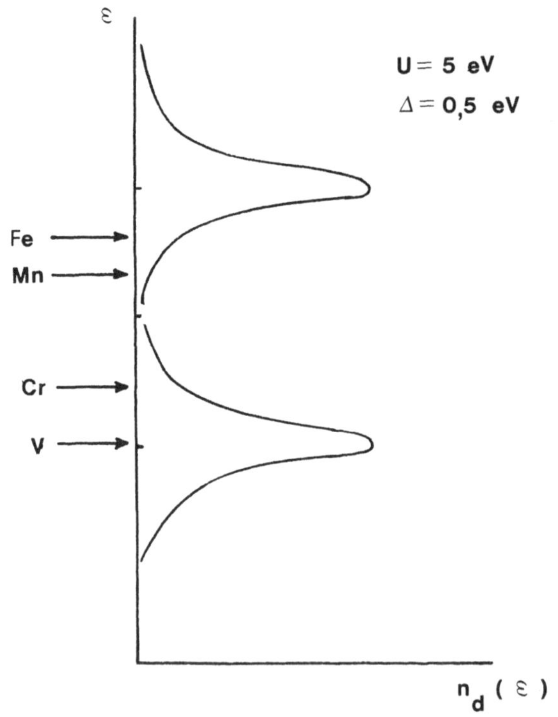

Figure 2. Positions of the virtual bound states relative to the Fermi energy as derived from optical absorption data on Cu-based alloys.

An experimental reflection of the double-peak structure in the density of states appears in the ordinary room temperature resistance of Cu-based magnetic alloys of the transition metal atoms Ti, V, Cn, Mn, etc.[14]

The basic physics of the Anderson model is vividly demonstrated by 1 % Fe impurities in alloys of Mo with Nb and independently with Re[15]. In effect, the Mo-Nb and Mo-Re hosts allow a wide range of Fermi level placements, thus admitting a favorable situation for a Fe magnetic moment over a prescribed range. This situation has been verified by susceptibility measurements over a wide range of Nb and Re content[15].

Finally a conclusive proof of the magnetic impurity involvement in the resistivity minimum was achieved by Sarachik[16], who found the anomalous temperature variation in only those alloys of Mo-Nb and Mo-Re which were known[15] to support a magnetic moment of an Fe impurity. Hence the one-to-one correspondence of a Curie-Weiss behavior of the susceptibility and the resistivity minimum in dilute alloys is firmly established.

II. KONDO EFFECT

A major breakthrough in understanding the resistivity minimum problem was attained by Kondo[17,18], who demonstrated that a single magnetic impurity yields a logarithmic divergence for the electron cross section at low temperatures. This surprising result appears in the third order perturbation theory as the many body effect depending on the distribution of the electron gas. The Kondo singularity immediately accounts for the resistivity minimum variation with impurity concentration, giving the established variation of the minimum temperature $T_{min} \propto x^{1/5}$.
However, the major excitement aroused by Kondo's calculation has been related to the obvious inadequacy of the perturbation theory at very low temperatures, namely the violation of the unitarity limit on the scattering. This issue has been studied by various techniques, culminating with a successful exact solution very recently.

It is convenient to consider the conduction electrons interacting with an impurity spin of magnitude $\underset{\sim}{S}$ according to the s-d Hamiltonian

$$H_{sd} = - J \, \underset{\sim}{S}_j \cdot \underset{\sim}{\sigma} \; \delta(\underset{\sim}{r} - \underset{\sim}{R}_j) \, , \tag{11}$$

where J is treated as a constant exchange interaction. The δ-function interaction was used in the earliest work of Fröhlich and Nabarro[19] and by Zener[20] in the study of nuclear spin relaxation. In the second quantization notation the Hamiltonian becomes

$$H_{sd} = -\frac{J}{2N}\sum_{\underset{\sim}{k},\underset{\sim}{q}} [S_z(c^+_{\underset{\sim}{k}+\underset{\sim}{q}\uparrow}c_{\underset{\sim}{k}\uparrow} - c^+_{\underset{\sim}{k}+\underset{\sim}{q}\downarrow}c_{\underset{\sim}{k}\downarrow}) + S_+c^+_{\underset{\sim}{k}+\underset{\sim}{q}\downarrow}c_{\underset{\sim}{k}\uparrow} + S_- c^+_{\underset{\sim}{k}+\underset{\sim}{q}\uparrow}c_{\underset{\sim}{k}\downarrow}] , \tag{12}$$

where the spin-flip operators S_+ and S_- are the essential ingredient in the Kondo scattering. It is interesting to express the effective s-d exchange constant J in terms of the Anderson model parameters, which are related by the Schrieffer-Wolff transformation[21] in the form

$$J \sim \frac{2|V_{km}|^2 U}{\varepsilon_d(\varepsilon_d + U)} < 0 . \tag{13}$$

It is important to note the antiferromagnetic character of the conduction electron coupling to the impurity spin, which is indeed prevalent in transition metal impurities. However, rare earth impurities may have ferromagnetic J values, suggesting the importance of electron correlations beyond the Hartree-Fock approximation in such cases. The Kondo effect yields a resistance minimum only in the antiferromagnetic case $J < 0$.

A divergence in the perturbation theory of the electron-impurity scattering is evident in the transition probability for an electron in initial state $\underset{\sim}{k}\uparrow$ scattering to a final state $\underset{\sim}{k}'\uparrow$ via the s-d interaction. The scattering probability may be expressed in powers of J according to

$$t_{\underset{\sim}{k}\uparrow,\underset{\sim}{k}'\uparrow} = t^{(1)}_{\underset{\sim}{k}\uparrow,\underset{\sim}{k}'\uparrow} + t^{(2)}_{\underset{\sim}{k}\uparrow,\underset{\sim}{k}'\uparrow} + \dots , \tag{14}$$

where the lowest order term does not involve spin flip terms and is given by

$$t^{(1)}_{\underset{\sim}{k}\uparrow,\underset{\sim}{k}'\uparrow} = -\frac{J}{2N}\langle S_z\rangle . \tag{15}$$

Hence this first Born Approximation gives the transition rate

$$W_{\underset{\sim}{k}\uparrow,\underset{\sim}{k}'\uparrow} = \frac{2\pi}{h}|t^{(1)}_{\underset{\sim}{k}\uparrow,\underset{\sim}{k}'\uparrow}|^2 \,\delta(E_{\underset{\sim}{k}\uparrow} - E_{\underset{\sim}{k}'\uparrow}) \tag{16}$$

$$W_{\underset{\sim}{k}\uparrow,\underset{\sim}{k}'\uparrow} = \frac{2\pi}{h}(J/2N)^2 S_z^2 \,\delta(E_{\underset{\sim}{k}\uparrow} - E_{\underset{\sim}{k}'\uparrow}) . \tag{17}$$

The next higher order process involves the spin-flip terms of Eq.12 and yields a divergent low temperature result because of the non-commutating property of the spin operators.

The two processes to be considered are : (a) scattering to an intermediate electron state $\underset{\sim}{q}\sigma$ and subsequently to the final state $\underset{\sim}{k}'\uparrow$; (b) an electron from a filled state $\underset{\sim}{q}\sigma$ scatters to the final state $\underset{\sim}{k}'\uparrow$, and later an electron from the initial state $\underset{\sim}{k}\uparrow$ scatters into the empty state $\underset{\sim}{q}\sigma$. Diagrammatically these two scattering events are analogous to the Compton scattering of an x-ray by an electron.

Without the spin-flip process, the second order scattering gives

$$t^{(2)}_{\underset{\sim}{k}\uparrow,\underset{\sim}{k}'\uparrow}\Big|_{\sigma=\uparrow} = \left(\frac{J}{2N}\right)^2 S_z^2 \left[\sum_{\underset{\sim}{q}} \frac{1-f_q}{\varepsilon_k-\varepsilon_q} + \sum_q \frac{f_q}{\varepsilon_k-\varepsilon_q}\right]$$

$$t^{(2)}_{\underset{\sim}{k}\uparrow,\underset{\sim}{k}'\uparrow}\Big|_{\sigma=\uparrow} = \left(\frac{J}{2N}\right)^2 S_z^2 \sum_q \frac{1}{\varepsilon_k-\varepsilon_q} \,. \tag{18}$$

Performing the integral over $\underset{\sim}{q}$, we find only a constant correction to the lower order term $t^{(1)}$. Because of the commuting nature of S_z^2, this result is in a sense similar to scattering by ordinary nonmagnetic impurities.

Spin flip terms in the s-d interaction of Eq.12 are essential at low temperatures, because there is not the simple cancellation of the Fermi function f_q for the intermediate state. Furthermore, since the spin flip dynamics involve zeroenergy transfer by the conduction electron, a divergent scattering amplitude appears : The corresponding terms are

$$t^{(2)}_{\underset{\sim}{k}\uparrow,\underset{\sim}{k}'\uparrow}\Big|_{\sigma=\downarrow} = \left(\frac{J}{2N}\right)^2 \sum_{\underset{\sim}{q}} \frac{1}{\varepsilon_k-\varepsilon_q}\left[S_-S_+\,(1-f_q) + S_+S_-f_q\right]$$

$$\cong 2\,(J/2N)^2\, S_z \sum_{\underset{\sim}{q}} \frac{f_q}{\varepsilon_k-\varepsilon_q} \,. \tag{19}$$

Here another term without f_q again yields an insignificant constant correction. Following Kondo, we transform the momentum integration to an energy variable ε by introducing a constant density of states $N(\varepsilon) \cong N(0)$ near the Fermi energy. Then Eq.19 can be integrated by parts to give

$$t^{(2)}_{\underset{\sim}{k}\uparrow,\underset{\sim}{k}'\uparrow}\Big|_{\sigma=\downarrow} = 2\,(J/2N)^2\, S_z \left[-N(0)\int d\varepsilon\, \frac{df}{d\varepsilon}\, \ln\left|\frac{\varepsilon}{D}\right|\right] \tag{20}$$

where a cut-off D is introduced which corresponds in actuality to the required momentum cut-off for a realistic exchange interaction $J_{\underset{\sim}{kq}}$. For a well-defined Fermi surface, Eq.20 immediately gives a term with $\ell n\, T$ and a total scattering amplitude (Eq.15 & Eq.20) :

$$t_{\underset{\sim}{k}\uparrow,\underset{\sim}{k}'\uparrow} = -\frac{J}{2N} S_z + 2\,(J/2N)^2 (S_z) N(0)\, \ell n\, \frac{T}{D}\,, \tag{21}$$

and, from the Golden rule of Eq.17, we find the transition probability to order J^3 as

$$W_{\underset{\sim}{k}\uparrow,\underset{\sim}{k}'\uparrow} = \frac{2\pi}{\hbar}\,[(J/2N)^2 S_z^2 + (J^3/2N^3) S_z^2 N(0)\ \ell n\ (\tfrac{T}{D}) + \ldots) \tag{22}$$

Finally the electrical resistivity becomes

$$R(T) = x\, R_m [1 + 2(J/N)N(0)\ell n(T/D) + \ldots]\,, \tag{23}$$

which is the well known Kondo effect formula with $R_m = 2\pi m x N(0) J^2 S(S+1)/ne^2\hbar$, where n gives the electron density. This result yields a very good fit to the resistivity minima of dilute alloys and, in tandem with the usual phonon contribution BT^5, accounts for the concentration dependence of the minimum temperature $T_{min} \propto x^{1/5}$.

It should be emphasized that the Kondo scattering is rapidly quenched by either external magnetic fields[22], or by interactions with other impurities[23]. Physically, both of these influences inhibit the spin-flip terms and thus quite naturally destroy the ℓnT singularity. As a general rule, the Kondo divergence is quenched at magnetic impurity concentrations exceeding one per cent.

At zero temperature the total scattering must of course be finite and less than the impurity limit

$$\Delta R = 3.8(2.5)\ \mu\Omega\text{-cm/at}\ \%\ , \tag{24}$$

which is also compatible with the vast data. In fact, the extreme low temperature resistivity is known to saturate by a power law of the type $R \simeq R_o - bT^2$ in Kondo systems, and this crossover from a logarithmic divergence to a power law has attracted the attention of numerous theoretical efforts[18,24].

Initially, higher order perturbation calculations were performed to leading order in the logarithmic terms. Sophisticated techniques were applied by various authors, including the evaluation of "parquet" diagrams which can be summed to yield a higher-order resistivity[18,24]

$$R = x R_m \frac{1}{[1 - 2 JN(0) \ell n(T/D)]^2} . \tag{25}$$

However, these calculations yield yet another type of divergence at the Kondo temperature defined by

$$T_k \cong D \exp \frac{1}{2 JN(0)} \tag{26}$$

for the antiferromagnetic case $J < 0$. This typical temperature, of order $T_k \sim 0.1°k$ for MnCu with $J = -1$ eV, provides a useful guide to the Kondo anomaly because it also appears in the magnetic susceptibility according to the relation[25]

$$\chi(T) = \frac{c}{T + 4.5 T_k} . \tag{27}$$

Thus the original association of a resistance minimum with a Curie-Weiss behavior of the susceptibility is now also established theoretically. Furthermore, Eq. 27 provides a useful guideline for the impurity concentration limitation on the Kondo effect, namely $T_k > T_N$, where T_N represents the magnetic ordering temperature for the impurities.

Finally, we mention the essential obstacles to perturbation theory methods at low temperature, namely the similarity to the problem of phase transitions which exhibit a critical behavior near the transition temperature. Hence, the low temperature Kondo regime has been considered using the renormalization group techniques[26,27]. In particular the approach of Wilson[28] has yielded low T limiting results which are numerically exact in the sense of the computer capability.

A proper conclusion to the survey of the Kondo effect is the recent exact solution derived recently by Andrei and Lowenstein, and independently by Wiegmann[29]. As in many other cases, their methods transcend the narrow boundaries of the resistance minimum anomaly, and have important implications for many body problems extending into the realm of high energy physics as well.

III. RESISTANCE MINIMUM FOR IMPURITY PAIRS

Recently it has been noted[30] that coupled magnetic impurities can yield a new type of resistivity divergence at the intermediate concentrations ranging from 1 to 10 %, where the Kondo scattering is inoperative. The anomalous many body scattering gives a resistivity of the form

$$\rho = \alpha x^2/T , \tag{28}$$

where $\alpha \propto J^4$ is a constant whose sign and magnitude is determined by the RKKY coupling between impurities. The effect will be present irrespective of the sign of J, but will yield a resistivity minimum only for ferromagnetic alignment of the impurity spins. This situation contrasts vividly for the Kondo minimum which requires $J < 0$ and is quenched by strong RKKY fields.

The physical origin of the 1/T divergence in ρ can be visualized in the process of an electron in state $\underset{\sim}{k}\uparrow$ scattering from an impurity by a spin-flip process to state $\underset{\sim}{q}\downarrow$, and subsequently scattering from a second impurity to the final state $\underset{\sim}{k}'\uparrow$. However, the two impurity spins are coupled via the indirect exchange of another electron, which suggests an analogy to the effective electron-electron interaction. Because the spin-flip process requires zero energy transfer by the scattering electrons, the cross section is divergent in the perturbation theory. A useful representation of the situation is possible by writing the electron-electron interaction arising from the intermediate-spin-flip of the impurity which takes the form[14]

$$H_{d-d} = -\left(\frac{J}{2N}\right)^2 S_z \sum_{\substack{\underset{\sim}{k},\underset{\sim}{q}\\ \underset{\sim}{p},\underset{\sim}{p}'}} \left\{\frac{1}{\varepsilon_p-\varepsilon_{p'}} - \frac{1}{\varepsilon_k-\varepsilon_q}\right\} c^+_{\underset{\sim}{p}'\uparrow} c_{\underset{\sim}{p}\downarrow} c^+_{\underset{\sim}{q}\downarrow} c_{\underset{\sim}{k}\uparrow} ; \tag{29}$$

This result follows by a direct canonical transformation of the s-d Hamiltonian of Eq.11. On the basis of this interaction, the sequential scattering by two impurities has a transition probability with a factor

$$W_{\underset{\sim}{k}\uparrow,\underset{\sim}{k}\uparrow} \propto x^2 J^4 \sum_q \frac{f_q}{(\varepsilon_k-\varepsilon_q)^2} \sim \frac{x^2 J^4}{\varepsilon_k} \sim \frac{x^2 J^4}{T} . \tag{30}$$

Note that the impurity pair resistance divergence is strongly enhanced at higher impurity content to this order of perturbation theory.

To achieve a quantitative description of the impurity pair scattering terms, and also to examine higher order processes it is convenient to employ the pseudo-Fermion method introduced by Abrikosov[31]. The motivation for the new approach is the fact that the dynamical spin operator S does not obey simple boson or fermion commutation relations. Abrikosov introduced the following representation for the impurity spin operators

$$S = a^+_{\beta'} S_{\beta'\beta} a_\beta , \tag{31}$$

where $S_{\beta'\beta} \equiv \langle\beta'|S|\beta\rangle$ are the standard matrix elements of the spin matrices, and repeated indices imply summation over the Zeeman β indices. The a_β^+ and a_β are creation and annihilation operators for the pseudofermion field, and obey the anti-commutation relations

$$\{a_\beta, a_{\beta'}^+\} = \delta_{\beta\beta'}, \ \{a_\beta, a_{\beta'}\} = 0 \ . \tag{32}$$

The drawback of the method is the presence of spurious states with unphysical multiple occupation. Fortunately, these are readily discarded by assigning a "kinetic energy" λ to each state and then normalizing the results to the probability of single occupation (2S+1) exp $(-\lambda/T)$ with $\lambda \to \infty$. We follow the usual diagram convention with solid lines representing electron propagators and the broken lines for pseudofermion Green's functions. In terms of these operators, the s-d Hamiltonian takes the form

$$H = \sum_{k,\alpha} \varepsilon_k c_{k\alpha}^+ c_{k\alpha} - J \sum_{\substack{\underset{\sim}{k},\underset{\sim}{k}',j \\ \alpha,\alpha',\beta,\beta'}} e^{i(\underset{\sim}{k}-\underset{\sim}{k}')\cdot \underset{\sim}{R}_j}$$

$$\times\ (\sigma_{\alpha'\alpha}\cdot S_{\beta'\beta})\ a_{j\beta'}^+\ a_{j\beta}\ c_{k\alpha'}^+\ c_{k\alpha'} \tag{33}$$

where j refers to the impurity located at site R_j.

A straight forward application of the s-d Hamiltonian is the derivation of the RKKY interaction between impurity spins arising from the indirect exchange of a conduction electron : The result is

$$H_{RKKY} = \frac{J^2 m k_F^4}{2\pi^3} F(2k_F\Delta) \underset{\sim}{S}_i \cdot \underset{\sim}{S}_j \ , \tag{34}$$

where k_F is the Fermi momentum, Δ is the impurity separation, and

$$F(y) \cong A_d \frac{\cos(y)}{y^d} e^{-\Delta/\ell} \ , \tag{35}$$

where d denotes the effective dimensionality of the system and ℓ is the electron mean free path. In the usual three dimensional, (d=3), "pure" metal, $\ell \to \infty$ and $A_3 = 9\pi/4$. However, in general it is important to consider the one-dimensional case d = 1, $A_d = \pi$ as well, because parallel sections of the electron Fermi surface may yield an effective d = 1 situation for the electron response.

Before proceeding to the actual calculation of the electron scattering amplitude, it is instructive to map out the phase diagram for ferromagnetic ordering determined by the criterion[32]

$$T_N = -xJ^2 S(S+1) \frac{Ad \cos(2k_F\Delta)}{E_F} . \tag{36}$$

This phase boundary is mapped in figure 3 using a typical value of the temperature $T_N = 10°k$.
This diagram illustrates again the extremely dilute concentration limit for the Kondo effect, i.e. $x < 1\%$.

At the intermediate concentrations of $1\ \% < x < 10\ \%$, we are also restricted to rather small values of the exchange coupling $J \simeq 0.1$ eV, which happens to be realized in a rather large variety of cases.

We now proceed to compute the interacting impurity pair contribution to the resistivity. The perturbation theory for the electron scattering amplitude can be constructed in terms of the Green's functions of the electron

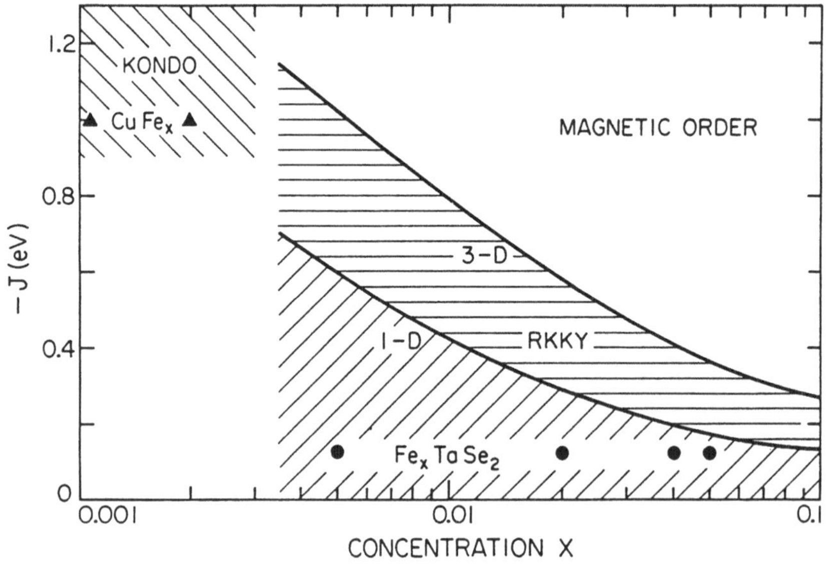

Figure 3. Separation of the ferromagnetic ordering phase as a function of J and x for both d = 1, and d = 3 dimensional cases. Typical data points for $CuFe_x$ and $Fe_x\ TaSe_2$ are indicated.

$$G = \frac{1}{i\omega_n - \varepsilon_k} \quad , \tag{37}$$

and the dispersionless impurity spin

$$\mathcal{G} = \frac{1}{i\omega_m} \quad . \tag{38}$$

Then the anomalous pair contribution is obtained from the electron self energy diagrams shown in fig.4.

The lowest order Born approximation gives the relaxation time used by Abrikosov and Gorkov in their classic work[33] on superconductivity :

$$\frac{1}{\tau_{AG}} = - \operatorname{Im} \Sigma_{AG} = 2\pi x N(0) J^2 S(S+1) \quad , \tag{39}$$

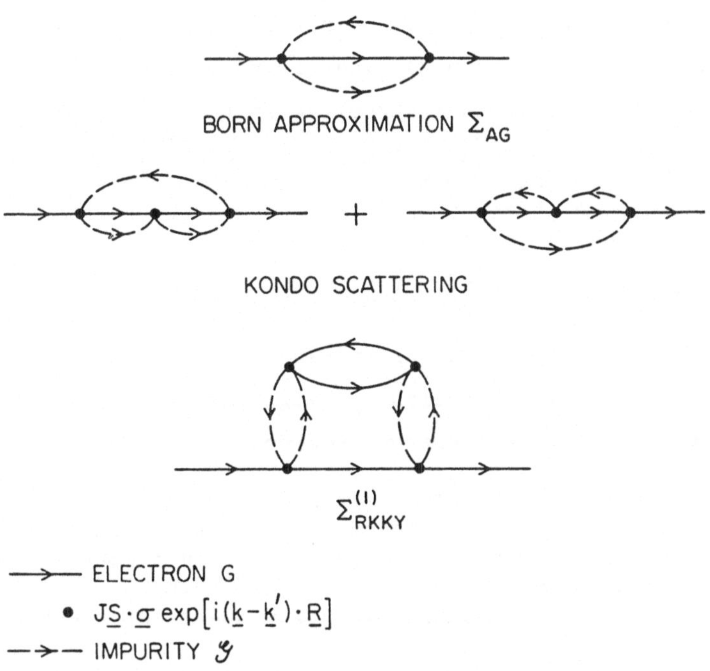

Figure 4. Diagrammatic representation of the scattering of an electron (solid line) by a magnetic impurity (dotted curve). Note that the divergent term $\Sigma_{RKKY}(1)$ involves spin-flop processes by two magnetic impurities and is fourth order in the exchange coupling J.

and is part of the constant resistivity contribution R_m in Eq.23. Thus a comparison of normal state resistivity and the superconducting transition sometimes allows a separation of the spin-flip amplitude of Eq.39 from ordinary potential scattering.

The lowest order Kondo diagrams are also shown in fig.4, and these contribute the J^3 term to the resistivity of Eq.23 which has the well-known $x \ln T$ behavior. Higher order scattering terms are clearly warranted at low temperatures and the Abrikosov pseudofermion technique provides a systematic method to collect the leading divergent terms. Furthermore, these Kondo-like diagrams[24] are analogous to the problem of x-ray absorption thresholds in simple metals, where the higher-order "parquet" diagrams can be summed to give an explicit power law for the frequency dependence of the absorption.

The magnetic impurity pair contribution to the electron self energy is designated by $\Sigma^{(1)}_{RKKY}$ in figure 4. It is of order J^4, proportional to x^2, and diverges as $1/T$ at low temperatures. Physically the process consists of a spin-flip scattering of an electron by the first impurity from initial state $\underset{\sim}{k}\uparrow$ to intermediate state $\underset{\sim}{q}\downarrow$ and subsequently scattering by the next impurity to the final state $\underset{\sim}{k}\uparrow$. It is instructive to trace the $1/T$ divergence in the self energy using the pseudofermion technique. A spin state "bubble" diagram has a double pole resulting from the localized nature of the impurity level, and thus the dotted line "bubble" graph is

$$\lim_{\lambda\to\infty} e^{\lambda/T} \sum_{\omega} (i\omega-\lambda)^{-2} = -\frac{1}{T} . \qquad (40)$$

Since $\Sigma^{(1)}_{RKKY}$ involves two "bubbles" giving $1/T^2$ and the intermediate state occupation function $f_q(T)$ giving one factor of T, the net result for the self energy is proportional to $1/T$.

Notice that the Kondo diagrams always have one electron line intersecting the "bubble" graphs, hence preventing terms proportional to $1/T$. Furthermore, the higher order vertex corrections to the Kondo series diverge as $(J)^n(\ln T)^m$, emphasizing once again the single impurity source of the Kondo scattering.

An explicit evaluation of the pair scattering in figure 4 gives

$$\Sigma^{(1)}_{RKKY} = - \frac{J^4[S(S+1)]^2}{T} \sum_{\omega_2} \frac{T}{(2\pi)^6} d\underset{\sim}{p}\, d\underset{\sim}{q}\, dq'$$

$$\times\ e^{i[\underset{\sim}{k}-\underset{\sim}{q}+\underset{\sim}{q}']\cdot\underset{\sim}{R}_j}\ e^{i[\underset{\sim}{q}-\underset{\sim}{k}-\underset{\sim}{q}']\cdot\underset{\sim}{R}_\ell}$$

$$\times \frac{1}{i\omega_2 - \varepsilon_p} \frac{1}{i\omega_2 - \varepsilon_{\underset{\sim}{p}-\underset{\sim}{q}'}} \frac{1}{i\omega - \varepsilon_q} \quad . \tag{41}$$

Taking the standard average over impurity sites and evaluating the integrals corresponding to the RKKY interaction, we find an electron relaxation time

$$\frac{1}{\tau^{(1)}_{RKKY}} \simeq - \text{Im}\ \Sigma^{(1)}_{RKKY} = \frac{2A_d x(n.n) J^4 \pi^2 N(E_F)[S(S+1)]^2 \cos(2k_F\Delta)}{T\ E_F\ (2k_F\Delta)^d} \tag{42}$$

This corresponds nicely with the standard RKKY expression of Eqs. 34, 35 when it is limited to nearest neighbor interactions with n.n being the number of nearest neighbors. This approximation is justified only in cases where the electron mean free path is short and consequently the RKKY interaction is restricted to have a short range. Of course an extension to further neighbors is quite straightforward.

Layered compounds provide an ideal testing ground for our theory since the magnetic impurities are constrained to specific sites in contrast to the usual random distributions. Also variations in the Fermi wavevector k_F allow a change in sign of the RKKY coupling, which permits a resistivity minimum only for ferromagnetic impurity spin alignment.

A resistance minimum can be described by adding the usual phonon scattering term BT^5 to yield a resistivity

$$\rho = \frac{\alpha x^2}{T} + BT^5 \ , \tag{43}$$

with α defined by the constants in Eq. 42, and with the standard resistivity formula $\rho = m/(ne^2\tau)$. Clearly the resistivity minimum temperature will follow a dependence $T_{min} \propto x^{1/3}$ from Eq. 43.

These general features of the RKKY-induced resistivity minimum apply particularly well to the data on Fe_xTaSe_2 obtained by the group of R.V. Coleman[34,35]. First of all, their measurements of the magnetic susceptibility rule out the Kondo effect for the range of Fe content in their samples. This test follows from the Curie-Weiss law $\chi = c/(T-\theta)$, where the constant $\theta = 4.5\ T_k$ yields a direct measure of the Kondo temperature T_K. At the lowest concentration examined (x = .005), $\theta = 1 \pm 1°k$, indicating a dubious Kondo temperature of $T_k \simeq 0.2 \pm 0.2°k$. However, at x = .08, the data gives $\theta \simeq -8°k$, indicating a dominant ferromagnetic RKKY interaction. Thus the magnetic ordering temperature is roughly two orders of magnitude larger than T_k, and the Kondo scattering is inoperative for Fe concentrations exceeding x ≃ .001 in these compounds.

A good fit of the RKKY theory is achieved with parameters estimated from independent measurements,[34,35] as shown in fig.5. The changes in the theory curves of fig.5 are due only to variations in the concentration x.

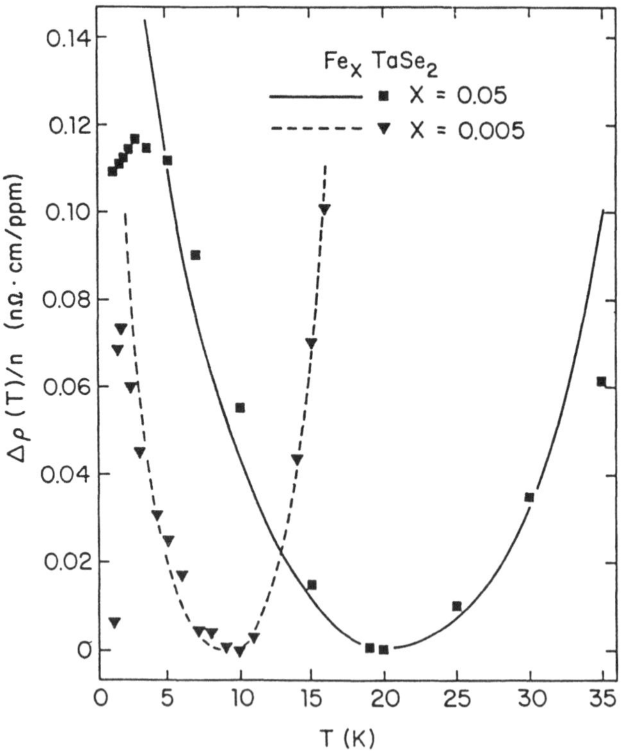

Figure 5. Excess resistivity of Fe_x in $TaSe_2$ as a function of temperature. The RKKY parameters are : J = −0.12 eV, $2k_F\Delta = 2.1$, $E_F = 0.4$ eV, $\alpha \simeq 1$, n.n. = 12x, and $A_d = 2.4$.

It is interesting to compare the dependence of the resistivity minimum temperature T_{min} as a function of x for various compounds[34,35] as shown in figure 6. Evidently the $T_{min} \propto x^{1/3}$ behavior works very well for the layered compounds at high impurity content. The contrasting Kondo behavior of $x^{1/5}$ is also shown for comparison in the dilute alloy regime.

An alternate example with no resistivity minimum is provided by Fe_xTaS_2 data,[34,35] which can be explained assuming antiferromagnetic impurity spin alignment. In fact the antiferromagnetic tendency is confirmed by susceptibility measurements on the concentrated alloys. A fit to the resistivity using Eq. 43 is shown in figure 7; using several values of the exchange constant J. Evidently the known small J values of order 0.1 eV have very little effect on the resistivity for the antiferromagnetic case. Also, it is instructive to note that the addition of Fe impurities causes the sample to become superconducting at low temperatures.

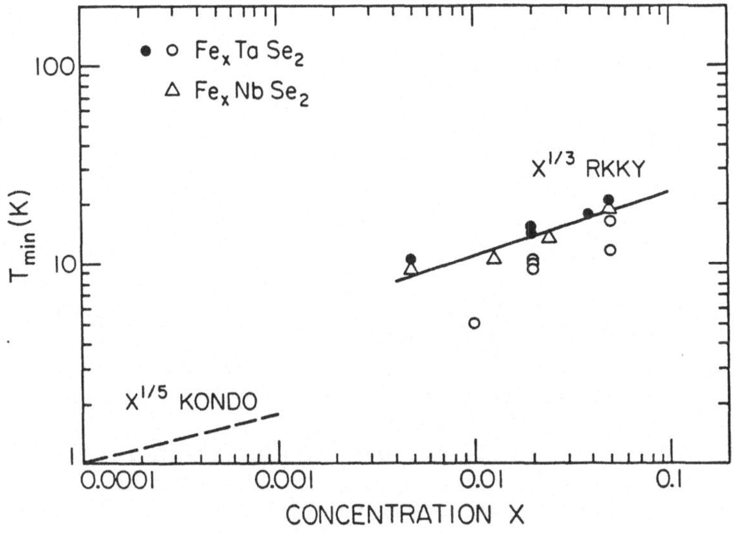

Figure 6. Resistivity minimum temperature showing the $x^{1/3}$ variation of the RKKY induced divergence in comparison to data on layered compounds.

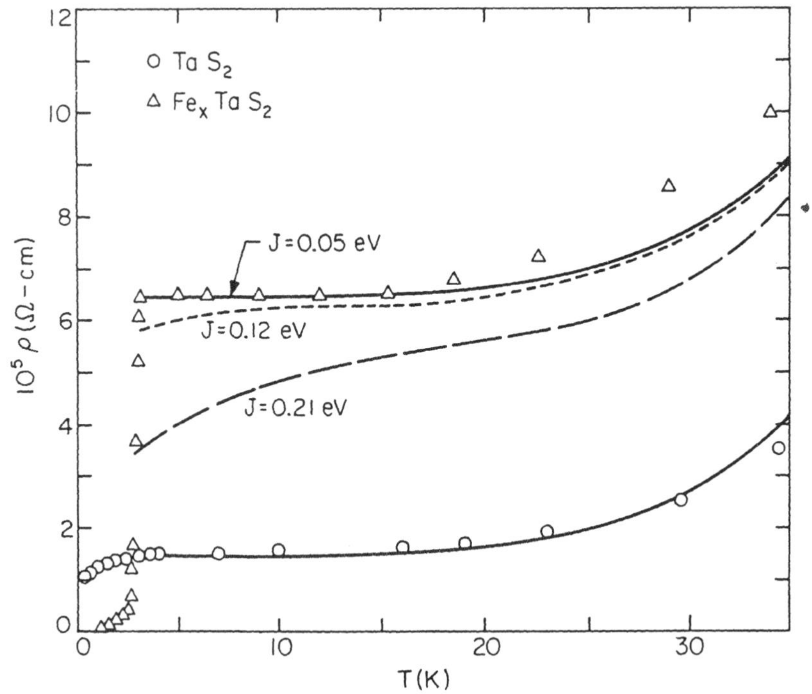

Figure 7. Resistivity of Fe_xTaS_2 using a phonon contribution of $B = 5 \times 10^{-7}$ of pure TaS_2, and $E_F = 0.4$ eV, $2\ k_F\Delta = 4.2$, $A_d = 2.4$, and n.n. = 12x.

A primary theoretical issue is the removal of the resistivity divergence at the lowest temperatures which must be done to comply with the unitarity limit on the scattering cross section. The measured resistivity naturally shows a saturation at low temperatures as seen for example in figure 5. As a first attempt to study the $T \to 0$ limit, we consider the multiple scattering processes which are obtained by interacting the electron-impurity scattering vertex with higher order "bubble" terms. Because the impurity spin "bubble" diagrams require elastic scattering in the sense of momentum and energy transfer, the sequential impurity terms in the self energy diverge as $T^{-\ell+1}$, where ℓ is the number of "bubbles". Thus these contributions are readily summed as a geometric series to yield a total relaxation time

$$\frac{1}{\tau_{RKKY}} = - \mathrm{Im} \frac{\Sigma^{(1)}_{RKKY}}{1 - \Sigma^{(1)}_{RKKY}/[\pi S(S+1)(n.n.)A_d J^2]} \quad . \tag{44}$$

This result is of course finite at T = 0, but now reveals a new divergence at a temperature T_N corresponding to the vanishing of the denominator in Eq.44. This is precisely the well-known condition for impurity ferromagnetism and the transition temperature T_N is identical to Eq.36, in agreement with previous work. As the phase transition is approached, the multiple scattering terms yield

$$\rho = \frac{\alpha x^2}{T - T_N} + BT^5 \ , \tag{45}$$

which indicates the mean field nature of the RKKY theory. Formula 45 is similar in structure to the result of Friedel and De Gennes for the resistivity of a system with long range magnetic ordering[36].

The influence of the higher order scattering on the resistivity is shown by the comparison of Eqs.43 and 45, using different J values as shown in figure 8.

The correspondence of the 1/T divergence in the transport properties to analogous terms in the thermodynamic free energy[32] provides an independent check on the RKKY analysis, since the respective terms have similar analytic structure. In this connection, we mention the Kondo-Miwa controversy[37,38] regarding the existence of a 1/T term in the <u>one impurity</u> scattering problem : By now it is established that proper account of the scattering processes rules out such a term in the electron t-matrix as well as in the corresponding thermodynamic potential[32].

At this point it may be worthwhile to summarize the key issues in the study of resistivity minima. We have learned that the Kondo minimum occurs only at very dilute impurity concentrations of well-defined magnetic impurities with <u>strong</u> and <u>antiferromagnetic</u> exchange coupling, typically J = -1 eV. Many examples of this situation are found and understood for transition metal impurities in metals. However, one class of "exceptional" compounds whose magnetic origin is characterized by very localized f-electrons is provided by rare earth impurities. These generally do not show a Kondo anomaly in the dilute limit! This may be due to a weak exchange interaction $|J| \simeq 0.1$ eV, as estimated from various experiments, or perhaps J is positive in these cases.

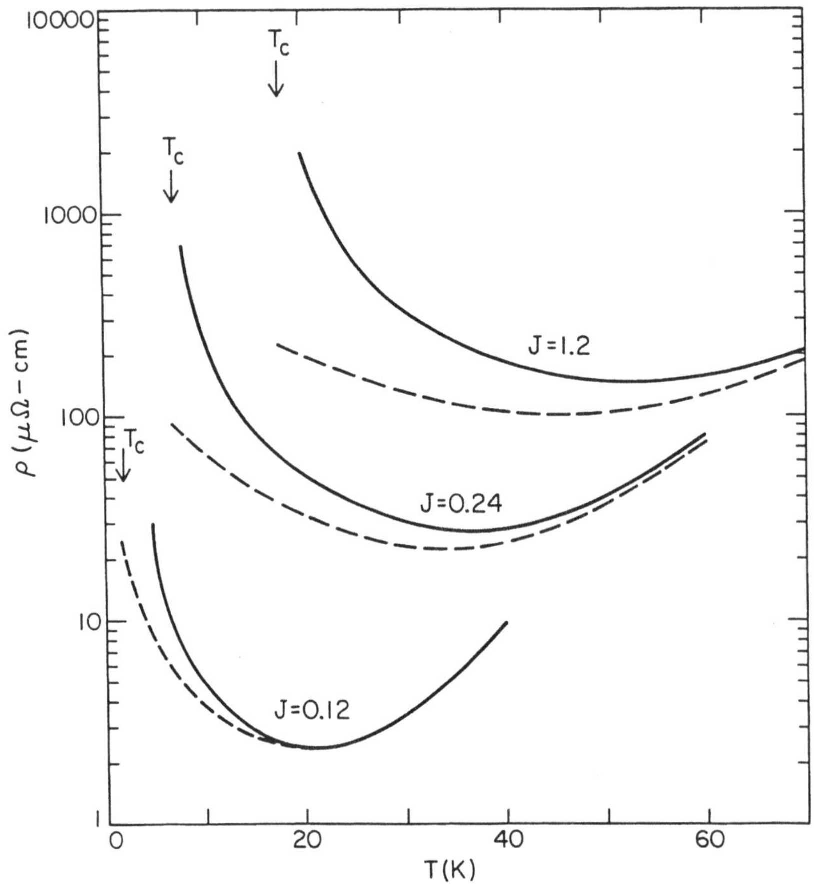

Figure 8. Comparison of the leading RKKY contribution of Eq. 43 (dotted curves) to the multiple scattering effects of Eq.44 (solid curves). Arrows point to the ordering temperature T_N.

Now an interesting possibility arises with rare earth impurities which show an anomalous resistivity only at higher concentrations as expected for the RKKY induced resistivity, which happens to be independent of the sign of J, but rather is sensitive to the impurity-impurity coupling. An example of this type is provided by the data on Ce_xLa alloys, whose resistivity is "normal" for $x \simeq 1$ %, but a well-defined minimum sets in at $x \simeq 3$ %[39]. Thus the rare earth impurities provide a stimulating challenge for future work, with particular attention to the numerous cases of mixed valence compounds and their assorted anomalies in the resistivity, magnetic susceptibility, and thermodynamic properties.

IV. SUPERCONDUCTIVITY

An impurity with a well defined magnetic moment has a strong and generally destructive effect on the superconducting transition temperature T_c, because of its tendency to break superconducting electron pairs by the spin-flip scattering. However, the divergent single electron cross sections discussed in the previous sections on resistivity, are also manifested in the electron-electron scattering amplitude which is responsible for the electron pairing in the BCS theory.

We consider three situations of interest : First the lowest order Abrikosov-Gorkov theory is presented as a guide to the behavior of T_c with impurity content in the dilute limit; for low concentrations $x \leqslant 1$ %, the Kondo scattering terms may alter the superconductivity near the lowest temperatures; and finally the concentrated impurity regime is dominated by the RKKY interactions which give a T_c enhancement for antiferromagnetic spin alignment, and the recently discovered re-entrant superconductivity for ferromagnetic alignment.

First we recall the classic work of Abrikosov and Gorkov[33] based on the Born approximation for the scattering of electrons by isolated magnetic impurities. The spin-flip scattering breaks the superconducting electron pairs and thus depresses the transition temperature T_c with increasing impurity content x. Their result gives the (AG) formula

$$\ln \frac{T_{co}}{T_c} = \psi(\tfrac{1}{2} + \tfrac{\xi}{2}) - \psi(\tfrac{1}{2}) \quad , \tag{46}$$

where ψ is the digamma function and

$$\xi^{-1} = \pi T_c \tau_s \quad . \tag{47}$$

Here the electron relaxation time is simply $\tau_s^{-1} = 2\pi X\, N(0) J^2 S(S+1)$. The functional dependence of T_c on impurity concentration is shown in figure 9.

Kondo scattering modifications of T_c have also been computed with good agreement in many dilute alloys[40]. Basically the corrections scale with the ratio of the Kondo temperature T_K

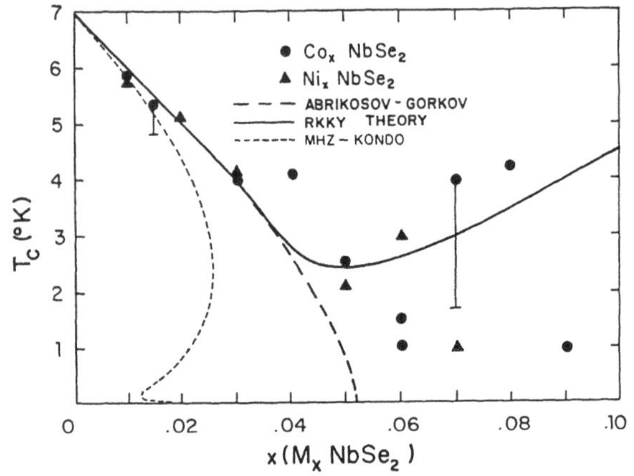

Figure 9. Superconducting temperature of $NbSe_2$ doped with Co and Ni impurities. The various theories are compared, using parameters J = -0.1 eV, E_F = 0.8 eV, $2k_F\Delta$ = 2.1, and A_d = 2.4. The error bars on the data points indicate the Tc transition widths.

to T_c. For small T_K, i.e. $T_K/T_c \ll 1$, the Kondo formulation of Müeller-Hartmann and Zittartz gives an interesting re-entrant form of T_c vs X as displayed in figure 9. For the opposite limit $T_K/T_c \gg 1$ the pair-breaking effect of impurities is reduced, with a consequent enhancement of T_c above the AG results. Re-entrant superconductivity has been discovered[41] in various compounds, and has an interesting connection to the general problem of the coexistence of magnetism and superconductivity.

The T_c data for the layered compound Co_xNbSe_2 [42] shown in figure 9 is of particular relevance to our work since it demonstrates a persistence of T_c at very high concentrations (x = .08), in contrast to previous theories. As we noted in the Introduction, the Kondo effect is inoperative at these high impurity compositions, because of the small Kondo temperature $T_K = 0.2 \pm 0.2°k$.

Recently we have shown[43] that the RKKY coupling of magnetic impurities accounts for the anomalous T_c behavior shown in figure 9, providing that the impurity spin interactions are antiferromagnetic in nature. These corrections to the AG formulation involve a modified relaxation time defined by

$$\hat{\xi} = \frac{1}{\pi T_c} \left(\frac{1}{\tau_s} + \frac{1}{\tau_{RKKY}}\right) , \tag{48}$$

where

$$\frac{1}{\tau_{RKKY}} = \frac{2\pi x \; B_d N(0)\,(n.n.)\, J^4 S(S+1) \cos(2k_F\Delta)}{T \; E_F \; (2k_F\Delta)^d} \; \frac{\chi_s}{\chi_N} \; . \tag{49}$$

Here the RKKY parameters are defined as in Eq.34, with the additional factor χ_s/χ_N which is the ratio of static susceptibilities in the superconducting (s) and normal (N) states. This ratio is non-vanishing by virtue of the finite electron spin-flip lifetime which gives $\chi_N/\chi_s = 12\, T_c \tau_s$ [44]. Again B_d is a constant of order unity depending on the dimensionality and the Fermi surface topology. As seen in figure 9, the RKKY theory gives a good fit to the behavior of $T_c(x)$ for magnetic impurities in $NbSe_2$.

The crucial distinction of our calculations is the sign of the RKKY coupling. For antiferromagnetic impurity alignment an enhanced superconducting state is predicted and observed, and by the same token there should not be a resistivity minimum. It is gratifying that in fact the resistivity measurements on Co_xNbSe_2 and Ni_xNbSe_2 do not show a minimum, and furthermore the susceptibility data support the antiferromagnetic RKKY coupling.

Lastly we mention another manifestation of the 1/T divergence in the electron-impurity-pair scattering amplitude for <u>ferromagnetic</u> RKKY interactions. From Eqs.47-49, it is readily shown that a <u>re-entrant</u> form of the superconductivity is expected along with a <u>minimum</u> in the normal resistivity. This situation is discussed in more detail in a previous publication[43]. Here we shall only demonstrate the RKKY induced re-entrant T_c in comparison to data on $Ce_xLaA\ell_2$ as shown in figure 10.

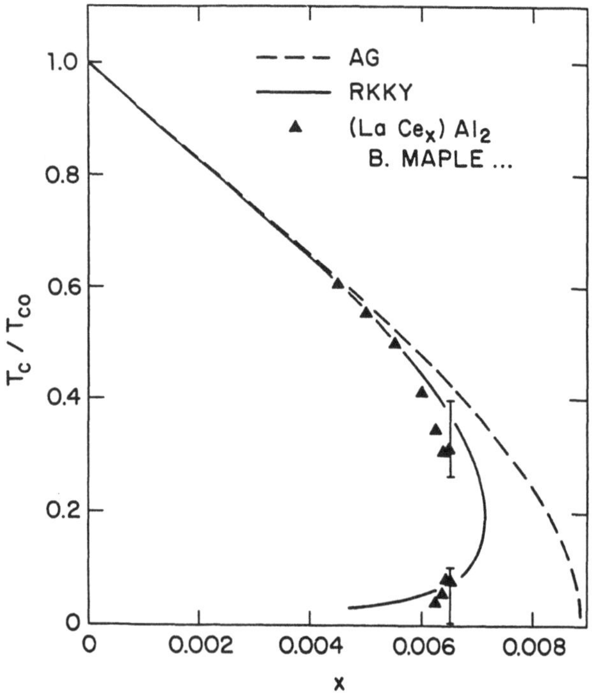

Figure 10. Superconducting transition temperature of $LaA\ell_2$ doped with Ce. The data of Maple (Ref.41) falls below the Abrikosov-Gorkov curve and shows two narrow transitions at a given concentration x = 0.006. This re-entrant behavior is reproduced by the RKKY theory with parameters $k_F = 0.6$ Å^{-1}, $\Delta = 3.5$ Å, $E_F = 2$ eV, N(0) = 0.5 states/eV-atom, and $|J| = 0.11$ eV.

V. CONCLUSIONS

We have examined the influence of coupled magnetic impurities on the transport properties of intermediate concentration alloys. Providing that the indirect exchange interaction between impurities is ferromagnetic in nature, a

1/T divergence in the resistivity is found, which combines with the phonon contribution to form a new type of resistivity minimum. At the lowest temperatures, this anomalous scattering suggests the need for further theoretical calculations, including for example, the higher order processes associated with broadening of the impurity state.

Although the 1/T term in the resistivity generally occurs in a concentration regime where the Kondo effect is quenched, it would nevertheless be interesting to examine experimental probes which sample only the RKKY-induced anomaly and not the traditional ℓn T divergence of the Kondo effect. One such case may be the nuclear spin relaxation rate which follows a smooth temperature variation in the Kondo case[4], but is likely to exhibit the low temperature contributions of the RKKY interaction. Furthermore, rare earth compounds which do not exhibit the Kondo effect (presumably because $J > 0$), may provide another distinguishing test at high concentrations since the RKKY anomaly is insensitive to the sign of J.

Similarly further calculations of the specific heat, magnetic susceptibility, and other properties of appropriate magnetic alloys may reveal new phenomena accessible to direct experimental tests.

As in the case of the Kondo effect, the formal description of the 1/T divergence in the electron scattering may have interesting analogies in other many body problems. The case of impurity ferromagnetics has already been widely considered, and the connection with the superconducting state has been mentioned in the present work. However, since the physical origin of the RKKY-induced divergence is related to the localized character of the magnetic impurity energy level, it may be worthwhile to examine analogous conditions in non-magnetic many body electronic systems.

This article is based on research performed with Dr. Fu-Sui Liu, and incorporates many of his valuable contributions. It is a pleasure to thank H. Keiter for a critical study of our analysis and his valuable comments. We have greatly benefited from stimulating discussions with various colleagues at the Institute for Theoretical Physics in Santa Barbara, and with R.V. Coleman, M. Fowler, S. Hillenius, R.C. Morris, and J. Poon. This research was supported by the Center for Advanced Studies at the University of Virginia, and by National Science Foundation Grants Nos. DMR 77-13167 and PHY 77-27084, and by the Department of Energy Grant No. DE-AS05-81ER10959.

REFERENCES

1. W. Meissner and G. Voight, Ann. Phys. 7, 761 (1930).

2. W.J. De Haas en J. De Boer, Physica 1, 1115 (1933).

3. J.O. Linde, Ann. Phys. 10, 52 (1931).

4. A.N. Gerritsen and J.O. Linde, Physica 17, 537 (1951).

5. J. Owen, M.E. Browne, W.D. Knight, and C. Kittel, Phys. Rev. 102, 1501 (1956).

6. G.K. White, Can. J. Phys. 33, 119 (1955).

7. G.J. Van Den Berg, in "Progress in Low Temperature Physics", (G.J. Gorter, ed.) Vol. 4, Ch. 4, North Holland, Amsterdam, 1964.

8. J. Korringa and A.N. Gerritsen, Physica 19, 457 (1953).

9. R.J. Elliott, Phys. Rev. 94, 564 (1954).

10. R.W. Schmitt, Phys. Rev. 103, 83 (1956).

11. K. Yosida, Phys. Rev. 107, 396 (1957).

12. A.D. Brailsford and A.W. Overhauser, J. Phys. Chem. Sol. 15, 140 (1960).

13. P.W. Anderson, Phys. Rev. 124, 1030 (1961).

14. A.J. Heeger, in Solid State Physics (ed. H. Ehrenreich, F. Seitz and D. Turnbull, Acad. Press London, 1969), Vol. 23, p.283 and references cited therein.

15. B.T. Matthias, M. Peter, H.J. Williams, A.M. Clogston, E. Corenzwit and R.C. Sherwood, Phys. Rev. Lett. 5, 542 (1960).

16. M.P. Sarachik, E. Corenzwit and L.D. Longinotti, Phys. Rev. 135, A1041 (1964).

17. J. Kondo, Prog. Theor. Phys. (Kyoto) 32, 37 (1964).

18. J. Kondo, in Solid State Physics (ed. H. Ehrenreich, F. Seitz and D. Turnbull, Acad. Press. London, 1969), Vol. 23, p. 183 and references cited therein.

19. H. Fröhlich and F.R.N. Nabarro, Proc. Roy. Soc. A175, 382 (1940).

20. C. Zener, Phys. Rev. 81, 440 (1951).

21. J.R. Schrieffer and P.A. Wolff, Phys. Rev. 149, 491 (1966).

22. P. Monod, Phys. Rev. Lett. 19, 113 (1967).

23. S.D. Silverstein, Phys. Rev. Lett. 16, 466 (1966); R.J. Harrison and M. Klein, Phys. Rev. 154, 540 (1967).

24. G. Grüner and A. Zawadowski, Reports on Prog. in Physics 37, 1497 (1974).

25. D.J. Scalapino, Phys. Rev. Lett. 16, 937 (1966).

26. P.W. Anderson, G. Yuval, and D.R. Hamann, Phys. Rev. B1, 4464 (1970).

27. M. Fowler and A. Zawadowski, Sol. St. Comm. 9, 471 (1971).

28. K.G. Wilson, Rev. Mod. Phys. 67, 773 (1975).

29. N. Andrei and J.H. Lowenstein, Phys. Rev. Lett. 46, 356 (1981); N. Andrei, Phys. Rev. Lett. 45, 379 (1980); P.B. Wiegmann, Phys. Lett. 80A, 163 (1980).

30. F.S. Liu and J. Ruvalds, Phys. Rev. Lett. (submitted for publication).

31. A.A. Abrikosov, Physics 2, 5 (1965); 2, 61 (1965).

32. R. Abe, Prog. of Theor. Phys. 36, 454 (1966).

33. A.A. Abrikosov and L.P. Gorkov, Zh. Eksp. Teor. Fiz. 39, 1781 (1960), [Sov. Phys. JETP 12, 1243 (1961)].

34. D.A. Whitney, R.M. Fleming and R.V. Coleman, Phys. Rev. B15, 3405 (1977).

35. S.J. Hillenius, R.V. Coleman, E.R. Domb and D.J. Sellmyer, Phys. Rev. B19, 4711 (1979).

36. P.G. De Gennes and J. Friedel, J. Phys. Chem. Sol., 4, 71 (1958).

37. J. Kondo, Progr. Theor. Phys. 34, 523 (1965).

38. K. Yosida and H. Miwa, Phys. Rev. 144, 375 (1966).

39. T. Sugawara, I. Yamase, and R. Soga, J. Phys. Soc. Japan 20, 618 (1965).

40. E. Müller-Hartman and J. Zittartz, Phys. Rev. Lett. 26, 428 (1971).

41. For an excellent review of re-entrant superconductivity, see, M. Brian Maple, in Magnetism, Vol. V, ed. G. Rado and H. Suhl (Acad. Press, New York, 1973), p.289.

42. J.J. Hauser, M. Robbins and F.J. Di Salvo, Phys. Rev. B8, 1038 (1973).

43. J. Ruvalds and Fu-sui Liu, Sol. State Comm. (1981); Fu-sui Liu and J. Ruvalds (to be published).

44. L.P. Gorkov and A.I. Rusinov, Sov. Phys. JETP 19, 922 (1964).

THEORY OF EXCHANGE-CORRELATION EFFECTS IN THE ELECTRONIC SINGLE- AND TWO-PARTICLE EXCITATIONS OF COVALENT CRYSTALS

W. Hanke, H.J. Mattausch and G. Strinati†

Max-Planck-Institut für Festkörperforschung
Heisenbergstr. 1, D-7000 Stuttgart 80
Federal Republic of Germany
†Istituto di Fisica "Guglielmo Marconi"
Università di Roma
00185 Roma, Italy

ABSTRACT

We summarize recent investigations on the importance of many-body effects for the single-particle excitation spectrum, the optical response and the impurity screening in covalent crystals which we have performed. Key tools in these investigations are the Green's function formalism and a local one-electron basis which facilitates an explicit solution of the coupled Dyson equations for the one- and two-particle Green's functions. We discuss a first-principle calculation of the single-particle excitations in diamond which rests on an energy-dependent non-local self-energy operator obtained by replacing the Coulomb potential in the exchange operator by a dynamically screened interaction. To be consistent with a variety of experiments on two-particle excitations, i.e. the optical absorption (studied in detail for diamond and silicon), the dielectric matrix of the medium was taken within the time-dependent screened Hartree-Fock approximation (TDSHF), thereby including both local-field and electron-hole (excitonic) effects. Previous calculations along similar lines have been restricted either to a RPA frequency-independent dielectric function or to a plasmon-pole approximation. For the first time the role of a realistic frequency and wave-vector dependent dielectric matrix was investigated and the relative importance of the electron-hole excitations and of the plasma resonance across the range of the valence and conduction bands was examined. Electron-hole mediated dynamical correlation

effects entirely determine the quasi-particle renormalization near the energy gap. On the other hand, the plasma-resonance does not contribute appreciably in the energy range about the band-gap while it contributes significantly to the valence band-width. Our values for the band-gap and the valence bandwidth are in good agreement with reflectivity and photoemission experiments (XPS). Implications for the local density and the energy-independent correlation approximations are discussed. In addition, our method, by utilizing an energy-dependent self-energy, has also enabled us to calculate quasi-particle damping times (specifically, intra-band Auger decay rates) that are consistent with photoemission spectra.
Finally, our detailed studies of both substitutional and interstitial impurity screening in diamond and silicon demonstrate again the necessity of including local-field and excitonic many-body effects. From these effects significant corrections to binding energies of impurities are to be expected.

I. INTRODUCTION

Simplifying and yet accurate descriptions of the many-body physics of the electronic system of crystals are still one of the main objects of contemporary solid state physics. For the treatment of ground- as well as excited-states of the system of interacting electrons a particle concept has proved to be a very successful working scheme [1,2]. The basic hypothesis consists in the conjecture to treat a complicated and highly correlated ground and excited state of an N electron system in an approximate way as a superposition of certain fundamental states which are assumed to have the character of only weakly interacting particles. Meaningful fundamental states can be single-particle (quasi-electron, hole) as well as two (or more)-particle (electron-hole pair, plasmon, etc.) excitations.

In this work we present a summary of recent investigations which we have performed on a quantitative description of the coupled single- and two-particle excitations in the electronic system of covalent crystals. In the many-body framework of the Feynman-Dyson perturbation formalism we have to solve the coupled set of integral equations for the one- and two-particle Green's function. From the one-particle Green's function we extract the spectrum of the complex quasi-particle (electron, hole) energies. From the two-particle Green's function, we derive the density-density correlation function from which we extract the optical response (long-wavelength limit) and the screening of impurities (static limit).

As a basis for our investigations serves a theory of elementary excitations in crystals with more or less localized electronic states (transition metals, semi-conductors, insulators), which has been developed in recent years [3,4,5]. In a variety of applications of this theory to different physical problems (optical properties, electron-phonon interaction, superconductivity) a direct interrelation between the importance of many-body effects and the degree of localization and thus inhomogeneity of the electronic charge density has emerged. This interplay also holds for the topics discussed in this article, i.e. the quasi-particle spectrum (one-particle Green's function) and the screening of impurities. One can make practical use of this interrelation and employ a local representation of the electronic wave functions. Very much as in our previous work on the two-particle Green's function [3,4,5] this leads to important simplifications for the treatment of the single-particle excitations.

The fundamental quantity for the determination of the electronic quasi-particle states is the selfenergy which includes exchange and correlation effects. Previous work on the interaction effects in the single-particle spectrum can be divided into two groups : on the one hand are the bandstructure calculations which are based on a practical ansatz for exchange and correlation, like the empirical (pseudo-) potential or the local density approaches (LDA) [7,8]. They have contributed enormously to our understanding of electronic and structural properties of solids and thus have given a kind of pragmatic answer as to how to include many-body effects into the single-particle excitations. On the other hand, we have the formal results of the many-particle theory [2,9-12]. In the case of metals, especially simple metals, which are near to the ideal case of a homogeneous electron gas, there has been a good deal of convergence between the pragmatic bandstructure and the many-body approaches. Here the results of the many-body description can be regarded as a justification and a formal basis of the pragmatic one-particle theories [9,10]. In the case of semiconductors and insulators, however, the situation is much less favourable. Because of the relatively strong localization of the electronic states and the resulting inhomogeneity of the electronic charge density it is rather more difficult to apply the formal equations of the many-body theory to a given special case. The few existing numerical evaluations of the quasi-particle properties could be carried out only through the use of rather crude screening models (polaron-[13], plasmon-model [14] and self-energy approximations [9] (COHSEX ≙ Coulomb hole plus screened exchange).

Table 1. Comparison of Hartree-Fock (HF) and local density (LDA) calculations with experiment.

crystal	direct band-gap HF	(eV) LDA	Exp.	valence band-width HF	(eV) LDA	Exp.
diamond	15 (15)	6.3 (16)	7.3 (17)	29 (15)	20.4 (16)	24.2 (18)
silicon	-	2.5 (19)	3.4 (20)	-	12.0 (19)	14.3 (21)

The LDA is, at least in principle, not applicable to a strongly inhomogeneous electronic system. Furthermore, it has been deviced for ground-state calculations only, although Sham and Kohn [10] have given an energy-dependent LDA version of the selfenergy potential. However, in the common LDA bandstructure determinations the ground-state and energy-independent functional for exchange and correlation is used. In a simplified physical picture this amounts to neglecting the "dynamic" aspects of the correlation hole surrounding an electron, i.e. that this hole depends on the energy or velocity of the particle under consideration. In view of this fact significant deviations of LDA bandstructures from experiment are not surprising as exemplified for the band-gap and valence-band width in Table 1. One essential question which we try to resolve for the prototype of a covalent crystal diamond is, whether it is the "locality" of the LDA or more the energy-independence of the exchange-correlation potential which is responsible for these deviations.

In the band-structure calculations with empirical (pseudo) potentials interband transitions are directly fitted to optical data [6] . Thereby RPA local-field and excitonical (electron-hole) many-body effects are included in an arbitrary an uncontrolled way into the single-particle excitations [5] .

In order to determine the quasi-particle Bloch-states we solve the equation of motion for the one-particle Green's function [9] (see chapter III). The crucial quantity in this equation is the self-energy or mass operator [10] . Since it is not possible to determine the self-energy exactly, one of the important problems is to find an adequate approximation for this quantity. The first non-trivial approximation is the Hartree-Fock (HF) approximation, where it is given by a product between an unscreened Coulomb potential and the one-particle Green's function. The HF approximation still is insufficient for the quasi-particle states, since correlation effects are neglected (see table 1). In chapter III we are going to investigate the time-dependent screened Hartree-Fock (TDSHF) approximation [9] which amounts to replacing the unscreened Coulomb potential by a dynamically screened potential. This is the lowest order process including dynamical correlation effects. The physical picture behind this approximation visualizes the quasi-particle as being surrounded by an energy- and wavevector-dependent cloud of electronic charge density (dynamical screening). Thus the dynamical reactions of the electronic system on a moving particle are included. From the considerations made above, it is clear that an accurate theoretical evaluation of the frequency and wave-vector-dependent two-particle Green's function or the related density response function of the crystal (chapter II) is of basic importance for the quality of the calculated quasi-particle states. It is precicely through the screening effects that the two-particle excitations couple to the single-particle excitations.

In a quantitative calculation of quasi-particle states of our prototype substance diamond we start from a Slater-Koster representation [22] of a self-consistent Hartree-Fock calculation [23] . The dynamical correlation corrections (real and imaginary part) are then determined for valence- and conduction-bands. In particular, we extract the influence of elementary excitations (electron-hole, plasmons) which are embedded in the dynamical screening over this energy range. Our findings reveal important differences for the microscopic dynamical correlations :

a) Near the band-gap correlation corrections are exclusively due to the excitation of electron-hole pairs and are practically not influenced by density fluctuations (plasmons). This has the important consequence that the corresponding correlation effect is predominantly of local nature and short-ranged (see chapter 3). This result still does not necessarily imply the validity of an energy-dependent local-density approximation for the self-energy which makes use of the homogeneous electron gas model [10]. But it unambiguously implies that the neglect of the energy-dependence (dynamical effect) in the commonly used LDA is a more serious approximation for single-particle like excitations near the band-gap than the local assumption.

b) Deeper into the valence and conduction bands collective excitations in the form of plasmons and therefore longer-ranged, non-local effects become increasingly important.

In summary, the calculations of the quasi-particle states of diamond lead to a quantitative agreement with experiment (reflectance [17], photoemission [18]) and allow detailed conclusions on the microscopic nature of the correlation effects.

Up to now we have restricted the discussion to the case of perfect crystals. Real crystals however are often far from being perfect and contain defects, dislocations etc. Of course, it is still possible to describe the electronic system of imperfect crystals within the framework of the quasi-particle and elementary excitation concept. The defects of the crystal show up in a modification of the spectrum of the quasi-particle states. New electronic quasi-particle states appear which are localized in the surroundings of the defects [24-26]. Since the defects are accompanied with a defect potential, the electronic system reacts also with a rearrangement of the electronic charge density leading to a screening of the defects. This process can be described within the framework of a perturbation treatment (linear response) through the creation of elementary excitations of the perfect crystal [2,5]. The exact determination of the screened potentials of defects (including many-body effects) is on one hand an interesting problem by itself and on the other hand a necessary prerequisite for the calculation of the localized electronic states and their binding properties [24-26]. We will investigate in this context the reaction of the electronic system to point defects on substitutional and interstitial sites in the lattices of diamond and silicon (chapter IV). Our special interest will focus on the influence of local-field (lattice effects) and excitonic (electron-hole interaction) effects. It turns out that both effects are of comparable magnitude and especially in the more localized (than silicon) electronic system of diamond are of the same magnitude as the zero-order RPA screening without local-field effects. Both many-body effects create an anisotropy in the induced potential and a dependence on the position (substitutional or interstitial) of the defect in the lattice.

II. MANY-BODY DESCRIPTION OF THE TWO-PARTICLE EXCITATIONS IN PERIODIC CRYSTALS : EXAMPLES DIAMOND AND SILICON

As already stated in the introduction, a detailed theoretical understanding of the frequency- and wavevector-dependent dielectric function or the density-density channel of the two-particle Green's function is a prerequisite for our investigations aiming at the single-particle states and the static screening of impurities. On the other hand, the knowledge of the quasi-particle states is also a prerequisite for the determination of the dielectric function [27] . This is formally expressed in the coupled set of integral equations of one- and two-particle Green's functions. In this chapter we will shortly review the method of Hanke and Sham [3-5] for the determination of the two-particle Green's function in periodic crystals. The method assumes that the quasi-particle states are known and includes local-field and electron-hole effects in the theoretical treatment. The scheme is then applied to an investigation of the frequency- and wave-vector-dependence of the dielectric function of diamond and silicon, extending earlier results on the optical response of covalent crystals [3,4] .

A. Many-body framework

For the applications we have in mind, it is sufficient to treat the longitudinal response function or the density-density correlation function [28] . In cubic crystals the dielectric function is a scalar [29] and in the long-wavelength limit the transverse dielectric function equals the longitudinal dielectric function [30] .

A1. Field-theoretical formulation

In the framework of the general many-body formulation of the dielectric response theory [31] we start from the interrelation

$$\varepsilon^{-1}(1,2) = \delta(1,2) + v(1,1')\ \chi(1',2) \tag{II.1}$$

between the inverse dielectric function ε^{-1} and the density-density correlation function χ. Each number in (II.1) represents a set of coordinates (space ($\vec{r}$), time (t), spin (δ)). Repeated numbers imply integration (or summation) over the corresponding sets of coordinates. $v(1,1')$ is the Coulomb interaction between unit charges. The density-density correlation function is connected via the relation [4]

$$\chi(1,2) = G(1,1;2,2) \tag{II.2}$$

to the two-particle Green's function G and is thus representing a specific channel. The equation of motion of the two-particle Green's function (for the diagrammatic representation : see fig.1)

$$G(1,1';2,2') = G^{\circ}(1,1';2,2') + G^{\circ}(1,1';3,3')\, I(3,3';4,4')\, G(4,4';2,2') \tag{II.3}$$

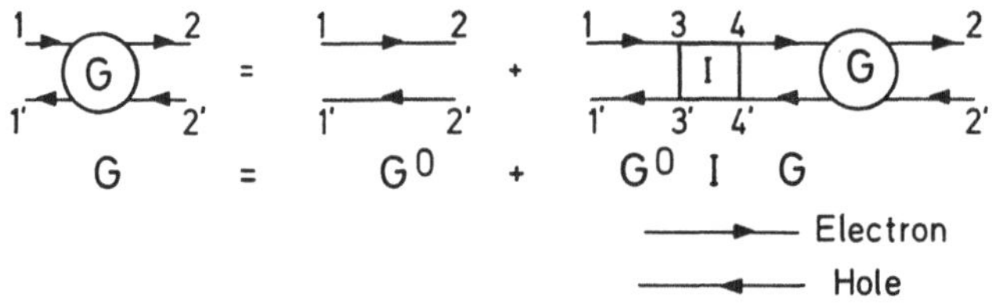

Figure 1. Graphical representation of the Bethe-Saltpeter equation for the two-particle Green's function.

is known as the Bethe-Salpeter equation [27,31] . In this equation

$$G^{\circ}(1,1';2,2') = g(1',2')\, g(2,1) \tag{II.4}$$

is the propagator of a non-interacting electron-hole pair. The one-particle Green's function which appears in (II.4) is in principle expected to contain the entire information about the quasi-particle states (bandstructure) of the crystal.

The term I in (II.3) represents the irreducible interaction between electrons and holes. In this article we will consider a screened electron-hole attraction Y and an unscreened exchange Z between electron-hole pairs (see fig.2).

$$I(3,3';4,4') = \delta(3,4)\delta(3',4')\,Y(3,3') + \delta(3,3')\,\delta(4,4')\,Z(3,4) \qquad (II.5)$$

It is quite clear that the exchange Z should not be screened, since it would not be irreducible otherwise. More complicated exchange terms, besides Z, are possible [4], but they are neglected here.

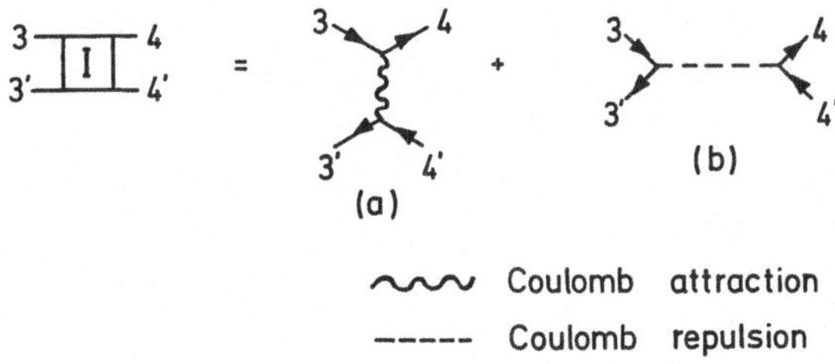

Figure 2. Irreducible interaction between electrons and holes
(a) electron-hole attraction (screened),
(b) electron-hole pair exchange (not screened).

In order to get a time-independent interaction term I, we will use a static screening for the electron-hole attraction Y. Then we have $t_1 = t_{1'}$, $t_2 = t_{2'}$ and the two-particle Green's function depends only on the time difference $t_1 - t_2$. In the following we will discuss the Fourier-transform of the above equations; thus, frequency-dependence (ω) instead of time-dependence appears.

A physical picture of the contents of the Bethe-Salpeter equation can be obtained through an expansion into a Born's series

$$G = G^\circ + G^\circ\, I\, G^\circ + G^\circ\, I\, G^\circ\, I\, G^\circ + \ldots\ldots$$
$$= G^\circ + G^\circ\, Y\, G^\circ + G^\circ\, Z\, G^\circ + \ldots\ldots \qquad \text{(II.6)}$$

For simplicity of notation arguments have been omitted in (II.6). Obviously, the interaction term I creates all couplings (attraction, exchange) between electrons and holes. In figure 3 the first two terms of the Born's series are represented graphically. The first term G° in this series represents the non-interacting electron-hole propagator. The second term $G^\circ\ I\ G^\circ$ shows in lowest order how the interaction term I introduces the screened attraction between electrons and holes and the unscreened electron-hole pair exchange (RPA Coulomb repulsion).

In this and in the following chapters we will use three different approximations for the two-particle Green's function or the dielectric function. To introduce these approximations, it is convenient to go to the wavevector representation $\varepsilon(\vec{q}+\vec{Q},\vec{q}+\vec{Q}';\omega)$ of the dielectric function.

a) The time-dependent Hartree approximation without local-field effects ($\overline{\text{RPA}}$) amounts to setting I = Z and taking the dielectric function diagonal in the reciprocal lattice-vectors ($\vec{Q} = \vec{Q}'$).

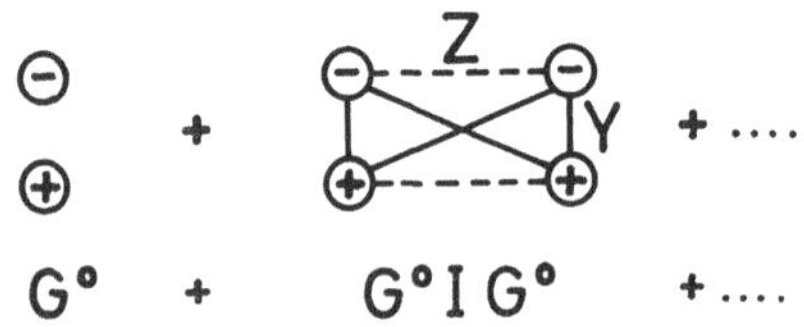

Figure 3. Graphical representation of the interaction between the electron-hole pairs for the first two terms of the Born's series.

b) In the time-dependent Hartree approximation with local-field effects (RPA) we put I = Z and in addition consider the non-diagonal elements ($\vec{Q} \neq \vec{Q}'$) in the wavevector representation of the dielectric function.
c) The time-dependent screened Hartree-Fock approximation with local-field effects (TDSHF) consists of putting I = Y + Z plus a non-diagonal wavevector representation of the dielectric function ($\vec{Q} \neq \vec{Q}'$).

A2. Local representation

Input quantities for the Bethe-Salpeter equation (II.3) are the irreducible interaction I and the electron-hole propagator G°. We will now shortly discuss a local representation of I and G°, which allows for a relatively simple matrix solution of the Bethe-Salpeter equation [3,4].

The electron-hole propagator G° is defined according to eq.(II.4) as a product of two one-particle Green's functions. The one-particle Green's function (energy representation)

$$g(1,2;\omega) = \sum_s \frac{\psi_s(1)\ \psi_s(2)}{\omega - E_s} \tag{II.7}$$

is determined through the spectrum of the quasi-particle states (ψ_s, E_s) of the crystal [2]. Here the index s ≙ ($\vec{k}$,n,σ) stands for the wavevector $\vec{k}$, the band-index n and the spin σ. The quasi-particle energies E_s are in general complex, with the imaginary part representing inverse life-times.

Since the wave-functions ψ_s are relatively well localized in covalent crystals (silicon, diamond), it is appropriate to utilize an expansion into suitable local orbitals ϕ (muff-tin or atomic orbitals, Wannier functions).
Thus, we take :

$$\psi_s(1) = N^{-1/2} \sum_d T_{ds}\ \phi_d(1)$$

with

$$\phi_d(1) = \phi_\nu(\vec{r}_1 - \vec{R}_t) \ ; \qquad T_{ds} = c_{n\nu}(\vec{k})\ e^{i\vec{k}\vec{R}_t} \tag{II.8}$$

The index d = (ν,t) identifies the different local orbitals ν and their positions $\vec{R}_t$ in the lattice. N is the number of lattice vectors (Wigner Seitz cells).

The local representation of the wave-functions allows for a representation of the electron-hole propagator G° and the irreducible interaction $I = Y + Z$ in terms of relatively small matrices, and thus for a matrix solution of the Bethe-Salpeter equation in the form [3,4]

$$G = G^\circ[1 - (Y+Z)G^\circ]^{-1} = [G^{\circ -1} - (Y+Z)]^{-1} \tag{II.9}$$

In the case of the density-density correlation function or the inverse dielectric function one finds the wave-vector representation [3,4]

$$\varepsilon^{-1}(\vec{q}+\vec{Q},\vec{q}+\vec{Q}';\omega) =$$
$$\delta_{\vec{Q},\vec{Q}'} + v(\vec{q}+\vec{Q}) \sum_{\delta\delta'} A_\delta(\vec{q}+\vec{Q})\, G_{\delta\delta'}(\vec{q};\omega)\, A^\star_{\delta'}(\vec{q}+\vec{Q}') \tag{II.10}$$

where $v(\vec{q}+\vec{Q})$ is the Coulomb potential in wave-vector representation and the

$$A_\delta(\vec{q}+\vec{Q}) = \int d^3r\, \phi^*_\nu(\vec{r})\, e^{-i(\vec{q}+\vec{Q})r}\, \phi_\mu(\vec{r}-\vec{R}_t) \tag{II.11}$$

are density form-factors. $\delta \mathrel{\hat{=}} (\nu,\mu,\vec{R}_t)$ stands for a set of indices with ν , μ representing local orbitals and $\vec{R}_t$ giving the distance between these orbitals.

We are now in the position to give a compact notation for the elements of the matrices which appear in (II.9). In the case of the electron-hole propagator we have

$$G^\circ_{\delta\delta'}(\vec{q},\omega) = N^{-1} \sum_{n,n',k} c^*_{n\nu}(\vec{k})\, c_{n'\mu}(\vec{k}+\vec{q})\, e^{i(\vec{k}+\vec{q})\vec{R}_t}$$
$$\frac{f_{n'}(\vec{k}+\vec{q}) - f_n(\vec{k})}{E_{n'}(\vec{k}+\vec{q}) - E_n(k) - \omega - i\eta} *$$
$$* \; e^{-i(\vec{k}+\vec{q})\vec{R}_t}\, c^*_{n'\mu'}(\vec{k}+\vec{q})\, c_{n\nu'}(\vec{k}) \tag{II.12}$$

Here $f_n(\vec{k})$ is the occupation number of the quasi-particle state $(n,\vec{k})$ and η is a positive convergence factor ($\eta \to 0^+$). For the electron-hole exchange one has

$$Z_{\delta\delta'}(\vec{q}) =$$

$$= \sum_m e^{-i\vec{q}\vec{R}_m} \int d^3r \int d^3r' \; \phi^*_\mu(\vec{r}-\vec{R}_t-\vec{R}_m) \; \phi_\nu(\vec{r}-\vec{R}_m) \; v(\vec{r}-\vec{r}') \; \phi^*_{\nu'}(\vec{r}') \; \phi_{\mu'}(\vec{r}'-\vec{R}_{t'})$$

$$= \sum_{\vec{Q}} A^*_\delta(\vec{q}+\vec{Q}) \; v(\vec{q}+\vec{Q}) \; A_{\delta'}(\vec{q}+\vec{Q}) \qquad \text{(II.13)}$$

The matrix elements of the electron-hole attraction are given by

$$Y_{\delta\delta'}(\vec{q}) =$$

$$= -\frac{1}{2}\sum_m e^{-i\vec{q}\vec{R}_m} \int d^3r \int d^3r' \phi^*_\mu(\vec{r}-\vec{R}_t-\vec{R}_m)\phi_\nu(\vec{r}'-\vec{R}_m)v_s(\vec{r}-\vec{r}')\phi^*_{\nu'}(\vec{r}')\phi_{\mu'}(\vec{r}-\vec{R}_{t'}) \qquad \text{(II.14)}$$

where $v_s(\vec{r}-\vec{r}')$ is the Coulomb potential with a static screening [4]. It should be noted that, in principle, the dielectric function, which we want to determine, enters also the screening in Y [31]. Static screening is necessary to make the procedure numerically feasible, and has been proved to be a good approximation for the description of the many-body effects in the optical properties of covalent crystals [3,4].

For the construction of the local (bonding- and anti-bonding) orbitals in our examples diamond and silicon (diamond structure see fig.4), we exploit the method of Hall [32]. In a first step sp^3-hybrids are formed from s- ($R_s(|\vec{T}|)$) and p-like ($R_p(|\vec{r}|)$) orbitals according to the prescription [33].

$$\zeta_{\vec{\nu}}(\vec{r}) = (4\sqrt{\pi})^{-1} \{R_s(|\vec{r}|) + \sqrt{3}(\vec{\nu}\vec{r}/|\vec{r}|) \, R_p(|\vec{r}|)\} \qquad \text{(II.15)}$$

Here $\vec{\nu} \mathrel{\hat{=}} (1,1,1); (1,-1,-1); (-1,1,-1); (-1,-1,1)$ is one of the four tetrahedral directions which in the following is referred to by the numbers $\nu = 1,2,3,4$. Pairs of hybrids from neighbouring atoms along the same tetrahedral direction are added or subtracted to form bonding (+) or antibonding (-) orbitals.

$$\phi_{\nu\pm}(\vec{r}) = K^{-1}_\pm \{\zeta_{\vec{\nu}}(\vec{r}) \pm \zeta_{-\vec{\nu}}(\vec{r} - b\vec{\nu})\} \qquad \text{(II.16)}$$

In (II.16) $K_\pm$ is a normalization constant and $b = \frac{a_0}{4}$ with the

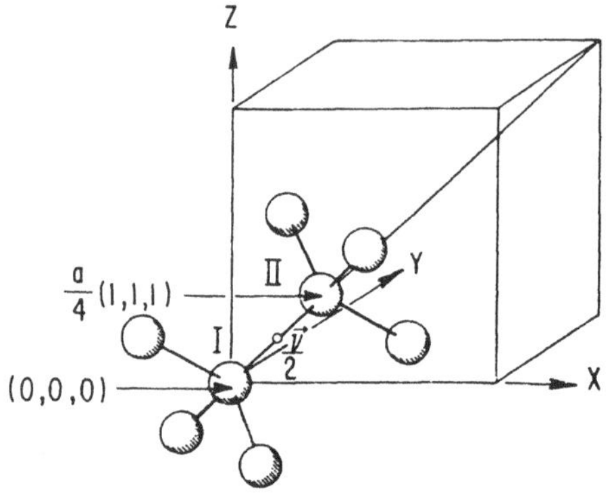

Figure 4. Lattice structure of diamond.

lattice constant a_o. The s- and p-like orbitals will be approximated by sums of Gaussian functions

$$R_s(r) = \sum_i a_i \, e^{-\alpha_i r^2} \; ; \quad R_p(r) = \sum_i b_i \, e^{-\beta_i r^2} \tag{II.17}$$

For the calculation of the electron-hole attraction we will make a further simplification and use a so-called "lobe"-representation of the p-like orbitals [34].

$$xR_p(r) = \sum_i \tilde{b}_i \, \{e^{-\beta_i(\vec{r}-\vec{d}_{ix})^2} - e^{-\beta_i(\vec{r}+\vec{d}_{ix})^2}\} \tag{II.18a}$$

with

$$\vec{d}_{ix} = \frac{x}{\sqrt{\beta_i}}(1,0,0) \; , \quad \tilde{b}_i = \frac{b_i}{4x\sqrt{\beta_i}} \tag{II.18b}$$

In diamond and silicon κ is taken as $\kappa = 0.09$, which reproduces well the angular dependence of the p-like orbitals [35]. We have now a complete representation of the local orbitals with the aid of Gaussian functions. This has the well-known advantage that the integration of all multi-center integrals A, Y and Z in eqs.(II.11) and (II.14) can be done analytically [4,36].

A3. Long-wavelength or optical limit

An important limit for the density-density correlation function or the dielectric function is the long-wavelength limit, since it can be directly compared with optical experiments. We will use this limit for two purposes. One is to investigate the importance of many-body effects in the optical properties of covalent crystals, and the other is to have a criterion for the quality of our theoretical description of the density response function.

In cubic crystals one can extract the macroscopic long-wavelength dielectric function including local-field effects from the prescription [37-39]

$$\varepsilon(\omega) = \lim_{\vec{q}\to 0} \frac{1}{\{\varepsilon^{-1}(\vec{q}+\vec{Q},\vec{q}+\vec{Q}')\}_{\vec{Q}=\vec{Q}'=0}} \qquad \text{(II.19)}$$

Here use is made of the previously defined longitudinal dielectric function which in cubic crystals and for $\vec{q} \to 0$ equals the actually needed transverse dielectric function [30].

With the help of (II.19) one derives the formula [3]

$$\varepsilon(\omega) = \varepsilon_1(\omega) + i\varepsilon_2(\omega) = 1 - \frac{8\pi}{\Omega_o} \sum_{\delta\delta'} f_\delta^\infty \, \overline{G}_{\delta\delta'}(\omega) \, f_{\delta'}^{\alpha *} \qquad \text{(II.20)}$$

for the macroscopic dielectric function. Here Ω_o is the volume of the Wigner-Seitz cell. $\overline{G}$ we get from (II.9) by setting $\overline{G^\circ}(\omega) = G^\circ(\vec{q}=0;\omega)$; $\overline{Y} = \overline{Y}(\vec{q}=0)$ and

$$\overline{Z}_{\delta\delta'} = \sum_{\vec{Q}\neq 0} A_\delta^*(\vec{Q}) \, v(\vec{Q}) \, A_{\delta'}(\vec{Q}) \qquad \text{(II.21)}$$

In the density form-factors

$$f_\delta^\alpha = \int d^3r \, \phi_\nu^*(\vec{r}) \, r_\alpha \, \phi_\mu(\vec{r}-\vec{R}_t) \qquad \text{(II.22)}$$

α denotes a cartesian coordinate.

The different approximations for the many-body effects which were introduced in paragraph A1, have the following meaning in eq.(II.20). Putting $\overline{G} = [\overline{G}^{\circ -1} - (\overline{Y} + \overline{Z})]^{-1}$, leads to the TDSHF, with $\overline{G} = (\overline{G}^{\circ -1} - \overline{Z})^{-1}$ or $\overline{G} = \overline{G}^\circ$ one gets the RPA or $\overline{\text{RPA}}$, respectively.

In the framework of a microscopic version of Maxwell's theory of electrondynamics the equation of continuity or the current conservation criterion leads to an important connection [40] between the density operator $\rho(\vec{q}) = e^{-i\vec{q}\vec{r}}$ and the current operator $\vec{j}(\vec{q}) = (\vec{p}\, e^{-i\vec{q}\vec{r}} + e^{-i\vec{q}\vec{r}}\, \vec{p})$

$$[\rho(\vec{q}),H] = \vec{q}\,\vec{j}\,(\vec{q}) \tag{II.23}$$

Here [,] denotes the commutator symbol, H is the Hamiltonian of the system and $\vec{p}$ is the electron momentum. Thus, one can transform the density-density correlation function into the equivalent current-current correlation function. All the previous formulas retain their structure, if we replace the density form-factors (II.11) and (II.22) by

$$A_{\delta}(\vec{q}+\vec{Q}) = (\vec{q}+\vec{Q}) \int d^3r\ \phi_{\nu}^{*}(\vec{r})\ \vec{j}(\vec{q}+\vec{Q})\ \phi_{\mu}(\vec{r}-\vec{R}_t) \tag{II.24}$$

and

$$f_{\delta}^{\alpha} = 2 \int d^3r\ \phi_{\nu}^{*}(\vec{r})\ \nabla_{\alpha}\ \phi_{\mu}(\vec{r}-\vec{R}_t)\ , \tag{II.25}$$

and in addition the electron-hole propagator (II.12) by

$$G^{\circ}_{\delta\delta'}(q,\omega) = N^{-1} \sum_{n,n'k} c^{*}_{n\nu}(\vec{k})c_{n'\mu}(\vec{k}+\vec{q})e^{i(\vec{k}+\vec{q})\vec{R}_t} \frac{f_{n'}(\vec{k}+\vec{q}) - f_n(\vec{k})}{E_{n'}(\vec{k}+\vec{q}) - E_n(\vec{k}) - \omega - i\eta} *$$

$$* \frac{1}{\{E_{n'}(\vec{k}+\vec{q}) - E_n(\vec{k})\}^2}\ e^{-i(\vec{k}+\vec{q})\vec{R}_{t'}}\ c^{*}_{n'\mu'}(\vec{k}+\vec{q})c_{n\nu'}(\vec{k}) \tag{II.26}$$

In the following, the current conservation criterion serves as a guide for construction and consistency of the choice of local orbitals ϕ.

B. Dielectric response of diamond

In this paragraph we closely follow the earlier investigations of Hanke and Sham [3] . Their work, mainly concerned with the long-wavelength limit, will be extended to include the wave-vector dependence of the dielectric function.

The X_{α} bandstructure of Painter et al [41] serves as a zeroth-order approximation to the actual quasi-particle states of diamond, in the numerical determination of the electron-hole propagator (see (II.12)). The band interpolation-scheme of Slater and Koster [42] is used to represent this bandstructure in the local basis of bonding and antibonding tetrahedral orbitals. Valence and conduction bands are treated independently [3,43]. Linear combinations of the four bonding orbitals (see (II.16)) are used to construct the wave-function of the four valence bands and linear combinations of the four antibonding orbitals are employed to construct the four conduction bands. The matrix-elements of the Hamiltonian between bonding (and correspondingly antibonding) orbitals are the parameters [42] for the interpolation of the bandstructure. One arrives at two eigenvalue equations with (4×4) matrices, one for the conduction bands and the other one for the valence bands. In total 10 independent hopping matrix-elements including interactions up to the third-nearest neighbouring bonds are considered. However, this may still not be enough to obtain full convergence of the wave-functions in the framework of the Slater-Koster interpolation scheme [44]. Therefore, a certain freedom in the construction of the local orbitals [45] is used to obtain a maximum agreement between the density and current representations of the dielectric function (in the $\overline{\text{RPA}}$) and therefore an internal consistency of our approximations.

To minimize the numerical effort, only nearest-neighbour overlaps (in bonding and antibonding orbitals) are used in (II.20) having found higher-order overlaps to introduce 5-10 % corrections [3].
The basic reason for this reduced overlap in the density response, when compared with the third-nearest neighbour Slater-Koster scheme, is that there the kinetic energy terms in the one-particle Hamiltonian are responsible for longer-range overlap effects.

The meaning of the indices ν, μ is restricted, so that ν refers to bonding and μ to antibonding orbitals. This limitation in the meaning of ν and μ can be strictly verified for the RPA and is an approximation for the TDSHF, since the symmetry properties under the interchange of ν and μ are different for the electron-hole attraction $\overline{Y}$ and the electron-hole exchange $\overline{Z}$. With nearest-neighbour overlap $\overline{G}^\circ$, $\overline{Y}$ and $\overline{Z}$ are matrices of dimension 28×28. $\overline{G}^\circ$ has 15, $\overline{Y}$ and $\overline{Z}$ have 17 independent elements.

The parameters (matrix-elements $\langle\phi_\nu(\vec{r})|H|\phi_\mu(\vec{r}-\vec{R}_t)\rangle$) of the Slater-Koster representation of the X_α bandstructure are given in Table 2. ζ_o to ζ_6 are determined in such a way that the X_α bandstructure is exactly reproduced at the symmetry points (Γ, X and L). ζ_7 tot ζ_9 serve for a fit in the Δ, Λ and Σ dimensions.

Table 2. Parameters of the Slater-Koster representation of the X_α band-structure of Painter et al [41] for diamond.

Parameter	ν	ν'	$R_t(\frac{a_o}{2})$	Valence bands (eV)	Conduction bands (eV)
ξ_o	1	1	(0,0,0)	-0.8844	9.1719
ξ_1	1	2	(0,0,0)	-1.4844	-0.0490
ξ_2	1	1	(1,1,0)	0.8203	-0.6724
ξ_3	1	2	(1,1,0)	-0.5844	0.5479
ξ_4	1	1	(1,-1,0)	-0.3172	0.2318
ξ_5	1	2	(0,1,-1)	-0.0622	-0.1172
ξ_6	1	1	(2,0,0)	0.0318	-0.0339
ξ_7	1	2	(1,-1,0)	0.1750	0.1495
ξ_8	1	2	(0,-1,-1)	-0.4000	-0.3000
ξ_9	1	2	(2,0,0)	-0.0550	-0.9890

Table 3. Parameters of the Gaussian representation of the local orbitals in diamond.

Parameter		Value (atomic units)
R_s	a_1	1.3570
	a_2	0.0887
	α_1	0.5020
	α_2	0.2130
R_p	b_1	0.3428
	b_2	0.3805
	β_1	0.2500
	β_2	0.4450

The parameters of the Gaussian representation of the local orbitals, obtained with the aid of the current-conservation criterion, are given in table 3. Two Gaussians have been used for the representation of R_s and R_p, respectively.

The density form-factors A_δ and f_δ^α can be calculated analytically. The independent matrix-elements [17] of the electron-hole exchange $\overline{Z}$ are evaluated fairly easy in wave-vector space.

More problematic is the calculation of the independent matrix-elements [17] of the electron-hole attraction $\overline{Y}$, since it has to be done in real space [3]. Because of the strongly localized orbitals it proves to be sufficient to consider only the $\vec{R}_m = 0$ term in the sum over lattice vectors. The utilization of a static screening

$$v_s(\vec{r}-\vec{r}\,') = \int d^3r'' |\vec{r}-\vec{r}\,''|^{-1} \varepsilon^{-1}(r'',r';\omega=0)$$
$$\simeq |r-r'|^{-1} \varepsilon^{-1}(\vec{r}-\vec{r}\,') \qquad \text{(II.27)}$$

is necessary. For this static and local screening we employ, similar to the local orbitals, an expansion into Gaussians [4].

$$\tilde{\varepsilon}^{-1}(\vec{r}) = \varepsilon_o^{-1} + (1-\varepsilon_o^{-1}) \sum_i T_i \, e^{-\tau_i r^2} \qquad \text{(II.28a)}$$

with

$$\sum_i T_i = 1 \qquad \text{(II.28b)}$$

ε_o is the static dielectric constant. For diamond a one Gaussian fit to numerical results [46], obtained with the simple isotropic Penn model [47], is used. Parameters are given in table 4.

The independent matrix-elements (15) of the electron-hole propagator $\overline{G}°$ are determined with the linear tetrahedron-method [49,50].

Table 4. Gaussian fit to Penn model results [40] for the spacial dispersion of the static dielectric function in diamond.

Parameter	value (atomic units)
ε_o	$5.85^{(48)}$
τ_1	1.2

Bl. Many-body effects in the optical properties

We shall now discuss the frequency dependence of the dielectric function of diamond in the $\overline{RPA}$, RPA and TDSHF approximations and compare the results to experimental data. Figure 5 shows the imaginary part.

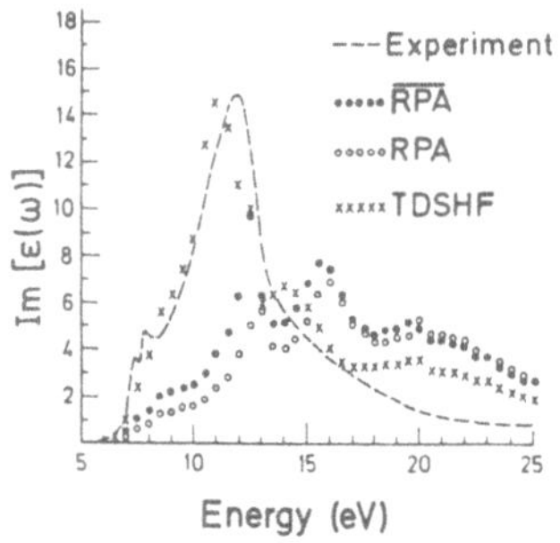

Figure 5. Imaginary part of the dielectric function of diamond. Comparison between theory and experiment.

Clearly, the agreement between theory and experiment is not good in the $\overline{RPA}$ calculation. For low energies the calculated values are too small and for high energies they are too big (turning point ∿14 eV). This is a general result of all numerical calculations of optical absorption spectra in semiconductors and insulators in the $\overline{RPA}$ [5]. Inclusion of local-field effects (RPA) furthers the discrepancy in comparison with experiment. This is in agreement with pseudo-potential calculation of Van Vechten and Martin [51]. Only the additional inclusion of the electron-hole attraction (excitonic effects) in the TDSHF leads to a now rather dramatic improvement. Oscillator strength is shifted to lower energies so that the local-field effects are overcompensated. Agreement with experiment is approximately within the numerical error of 10 %.

The dielectric constant is found to be 4.82 in the $\overline{\text{RPA}}$, 4.25 in the RPA and 6.10 in the TDSHF. The experimental value [48] is 5.85.

The sum-rules [1] for the imaginary part of the dielectric function

$$\int_0^\infty \mathrm{Im}[\varepsilon(\vec{q},\omega)]\,\omega\,d\omega = \frac{\pi}{2}\,\omega_p^2 \tag{II.29a}$$

and the imaginary part of the inverse dielectric function

$$\int_0^\infty \mathrm{Im}[\varepsilon^{-1}(\vec{q},\omega)]\,\omega\,d\omega = -\frac{\pi}{2}\,\omega_p^2 \tag{II.29b}$$

are fulfilled rather accurately. For the integrals (II.29a) and (II.29b) one gets (in eV^2) 1661 and -1652 respectively in the TDSHF. The experimental value [52] for the plasma resonance in diamond ω_p = 33.3 eV gives $\frac{\pi}{2}\,\omega_p^2$ = 1742 in comparison.

The theoretical value for the plasma resonance is given by the last zero of $\varepsilon_1(\omega)$, which is at 38 eV in the TDSHF. This discrepancy can be attributed to the X_α bandstructure [41] which serves as an approximation to the quasi-particle states in diamond. Here the width of the valence and conduction bands as well as the band-gap are too small in comparison with experiment, as will be discussed in detail in chapter III. Therefore, the combined density of states vanishes already at 34 eV and is too big over a broad energy range (18-30 eV). Thus the plasma resonance is shifted to higher energies.

The clear separation between the continuum of electron-hole excitations and the plasma resonance allows for an investigation of their relative contribution to the sum-rule for the imaginary part of the inverse dielectric function (II.29b). We find that the electron-hole continuum accounts for 10 % and the plasma resonance (delta-function character since ε_2 = 0 there) for 90 % of the sum-rule. This is different for metals, where the plasma resonance exhausts (at least for small $\vec{q}$) the whole sum-rule [1]. These 10 %, when weighted with the one-particle Green's function g play a dominant role in the selfenergy Σ, as discussed later.

B2. Wave-vector dependence

Since an accurate description of the wave-vector dependence of the density response function is necessary for the investigations in chapter III and IV, we now discuss this problem.

The wave-vector dependence of the dielectric function (see (II.10)) is mainly determined by the density form-factors $A_\delta(\vec{q}+\vec{Q})$ and the two-particle Green's function $G_{\delta\delta'}(\vec{q},\omega)$. The $A_\delta(\vec{q}+\vec{Q})$ can be easily calculated as a function of $\vec{q}$ and $\vec{Q}$, but the evaluation of the wave-vector dependence of $G_{\delta\delta'}(\vec{q},\omega)$ is a difficult numerical problem.

The two-particle Green's function has three wave-vector dependent ingredients, the electron-hole propagator $G°(\vec{q},\omega)$, the electron-hole exchange $Z(\vec{q})$ and the electron-hole attraction $Y(\vec{q})$. The electron-hole exchange can easily be obtained from the $A_\delta(\vec{q}+\vec{Q})$ and thus will be included exactly. For the electron-hole attraction (see (II.14)) terms with $R_m \neq 0$ can be neglected in the $\vec{q} = 0$ limit. Therefore the approximation $Y(\vec{q}) = Y(\vec{q}=0)$ is justified.

The crucial approximation, which we introduce here and then have to justify, is to assume that the electron-hole propagator is only weakly wave-vector dependent so that $G°(\vec{q},\omega) \cong G°(q=0,\omega)$. To elucidate and justify this approximation, we calculate the static dielectric function in the $\overline{\text{RPA}}$ for a special direction [53] in the irreducible Brillouin zone (see (III.22), later used for k-summations according to the relation

$$\overline{\varepsilon}(\vec{q}) = 1 - \frac{v(\vec{q})}{\Omega_o} \sum_{\delta\delta'} A_\delta(\vec{q})\ G^\circ_{\delta\delta'}(\vec{q}=0,\omega=0)\ A^*_{\delta'}(\vec{q}) \qquad \text{(II.30)}$$

The only carriers of the wave-vector dependence (besides $v(\vec{q})$) are the density form-factors $A_\delta(\vec{q})$. To reduce the influence of anisotropy we average (II.30) over the corresponding star of the wave-vector $\vec{q}$. Figure 6 displays the result in comparison with the interpolation formula of Penn [47], normalized to the value 4.82 of (II.30) for $\vec{q} = 0$. The good agreement confirms the assumption, i.e. that the wave-vector dependence of the electron-hole propagator stems entirely from its matrix-elements. On the other hand, one expects the approximation to become slightly worse for increasing $\omega \neq 0$, especially, if one wants to study details of the wave-vector and frequency-dependent dynamical structure factor. Nevertheless, the approximation should be sufficient for the investigations in chapter III (IV is restricted to $\omega = 0$), since in addition we have there to average over the Brillouin zone.

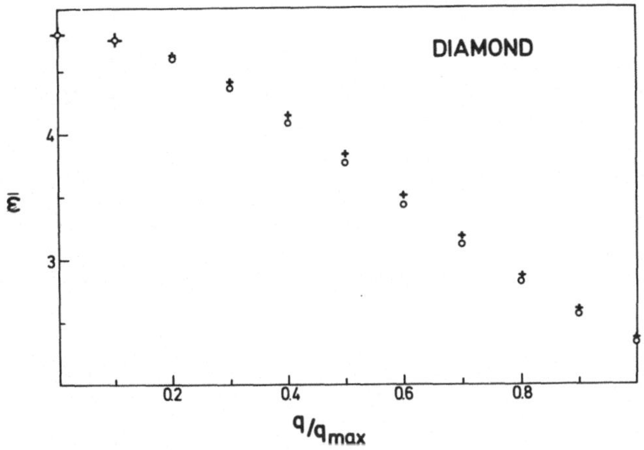

Figure 6. Wave-vector dependence of the dielectric function of diamond for $\omega = 0$. q_{max} denotes the edge of the Brillouin zone in the special direction. Displayed are the Penn model results (o) and the symmetrized RPA (+).

C. Dielectric response of silicon

Here we give an improvement investigation of the many-body effects in the dielectric function of silicon, which starts from the previous results of Hanke and Sham [41], but includes higher overlap and orthogonalization corrections.

As a basis for the determination of the electron-hole propagator serves again a Slater-Koster representation (see fig.7) of the OPW bandstructure of Ortenburger and Rudge [54]. This bandstructure is in quite good agreement with the pseudo-potential results of Chelikowsky and Cohen [55] for the upper part of the valence bands and the lower part of the conduction bands. For instance, we have for the energy-differences L_1-L_3 and X_1-X_4 in the Slater-Koster bandstructure 3.44 eV and 4.36 eV as compared to 3.37 eV and 4.21 eV in the pseudo-potential bandstructure.

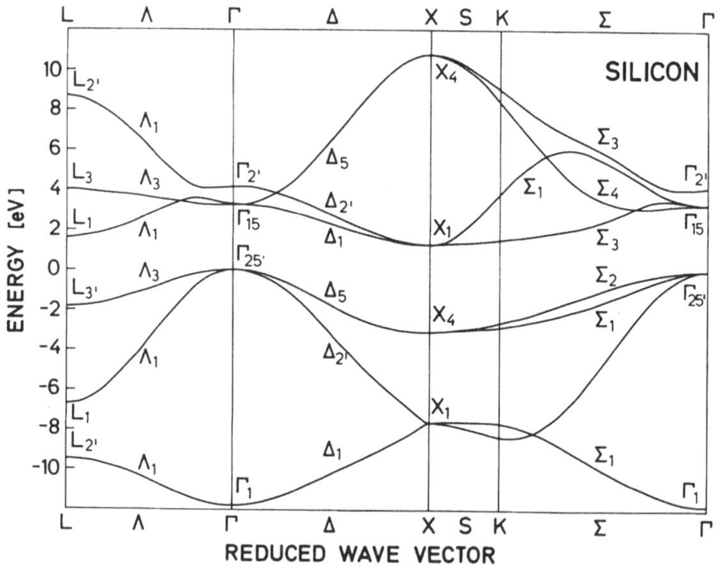

Figure 7. Slater-Koster representation of the OPW bandstructure of Ortenburger and Rudge [54] for silicon.

Table 5 shows the 10 independent matrix-elements of the Slater-Koster fit for the valence and conduction bands.
The parameters of the Gaussian representation of local orbitals in Si are optimized such that, on one hand, the valence-electron density [19,55] is fairly well reproduced (especially in the bond-region) and that, on the other hand, the current conservation criterion is fulfilled. Again, two Gaussians are used to represent R_s and R_p. The d-part of the local orbitals will not be included in our following discourse explicitly, but will be simulated by the overlap of the p-orbitals of neighbouring atoms [56]. However, we have also checked a representation of the local orbitals [57] which explicitly includes the d-part. This leads to an enormous increase of the numerical expense, but does not or only very slightly improve the optical response in comparison to experiment. The parameters of the Gaussian representation of the local orbitals are shown in table 6.

Since the band-gap in silicon is significantly smaller than in diamond, the orthogonalization of the conduction band to the valence band wave-functions gains importance. We include the orthogonalization between conduction and valence band functions

Table 5. Parameters of the Slater-Koster representation of the OPW bandstructure of Ortenburger and Rudge [54] for silicon.

Parameter	ν	ν'	$\vec{R}_t\ (\frac{a_o}{2})$	Valence bands (eV)	Conduction bands (eV)
ξ_0	1	1	(0,0,0)	-4.8600	4.9837
ξ_1	1	2	(0,0,0)	-1.1471	-0.6433
ξ_2	1	1	(1,1,0)	0.5279	-0.5015
ξ_3	1	2	(1,1,0)	0.0175	0.3400
ξ_4	1	1	(1,-1,0)	-0.2254	0.1877
ξ_5	1	2	(0,1,-1)	-0.1052	-0.1408
ξ_6	1	1	(2,0,0)	0.0167	0.0665
ξ_7	1	2	(1,-1,0)	-0.0500	0.1400
ξ_8	1	2	(0,-1,-1)	0.0900	-0.1200
ξ_9	1	2	(2,0,0)	-0.0200	-0.0400

Table 6. Parameters of the Gaussian representation of the local orbitals in silicon.

Parameter		Value (atomic units)
R_s	a_1	2.72710
	a_2	-2.43370
	α_1	0.24000
	α_2	0.32000
R_p	b_1	0.13438
	b_2	0.02868
	β_1	0.09000
	β_2	0.19150

or antibonding and bonding orbitals with the aid of Schmid's orthogonalization procedure [58]. For orthogonalization all the neighbouring bonding orbitals of a given antibonding orbital have to be considered :

$$\tilde{\phi}_{\mu-}(\vec{r}) = \frac{1}{K(+,-)} \{\phi_{\mu-}(\vec{r}) - \frac{1}{K_+} <\phi_{1+}|\phi_{2-}> \sum_{\nu\neq\mu}^{4} [\phi_{\nu+}(\vec{r}) - \phi_{\nu+}(\vec{r}-\vec{R}_{\mu\nu})]\} \tag{II.31}$$

Here K(+,-) is a normalization constant and $\vec{R}_{\mu\nu} = \frac{a_o(\vec{\mu}-\vec{\nu})}{4}$.
The new antibonding orbitals $\tilde{\phi}_{\mu-}(\vec{r})$, together with the old bonding orbitals $\phi_{\nu+}(\vec{r})$, are then used to construct the two-particle Green's function according to the formulas of paragraph IIA.

In comparison to diamond the local orbitals in silicon are much more extended and therefore it is necessary to include more than only next-nearest-neighbour orbitals. We include in silicon also the second-nearest orbitals which are directed parallel. The other second-nearest neighbour elements are neglected, since their density form-factors A_δ and f_δ^α are smaller by a factor 3 and the elements of the electron-hole propagator $G^o_{\delta\delta'}$ are smaller by a factor 2 on the average. Thus, $\overline{G^o}$, $\overline{Y}$ and $\overline{Z}$ in Si become matrices of dimension 52×52. $\overline{G^o}$ has 35 and $\overline{Y}$, $\overline{Z}$ each have 47 independent elements.

Because of the orthogonalization, the calculation of the density form-factors A_δ and f_δ^α as well as of the independent matrix-elements of the electron-hole exchange $\overline{Z}$ becomes more complicated. In comparison to diamond additional terms appear in (II.11), (II.21) and (II.22).

For the electron-hole attraction $\overline{Y}$ the 30 independent elements with second-nearest neighbours are negligibly small and are therefore set equal to zero. Because of the more extended wave-functions, we have to consider in the Fourier-sums for the remaining 17 elements terms with lattice-vectors $\vec{R}_m \neq 0$ (6 on the average). The orthogonalization again leads to additional terms, the most important of which we consider. For the static screening of the Coulomb potential a Gaussian fit to Penn-model results [46] is used. Parameters are shown in table 7. For the calculation of the independent elements [35] of the electron-hole propagator $\overline{G^o}$, the linear tetrahedron method [49,50] is used.

Table 7. Gaussian fit to Penn model results [46] for the spacial dispersion of the static dielectric function in silicon.

Parameter	Value (atomic units)
ε_o	11.7
T_1	0.7135
T_2	2.2865
τ_1	2.0100
τ_2	0.2485

Cl. Many-body effects in the optical properties

This section presents our theoretical results for the optical properties of silicon in comparison to experiment.

Figures 8 and 9 show a comparison of the theoretical results ($\overline{\text{RPA}}$, RPA, TDSHF) for the imaginary and real part of the dielectric function in the region of the E_1 and E_2 structures of the optical spectrum of silicon. The inclusion of the different many-body corrections leads to effects which in their trend are completely analogous to the case of diamond. However, they are smaller in size, because of the more extended electronic wave-functions. Pseudo-potential calculations of Louie et al. [59] display similar overall trends for the $\overline{\text{RPA}}$ and RPA.

The orthogonalization of conduction and valence band wave-functions leads to a reduction of the differences between the theoretical calculations ($\overline{\text{RPA}}$, RPA, TDSHF) also with respect to ref.4.

Figures 10 and 11 give the imaginary and real part of the experimental dielectric function of silicon [60] together with the TDSHF calculation. The agreement between theory and experiment is quite satisfactory in the region of the E_1 and E_2 structures. The E_2 structure is somewhat too small. The reason behind this fact is probably the origin of the E_2 structure, namely a lower conduction and an upper valence band which are parallel over a large part of the Brillouin zone [61]. This is very difficult to reproduce in a Slater-Koster bandstructure.

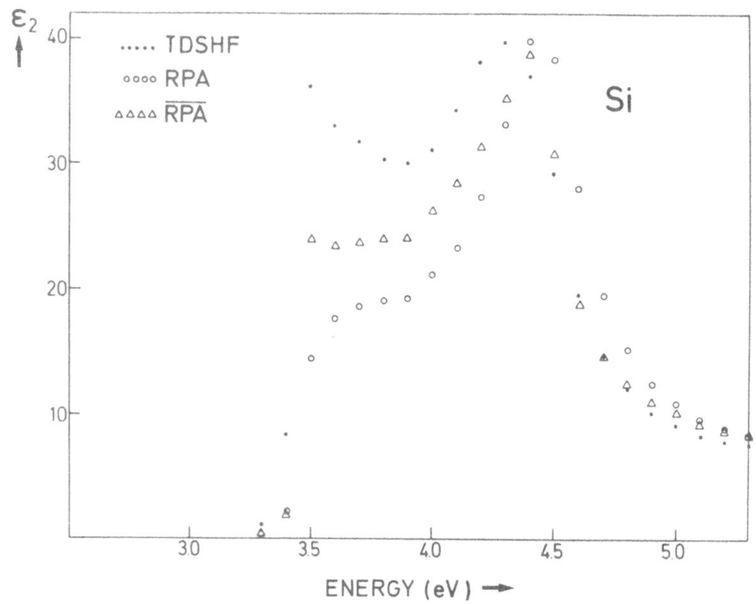

Figure 8. Comparison of the theoretical results for the imaginary part of the dielectric function in silicon.

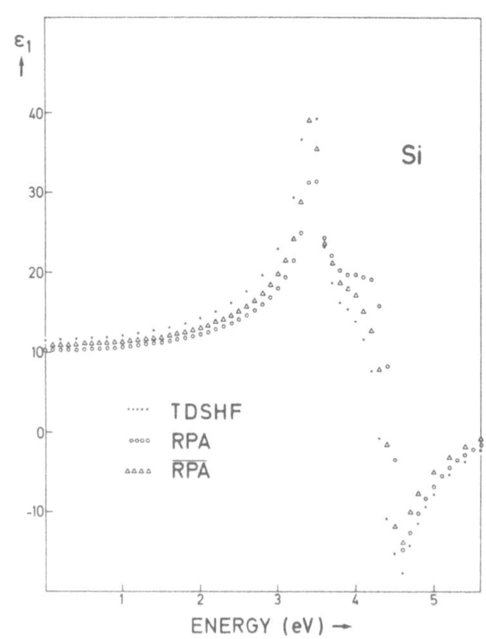

Figure 9. Comparison of the theoretical results for the real part of the dielectric function in silicon.

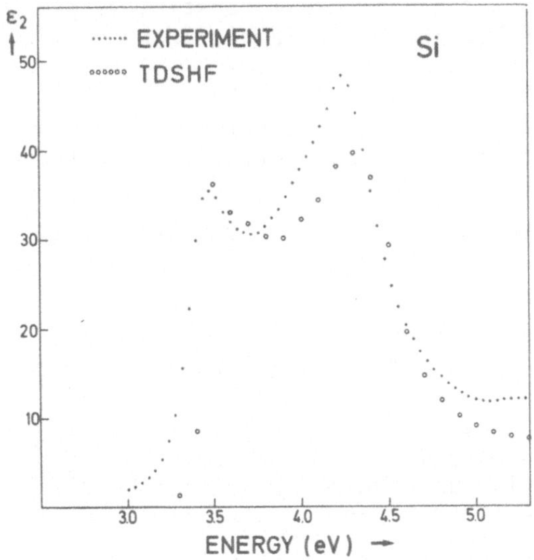

Figure 10. Comparison between theory (TDSHF) and experiment [60] for the imaginary part of the dielectric function of silicon.

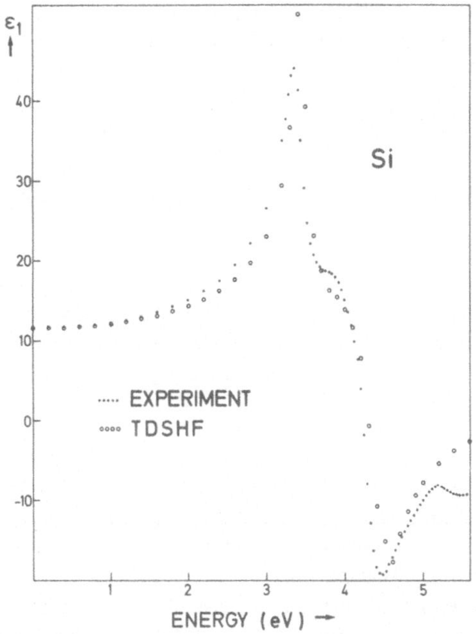

Figure 11. Comparison between theory (TDSHF) and experiment [60] for the real part of the dielectric function of silicon.

For the static dielectric constant we get 10.82 in the $\overline{\text{RPA}}$, 10.31 in the RPA and 11.62 in the TDSHF. The experimental value [60] is $\varepsilon_o = 11.7$.

The integrals in the sum-rules (II.29a,b) give in the TDSHF 695 and -680 (in eV^2) respectively. With the experimental plasma resonance [62] $\omega_p = 16.9$ eV we have on the other hand $\frac{\pi}{2}\omega_p^2 = 450$. The sum-rules are clearly not as well fulfilled as in diamond.

Also the position of the plasma resonance with approximately 22 eV is not satisfactorily reproduced in the TDSHF.

The reason for these difficulties is probably related to the decreasing quality of the underlying bandstructure [54] for the upper conduction and the lower valence bands. Using the theoretical value (TDSHF) for the plasma resonance, one finds $\frac{\pi}{2}\omega_p^2 = 760$ in good agreement with the integrals in eqs. (II.29a,b) pointing to an internal consistency of the calculation.

In conclusion, the calculated dielectric function is quite accurate for optical energies (< 8 eV) and becomes increasingly worse for higher energies. Already here the importance of a reliable quasi-particle description emerges.

C2. Wave-vector dependence

Our investigations are restricted to the wave-factor dependence in the static case ($\omega = 0$) which is needed in chapter IV.

For the components ($G°$,Y,Z) of the two-particle Green's function the same approximations as in the case of diamond will be made. Here the assumption of a wave-vector independent electron-hole attraction Y is less well justified, since the terms with $\vec{R}_m \neq 0$ are greater in silicon than in diamond.

To confirm the central approximation of a wave-vector independent electron-hole propagator $G°$ in eq. (II.12), we again calculate as in diamond the static dielectric function $\overline{\varepsilon}(\vec{q})$ (see (II.30)) in the $\overline{\text{RPA}}$ approximation along a special direction [53] in the irreducible Brillouin zone and average over the corresponding star. Figure 12 shows the result in comparison to the interpolation formula of Penn [47], fitted to the same value 10.87 for $\vec{q} = 0$. The good agreement between the two results is again (as in C) a strong hint that the wave-vector dependence

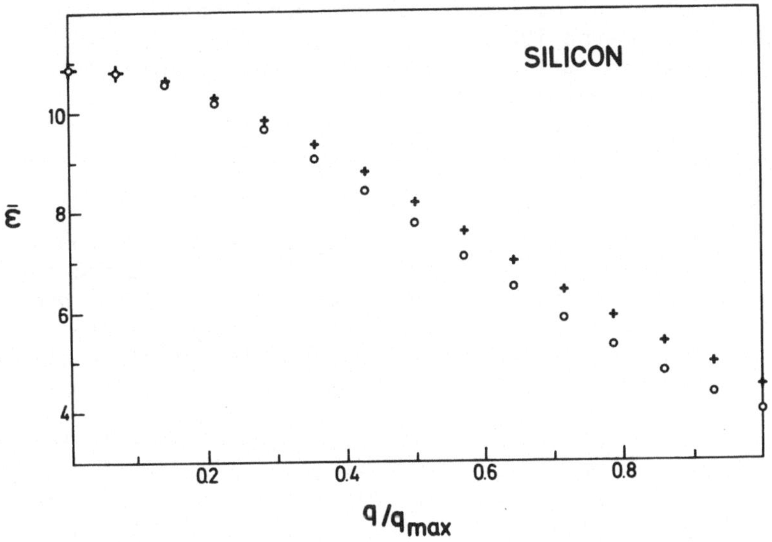

Figure 12. Wave-vector dependence of the dielectric function of silicon for $\omega = 0$. q_{max} denotes the edge of the Brillouin zone in the special direction. Displayed are the Penn model results (o) and the symmetrized RPA (+).

of the bare electron-hole propagator G° can be neglected in eqw. (II.9 and 10) and confined to the matrix-elements of $A_{\delta}(\vec{q})$ and Y and Z.

III. DYNAMICAL CORRELATION EFFECTS IN THE ELECTRONIC QUASI-PARTICLE STATES OF PERFECT CRYSTALS : EXAMPLE DIAMOND

In this chapter the equation of motion of the one-particle Green's function [9] will be reformulated in a local representation and applied to the calculation of the electronic quasi-particle states in the covalent crystal diamond [63] . Especially the importance of many-body effects (exchange and correlation) will be analysed with the use of a systematic Green's function expansion and the aid of consistency criteria. Thus, a practical way for the determination of quasi-particle states in a crystal with a strongly inhomogeneous electronic charge density is demonstrated.

A fundamental difficulty in the application of the Green's function method [9] lies in the fact that the detailed screening properties (ε^{-1}) of the crystal must be known beforehand, if reasonable quasi-particle states are to be obtained. Normally this leads to a self-consistency cycle which we will shorten making use of the results of chapter II.

A different systematic approach to the inclusion of many-body effects is given by the density-functional method [7,64]. From its theoretical justification (LDA) it is best suited for systems with slowly varying electronic densities (metals) and has given, though in principle only applicable to ground state properties, also encouraging results for quasi-particle states [8]. However, an application of the LDA to systems with a strongly varying density (semiconductors, insulators) can in principle not be justified on rigorous theoretical grounds. This is for instance clearly shown in the large deviations (bandgap, valence band width) of LDA calculations for diamond [16,41] and silicon [19] from experiment [17,18,20,21] (see table 1). On the other hand, the LDA provides such a simple, practical scheme for exchange and correlation that it seems very important to know the basic reasons for these deviations (energy dependence and/or nonlocality) and possibly extract still practical extensions.

A. Eigenvalue equation for the quasi-particle states

Quasi-electrons or holes denote one-particle excitations of a system with (N+1) or (N-1) electrons (in comparison to the ground-state of the N-electron system) respectively. Their excitation energies and energetic widths (related to the life-time by the uncertainty relation) are given by the real and imaginary parts of the poles of the one-particle Green's function g [65] and form a complex many-body bandstructure.

With the aid of the functional derivative formalism [9] one can derive an eigenvalue equation

$$[-\Delta + V_{av}(\vec{r})]\psi_i(\vec{r}) + \int d^3r' \, \Sigma(\vec{r},\vec{r}\,';E_i)\, \psi_i(\vec{r}\,') = E_i\psi_i(\vec{r}) \tag{III.1}$$

in which the self-energy or mass operator $\Sigma(\vec{r},\vec{r}\,';E_i)$ plays a crucial role. With its knowledge the (complex) energies E_i of the quasi-particle states ψ_i can be determined through a self-consistent solution of (III.1). $V_{av}(\vec{r})$ is an averaged local potential (Hartree potential) which represents the ions and the valence charge-density of the ground state. Relativistic and spin-dependent effects as well as lattice-dynamics effects are neglected here.

Of course, the exact expressions for the self-energy Σ, as well as for the one-particle Green's function are not known exactly. However, one can derive a set of coupled integral equations which interrelates the exact self-energy Σ to the exact one-particle Green's function g, the irreducible polarizability $\tilde{\chi}$, the vertex-function Γ and the dynamically screened Coulomb interaction W [9] :

$$\Sigma(12) = i \int d34\ W(1^+3)\ g(14)\ \Gamma(42,3) \tag{III.2a}$$

$$W(12) = v(12) + \int d34\ v(13)\ \tilde{\chi}(34)\ W(42) \tag{III.2b}$$

$$\tilde{\chi}(12) = -i \int d34\ g(23)\ g(42)\ \Gamma(34,1) \tag{III.2c}$$

$$\Gamma(12,3) = \delta(12)\ \delta(13) + \int d4567\ \frac{\delta\Gamma(12)}{\delta g(45)}\ g(46)\ g(75)\ \Gamma(67,3) \tag{III.2d}$$

In these equations each numeral represents a set of coordinates (space ($\vec{r}$), time (t) and spin (σ)). 1^+ means that the time-variable t_1 is accompanied by a positive infinitesimal. Equations (III.2) indicate, how the self-energy Σ is interrelated with the dielectric function ($\varepsilon = 1 - v\tilde{\chi}$) through the vertex-function Γ and how Γ on the other hand depends on Σ.

The basic problem in any application of the outlined formalism is the decoupling of the equations (III.2). Normally two separated approximations for Σ and Γ are employed. For the self-energy Σ often the first term in a Green's function expansion (GW approximation)

$$\Sigma_{GW} = iW(1^+2)\ g(12) \tag{III.3}$$

is chosen [9] . The vertex-function Γ is usually taken as

$$\Gamma(12,3) = \delta(12)\ \delta(13) \tag{III.4}$$

implying that vertex-corrections are neglected and that the screening (dielectric function) is treated in the RPA. There is, however, no criterion which gives a preference to this special decoupling.

Criteria for a special choice of Σ and Γ cannot be obtained from the equations (III.2) themselves, but only from reasonable physical constraints as for instance the symmetries of the system. In particular the invariance of the theory under local gauge-transformations (related to the current conservation) gives rise to a well-known connection between the self-energy Σ and the vertex-function Γ, i.e. Ward identity [65] .

$$i(t_1-t_2)\ \Sigma(12) = \delta(12) - \int d3\ \Gamma(12,3) = -\int d3\ \Lambda(12,3) \qquad \text{(III.5)}$$

(III.5) can be used as a criterion for extracting the approximations for Γ, which are consistent with the GW approximation for Σ. The vertex-function Γ thus consists of three terms which are the RPA term (III.4) and two terms from the functional derivative of Σ with respect to g. The derivative with respect to the explicitly in (III.3) appearing one-particle Green's function g leads to the TDSHF and the functional derivative of W (containing g implicitly) to a third and more complex approximation. This last contribution to Γ is numerically extremely difficult to handle and therefore, the investigations will be restricted to the RPA and TDSHF. Conclusions to the quality and the range of validity of these approximations must be reserved for the comparison of numerical calculations and experiment. Finally, the dynamically screened interaction W, appearing in the vertex-function Γ of the TDSHF will be further approximated by a statically screened interaction (as in chapter II).

The GW approximation itself is the lowest-order term in an expansion of the self-energy in powers of the dynamically screened interaction W. The neglect of higher terms cannot strictly be justified by smallness arguments. From the comparison of phase-factors and rough calculations of the integrals one finds that the next term in the expansion should be smaller by about a factor 1/30 for diamond. Nevertheless, the step to a non-local dynamical theory of correlations in semiconductors and insulators, containing in a quantitative way the most important two-particle excitations (electron-hole pairs, plasmons), is a major step forward. Higher-order terms will undoubtedly cause numerical changes, but will not change the basic physical conclusions.

B. The self-energy, properties and approximations

In this paragraph we discuss general properties of the self-energy in a periodic crystal and previously used approximations (in comparison to our approximation introduced in III.A).

The general properties of Σ will be especially useful for casting the eigenvalue equation (III.1) into a form suitable for numerical treatment.

A symmetry property of the self-energy (and (III.1) as a whole) is of course the invariance under the space-group operations $\{R|\vec{w}\}$ of the crystal

$$\Sigma(\{R|\vec{w}\}\vec{r},\{R|\vec{w}\}\vec{r}\,';E) = \Sigma(\vec{r},\ \vec{r}\,';E) \qquad \text{(III.6)}$$

Here $\{R|\vec{w}\}\vec{r} = R\vec{r} + \vec{w}$ with the rotation R and the translation $\vec{w}$. (III.6) can be proved by showing the property first for the one-particle Green's function g with the aid of its spectral representation and then by the use of the functional relation (Dyson equation) between Σ and g. (III.6) implies that Bloch's theorem is fulfilled for the wave-function ψ_i so that $i = (n,\vec{k})$ with the band index n and the wave-vector $\vec{k}$ (reduced to the first Brillouin zone).

The invariance under interchange of space-coordinates $\vec{r}$ and $\vec{r}'$

$$\Sigma(\vec{r},\vec{r}';E) = \Sigma(\vec{r}',\vec{r};E) \tag{III.7}$$

is a second general property of the self-energy which is verified similarly to (III.6).

A third property of Σ which is of fundamental importance for our numerical procedure is its short-rangedness in $|\vec{r}-\vec{r}'|$, as shown by Sham and Kohn [10] on the basis of general perturbation-theoretical considerations. This implies that Σ, viewed as a continuous matrix in $\vec{r}$ and $\vec{r}'$, is appreciably different from zero only near the diagonal $\vec{r} = \vec{r}'$.

Several approximations for the self-energy Σ and thereby for the quasi-particle solutions of (III.1) appear in literature. They will be discussed in the following.

The simplest possible ansatz is trivially $\Sigma = 0$. This leads to the Hartree approximation for the quasi-particle states, where the many-body interaction of one selected electron with all the remaining electrons is replaced by the interaction with an averaged local charge density of the electronic system in the ground state. The Hartree approximation neglects parallel spin correlations (Pauli principle) as well as Coulomb correlations and has therefore long been recognized as being insufficient [2]. An improvement is given by the Hartree-Fock approximation in which the representation of the self-energy is formally similar to the GW approximation of paragraph III.A. But instead of a dynamically screened Coulomb interaction W an unscreened "bare" Coulomb interaction v is used. The Hartree-Fock approximation includes exchange-interactions between electrons with parallel spin, but remains static (energy independent) like the Hartree approximation. From table 1 (diamond) we can see that this is still insufficient even for an approximate description of the quasi-particle states of a real system, especially if it has a strongly inhomogeneous electronic charge density. It is customary to regard the Hartree-Fock approximation as a reference level and to refer to many-electron interactions which go beyond it, as correlations. The representation (III.2b) of the screened Coulomb interaction W can be used to divide the GW approximation (III.3) into a Hartree-Fock part Σ_{HF} and a remainder Σ' which describes the correlation, i.e.

$$\Sigma_{GW}(12) = \Sigma_{HF}(12) + \Sigma'(12) \qquad \text{(III.8)}$$

In the picture of the static pair-correlation function the inclusion of exchange and correlations corresponds to the creation of an exchange-correlation hole around the particles of extension $\sim 1/\vec{k}_F(\vec{v})$, where $\vec{k}_F(\vec{v})$ is the local Fermi vector.

The short range of the combined effects of exchange and correlations which is reflected in the self-energy is also an important ingredient of the density-functional method of Hohenberg, Kohn and Sham [7,10,64]. It is based on the fact that the ground state energy of an interacting inhomogeneous electron-gas in a static external potential (for instance from the ions of the crystal) is a universal functional $E[\rho]$ of the ground state density $\rho(\vec{r})$ which moreover has its minimum value for the correct density. One finds that $\rho(\vec{r})$ can be determined from the self-consistent solution of a one-particle equation of the Schrödinger type. In this equation the functional derivative of $E[\rho]$ appears as the so-called exchange-correlation potential $v_{xc}(\rho(\vec{r}))$. Although the eigenstates of the one-particle equation cannot be interpreted as quasi-particle states of the system, right at the Fermi-energy μ, the LDA treatment of the self-energy converges to $v_{xc}(\rho(\vec{v}))$ [8]. For metals such a procedure has been quite successful [8]. For semiconductors and insulators one might argue that one is necessarily eV's away from the Fermi surface and the deviations in table 1 are to be expected. Since one knows that the self-energy is short-ranged in $|\vec{r}-\vec{r}'|$, it should still be possible to neglect the non-locality in first order. Within the LDA, the energy-dependence can be included by an expansion into powers of the deviation from the Fermi-energy. Sham and Kohn [10] showed that

$$\Sigma(\vec{r},\vec{r}';E) \cong \delta(\vec{r}-\vec{r}')[v_{xc}(\rho(\vec{r})) + (E-\mu)(1-m^*(\rho(\vec{r})) + \ldots] , \qquad \text{(III.9)}$$

where m^* is the density of states mass [67]. As we shall see in the investigations of the quasi-particle properties of diamond, developments along similar lines which emphasize the energy dependence of Σ are expected to lead to fruitful improvements of the LDA method of bandstructure calculations in semiconductors and insulators.

Many schemes for approximating (III.1) start from Hartree-Fock and include correlation via simplified polarization models. In this category belong the plasmon-model of Overhauser [68,14] and the polaron-model of Toyozawa and Kunz [13,69-71] . They use the so-called configuration-mixing treatment [70] , where the interaction is approximated in the form of a surrounding cloud of plasmons (in the plasmon-pole approximation) or (two-band) excitons, respectively.

Previous applications of the GW approximation with a screened interaction W all start from simplified forms of the screening and treat the energy-dependence of the dielectric sfreening with two rather extreme model assumptions : the dielectric function is either assumed to be completely energy-independent (static) or its imaginary part is replaced by a delta-function at the plasma-resonance. The first approximation is the so-called COHSEX (Coulomb hole plus screened exchange) approximation of Hedin [9] and was applied by several authors to the calculation of quasi-particle states in semiconductors and insulators [72-74] , using model descriptions (e.g. Penn) for the wave-vector dependence of the dielectric function. The plasmon-pole scheme has been mainly applied to metals [9] , but calculations for semiconductors have also been tried [75] .

The above discussions demonstrate the need for an a-priori study of correlation effects which rests on a quantitative determination of the two-particle excitations embedded in a non-local and energy-dependent screening calculated from first principles. Only then it is possible to judge the merits of a given approximation. Our calculations will demonstrate that the GW is indeed a very reasonable approximation to dynamical correlation, and has the additional virtue of being based on a systematic Green's function treatment. In particular, the two-particle excitations and the role of many-body effects are calibrated by experiment. An investigation of the relative importance of elementary excitations of electron-hole and plasmon type in the dynamically screened interaction W gives important insights into the long-range versus short-range character of correlation. Proper inclusion of the energy-dependence in Σ furthermore allows for an accurate determination of the life-times of the quasi-particle states [76] .

C. Local representation of the eigenvalue equation

With the help of a local-orbital representation (see eq. (II.8)) the eigenvalue equation (III.1) can be transformed into an equivalent algebraic eigenvalue problem for the expansion coefficients $c_{n\nu}(\vec{k})$.

$$\sum_{\nu'} \{\langle\nu|\varepsilon(\vec{k})|\nu'\rangle + \langle\nu|\Sigma(\vec{k};E_n(\vec{k}))|\nu'\rangle\} c_{n\nu'}(\vec{k}) = E_n(\vec{k}) c_{n\nu}(\vec{k}) \tag{III.10}$$

In eq.(III.10), which has to be solved separately at each point of the Brillouin zone, we use the notations ($\vec{R}_t \hat{=}$ lattice-vector, $H_{uv} \hat{=}$ Hartree part of (III.1))

$$\langle\nu|\varepsilon(\vec{k})|\nu'\rangle = \sum_{\vec{R}_t} e^{i\vec{k}\vec{R}_t} \langle\phi_\nu(\vec{r})|H_{uv}|\phi_{\nu'}(\vec{r}-\vec{R}_t)\rangle \tag{III.11a}$$

and

$$\langle\nu|\Sigma(\vec{k};E)|\nu'\rangle = \sum_{\vec{R}_t} e^{i\vec{k}\vec{R}_t} \langle\phi_\nu(\vec{r})|\Sigma(\vec{r},\vec{r}\,';E)|\phi_{\nu'}(\vec{r}\,'-\vec{R}_t)\rangle \tag{III.11b}$$

Because the self-energy is short-ranged in $|\vec{r}-\vec{r}\,'|$ and because of the localization of the orbitals $\phi_\nu(\vec{r})$, one can expect the sum over $\vec{R}_t$ in (III.11b) to be restricted similarly as in (III.11a) to a small number of lattice vectors.

The eigenvalues $E_n(\vec{k})$ of (III.10) which also appear in the self-energy, have to be determined self-consistently at each point of the Brillouin zone. An additional difficulty is given by the fact that the matrix $\langle\nu|\Sigma(\vec{k};E)|\nu'\rangle$ is in general non-hermitian. The eigenvalues $E_n(\vec{k})$ are thus complex, with the real part and the imaginary part furnishing the excitation energy and the energetic width (life-time) of the quasi-particles, respectively. An energy-independent approximation for the self-energy Σ leads to a hermitian matrix (III.11b) and thus prevents the determination of life-times [76] .

Starting point for the calculation of (III.11b) is the energy representation of the GW approximation (III.3)

$$\Sigma(\vec{r},\vec{r}\,';E) = i(2\pi)^{-1} \int_{-\infty}^{\infty} dE'\, e^{iE'\delta} g(\vec{r},\vec{r}\,';E+E')\, W(\vec{r},\vec{r}\,';E') \tag{III.12}$$

where δ is a positive infinitesimal ($\delta \to 0^+$) to insure convergence of the Hartree-Fock term. It takes the usual form of a convolution integral, if one considers that W is an even function of E'. The one-particle Green's function g which appears in (III.12) (also indirectly through W), should in principle be determined self-consistently from the related Dyson equation [9] . This means that one starts from an approximation for the quasi-particle states, calculates the self-energy after (III.12) and solves (III.1) for an improved quasi-particle spectrum, which afterwards is used in the same cycle and so forth, until convergence is reached. In the calculation for diamond we abbreviate this cycle by using for the determination of (III.12) a X_α bandstructure [41] which has been proved to give a very satisfactory description of the dielectric function (see (II.8)) and already includes exchange and correlation approximately.

For our later discourse it is also necessary to represent the one-particle Green's function in the local basis.

$$g(\vec{r},\vec{r}\,';E) = \sum_{\sigma\sigma'} \sum_{\vec{R}_s\vec{R}_{s'}} \phi_\sigma(\vec{r}-\vec{R}_s)\, g_{\sigma\sigma'}(\vec{R}_s-\vec{R}_{s'};E)\, \phi^*_{\sigma'}(\vec{r}\,'-\vec{R}_{s'}) \tag{III.13}$$

Here the matrix-elements $g_{\sigma\sigma'}(\vec{R}_m;E)$ are given by

$$g_{\sigma\sigma}(\vec{R}_m;E) = N^{-1} \sum_n \sum_{\vec{k}}^{BZ} \frac{c_{n\sigma}(\vec{k})\, e^{i\vec{k}\vec{R}_m}\, c^*_{n\sigma'}(\vec{k})}{E - E_n(\vec{k}) + i\eta\, \mathrm{sgn}(E_n(\vec{k}) - E_F)} \tag{III.14}$$

where η is again a positive infinitesimal ($\eta \rightarrow 0^+$) and E_F is the Fermi energy.

The screened interaction W is splitted into an unscreened Hartree-Fock part v and a correlation part W'.

$$W(\vec{r},\vec{r}\,';E) = v(\vec{r}-\vec{r}\,') + W'(\vec{r},\vec{r}\,';E) \tag{III.15}$$

The two parts in (III.15) directly correspond to Σ_{HF} and Σ' in (III.8). With the density-density correlation function χ we can write for the correlation part

$$W'(\vec{r},\vec{r}\,';E) = \int d^3r_1 d^3r_2\, v(\vec{r}-\vec{r}_1)\, \chi(\vec{r}_1,\vec{r}_2;E)\, v(\vec{r}_2-\vec{r}\,') \tag{III.16}$$

To obtain a local basis representation of W', we employ the local representation of χ

$$\chi(\vec{r},\vec{r}\,';E) = V^{-1} \sum_{\vec{q}}^{Bz} \sum_{\vec{Q}\vec{Q}'} e^{i(\vec{q}+\vec{Q})\vec{r}}\, \chi(\vec{q}+\vec{Q},\vec{q}+\vec{Q}';E)\, e^{-i(\vec{q}+\vec{Q}')\vec{r}\,'} \tag{III.17a}$$

or

$$\chi(\vec{q}+\vec{Q},\vec{q}+\vec{Q}';E) = \Omega_o^{-1} \sum_{\delta\delta'} A_\delta(\vec{q}+\vec{Q})\, G_{\delta\delta'}(\vec{q};E)\, A^*_{\delta'}(\vec{q}+\vec{Q}') \tag{III.17b}$$

where $V = N\Omega_o$ is the crystal volume. Because we want to restrict the integration in (III.12) to positive energies (χ is an even function of E), a modified one-particle Green's function

$$\bar{g}(\vec{r},\vec{r}\,';E,E') = g(\vec{r},\vec{r}\,';E+E') + g(\vec{r},\vec{r}\,';E-E') \qquad \text{(III.18)}$$

is introduced. For large E one can show $\chi(\vec{r},\vec{r}\,';E) \sim 0(E^{-2})$, so that the convergence factor $e^{iE\delta}$ can be left out in the correlation part of (III.13). Using equations (III.12) - (III.18), we get

$$\langle\phi_\tau(\vec{r})|\Sigma'(\vec{r},\vec{r}\,';E)|\phi_{\tau'}(\vec{r}-\vec{R}_t)\rangle =$$

$$= i(2\pi)^{-1} \int_0^\infty dE' \sum_{\sigma\sigma'} \sum_{\vec{R}_m\vec{R}_{m'}} \bar{g}_{\sigma\sigma'}(-\vec{R}_m;E,E') \cdot$$

$$\cdot\, N^{-1} \sum_{\vec{q}} e^{-i\vec{q}\vec{R}_m} W'_{\vec{R}_m+\vec{R}_{m'},\sigma\tau;\vec{R}_{m'}+\vec{R}_t\sigma'\tau'}(\vec{q},E') \qquad \text{(III.19)}$$

for the correlation part Σ' of the self-energy in the local representation. In (III.19)

$$W'_{\delta\delta'}(\vec{q};E) = \sum_{\delta_1\delta_2} Z_{\delta\delta_1}(\vec{q})\, G_{\delta_1\delta_1}(\vec{q};E)\, Z_{\delta_2\delta'}(\vec{q}) \qquad \text{(III.20)}$$

is the Fourier-transformed matrix of the potential $W'(\vec{r},\vec{r}\,';E)$ between pairs (δ,δ') of local orbitals. The representation (III.20) is especially convenient, since $W'(\vec{q};E)$ can be obtained from simple matrix multiplications of the two-particle Green's function G and the Coulomb matrix or electron-hole exchange Z.

D. Calculation of the quasi-particle states of diamond

In this paragraph the eigenvalue equation (III.10) is solved for the case of diamond. Since our starting point is a self-consistent Hartree-Fock bandstructure [15] , the matrix-elements of the self-energy (III.10b) are splitted into a Hartree-Fock part $\langle\nu|\Sigma_{HF}(\vec{k})|\nu'\rangle$ and a correlation part $\langle\nu|\Sigma'(\vec{k};E)|\nu'\rangle$. For the energy-independent Hartree-Fock part $\langle\nu|\varepsilon(\vec{k}) + \Sigma_{HF}(\vec{k})|\nu'\rangle$ of (III.10) we use a Slater-Koster representation [22] of an existing self-consistent Hartree-Fock bandstructure [15]. The energy-dependent correlation part is obtained from an explicit evaluation of the matrix-elements (III.19).

Having found higher-order overlaps rapidly falling off, we make the approximation $\vec{R}_m = \vec{R}_t = -\vec{R}_{m'}$ and σ parallel to τ

as well as σ' parallel to τ' in the matrix W' in (III.19). In the summation of (III.20) on the other hand all next-nearest neighbour terms are considered which is dictated by our results on the optical response (see chapter II). With these approximations one obtains

$$\langle\phi_{b_i}(\vec{r})|\Sigma'(\vec{r},\vec{r}';E)|\phi_{b_j}(\vec{r}'\vec{R}_t)\rangle =$$

$$= i(2\pi)^{-1}\int_0^\infty dE'\{\bar{g}^{(v)}_{b_ib_j}(-\vec{R}_t;E,E')\,W'_{b_ib_i,b_jb_j}(\vec{R}_t;E') +$$

$$+\bar{g}^{(c)}_{a_ia_j}(-\vec{R}_t;E,E')\,W'_{a_ib_i,a_jb_j}(\vec{R}_t;E')\} \qquad \text{(III.21a)}$$

for the valence bands (the indices a and b denote antibonding and bonding orbitals, respectively, (i,j) = (1,2,3,4) are indices for the four tetrahedral directions) and

$$\langle\phi_{a_i}(\vec{r})|\Sigma'(\vec{r},\vec{r}';E)|\phi_{a_j}(\vec{r}'-\vec{R}_t)\rangle =$$

$$= i(2\pi)^{-1}\int_0^\infty dE'\{\bar{g}^{(v)}_{b_ib_j}(-\vec{R}_t;E,E')\,W'_{b_ia_i,b_ja_j}(\vec{R}_t;E') +$$

$$+\bar{g}^{(c)}_{a_ia_j}(-\vec{R}_t;E,E')\,W'_{a_ia_i,a_ja_j}(\vec{R}_t;E')\} \qquad \text{(III.21b)}$$

for the conduction bands. Here the expression

$$W'_{\sigma_i\tau_i,\sigma_j\tau_j}(\vec{R}_t;E) = N^{-1}\sum_{\vec{q}}^{BZ} e^{-i\vec{q}\vec{R}_t}\,W'_{\vec{o}\sigma_i\tau_i,\vec{o}\sigma_j\tau_j}(\vec{q};E) \qquad \text{(III.21c)}$$

has been introduced. It should be noted that the summation over the band indices in the definition (III.14) of the one-particle Green's function is explicitly reduced for $\bar{g}^{(v)}$ and $\bar{g}^{(c)}$ to occupied (v) and unoccupied (c) bands. For the calculation in diamond only four valence and four conduction bands are considered.

The choice of diamond as a prototype of a covalent crystal has several reasons : Firstly, accurate Hartree-Fock band determinations exist and a local orbital representation of the wave-functions is appropriate. Moreover, the dielectric function including many-body effects (local-field, excitonic) is known (see II.B) and finally the clear separation of electron-hole from plasmon excitations allows to study their different effects on the self-energy and to extract a more short-ranged (e-h) or long-ranged (plasmon) correlation behaviour.

D1. The Hartree-Fock part of the eigenvalue equation

As already mentioned, the Hartree-Fock part of (III.9) is obtained from a Slater-Koster representation of the self-consistent Hartree-Fock bandstructure of Mauger and Lannoo [15]. The procedure is analogous to the description in chapter II. Valence and conduction bands are treated separately and are represented by linear combinations of bonding and antibonding tetrahedral orbitals, respectively. In the sums over lattice vectors matrix-elements between up to third-nearest orbitals are considered. The 10 independent matrix-elements are shown in table 8.

Table 8. Parameters of the Slater-Koster representation of the self-consistent Hartree-Fock bandstructure of Mauger and Lannoo [15] for diamond.

Parameter	ν	ν'	$R_t\ (\frac{a_o}{2})$	Valence bands (eV)	Conduction bands (eV)
ξ_0	1	1	(0,0,0)	-15.3687	18.2813
ξ_1	1	2	(0,0,0)	- 2.6646	-0.6296
ξ_2	1	1	(1,1,0)	1.2198	-1.1302
ξ_3	1	2	(1,1,0)	- 0.4463	1.5438
ξ_4	1	1	(1,-1,0)	- 0.5011	0.4740
ξ_5	1	2	(0,1,-1)	- 0.1640	-0.2315
ξ_6	1	1	(2,0,0)	- 0.0323	-0.0990
ξ_7	1	2	(1,-1,0)	0.0200	-0.1000
ξ_8	1	2	(0,-1,-1)	0.1000	0.1400
ξ_9	1	2	(2,0,0)	0.0400	-0.1800

The parameter ζ_o is directly related to an absolute energy scale. In particular, the calculation of Mauger and Lannoo [15] places the upper edge of the valence band complex at an energy value of -4 eV. This is an additional information which can be obtained from Hartree-Fock bandstructure determination.

D2. The one-particle Green's function

The calculation of the one-particle Green's function should be based on a bandstructure which is near to the expected result for the quasi-particle spectrum. Here we take the Slater-Koster representation of the X_α bandstructure of Painter et al [41] which we already used for the investigations of the dielectric function. The parameters of this representation in table 2 place the valence band edge at -1.5 eV. This choice corresponds to an expected displacement of the correlated valence bands (with respect to the Hartree-Fock calculation) by 2.5 eV to higher energies. The expected displacement is taken from an earlier calculation (COHSEX) of Brener [74]. The investigations which we describe here, locate the upper valence band edge at 0.25 eV. Thus an appropriate change of the energy-scale of the one-particle Green's function becomes necessary.

Since real functions (see (II.16) - (II.18)) are chosen for the local orbitals, the one-particle Green's function (III.14) can easily be separated formally into real and imaginary part. The real part is connected to the imaginary part by a Cauchy principle-value integral over the energy. In the present context the imaginary part (Brillouin zone integration) is calculated first with the aid of the linear tetrahedron-method [49,50]. Afterwards the principle-value integral is used to extract the real part.

In the energy-dependence of the matrix-elements with lattice vectors $\vec{R}_m \neq 0$ an oscillatory behaviour appears which increases with increasing $|\vec{R}_m|$. This oscillatory behaviour will be important for qualitative arguments later.

D3. The dynamically screened interaction (correlation part)

Three types of matrix-elements (III.21c) determining the screened interaction W remain to be calculated. The linear tetrahedron-method [49,59], used for calculating the one-particle Green's function g and the electron-hole propagator G°, cannot be used here, because of an enormous requirement of computer time. On the other hand, the special-point method [77] can also not be employed, because the integrant diverges for $\vec{q} \to 0$. Best suited for the integration over the Brillouin zone seems to be the method of Bausil [53]. This method allows, under certain conditions, to replace the three-dimensional integral over the Brillouin zone by a sum over one-dimensional integrals along specially chosen directions in the Brillouin zone. The angular dependence of the integrant, which can be singular at the Γ-point, is assumed to be not too strong. The special directions are found from the condition that an appropriate number of cubic harmonics, up to a given angular momentum l_{max}, vanish. For the calculation of the matrix-elements of the screened interaction W after (III.21c) the minimum set of special directions is used. This set consists of the line

$$(k_x, k_y, k_z) = \frac{2\pi}{a_o}(0.83539,\ 0.40747,\ 0.25714)\zeta \qquad (0 < \zeta < 1) \qquad \text{(III.22)}$$

in the irreducible Brillouin zone and the corresponding star (48 lines). Angular dependence up to totally symmetric cubic harmonics with l = 6 is thus included exactly. Along the special line several points (11) are chosen for the numerical integration.

Finally, the transformation properties of the matrix-elements $W_{\vec{o}\sigma_i\tau_i,\vec{o}\sigma_j\tau_j}(\vec{q},E)$ under the cubic point group, which are non-trivial and contain complex phase-factors, are exploited fully.

The electron-hole exchange $Z(\vec{q})$ and the two-particle Green's function $G(\vec{q};E)$ which are needed for calculating $W'(\vec{q};E)$, can be obtained from the results of chapter II.

We turn now to the discussion of the energy-dependence of $W'(\vec{R}_t;E)$. Since it is controlled by the two-particle Green's function, all matrix-elements are expected (and the calculations confirm this) to have a quite similar form. The problem of the numerical inclusion of the plasma-resonance which the theory places at 38 eV (last zero of $\varepsilon_1(\omega)$), while the imaginary part of the electron-hole propagator G° vanishes already at 34 eV, is solved by extending the last non-zero value of $Im(G^\circ)$ up to our maximum energy of 40.5 eV. As an example we show in figure 13 the most important matrix-element of W' (III.21c) with the index-combination (b $\hat{=}$ bonding, a $\hat{=}$ antibonding) bb-bb. Matrix-elements of the index-combination aa-aa and ba-ba look similar, but while the elements aa-aa are of the same order of magnitude, the elements ba-ba are smaller by more than an order of magnitude. Formally, this is related to the fact that for $|\vec{q}+\vec{Q}| \to 0$ the form-factors $A_\delta(\vec{q}+\vec{Q})$ (see (II.11)) have the limit 1 in the first

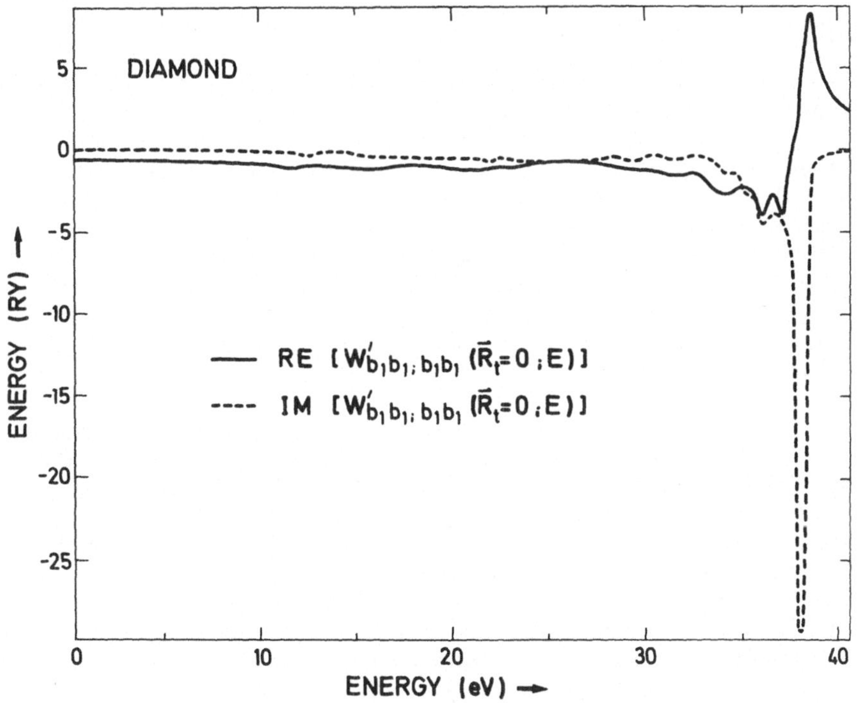

Figure 13. Energy-dependence of the most important matrix-element of the dynamically screened interaction (correlation part) between two pairs of bonding orbitals.

case (bb or aa) and 0 in the second case (ba). Physically it means that the self-energy corrections to the valence bands (see (II.21a)) are mainly determined by matrix-elements between bonding orbitals (1.term), while the coupling-term to the conduction bands (2.term) represents only a small correction. An equivalent conclusion is true for the conduction bands. Another important result is that the matrix-elements of W' are rather flat over a large energy-range (0-32 eV). Only near the plasma-resonance (∿38 eV) the corresponding distinct structures appear. Together with the increasingly oscillatory behaviour of the matrix-elements of the one-particle Green's function with increasing $|\vec{R}_t|$, the flatness of the matrix-elements of the dynamically screened interaction leads to the short-range properties of the self-energy corrections. The qualitative arguments are confirmed by the numerical calculations. Over the whole energy-range the matrix-elements (III.21a) and (III.21b) become clearly smaller with increasing $|\vec{R}_t|$ (up to several orders of magnitude).

D4. Results for the quasi-particle states and discussion

Armed with the knowledge of one- and two-particle Green's functions, we are now in the position to calculate the matrix-elements (III.11a), (III.11b) and to solve the non-hermitian eigenvalue-problem (III.9) for the electronic quasi-particle states. Valence and conduction bands are treated separately. The complex quasi-particle energies $E_n(\vec{k})$ at a given $\vec{k}$-vector are determined by diagonalizing the corresponding 4×4-matrices first over an appropriate energy-range with a point-separation of 0.25 eV (eigenvalues belonging to E are denoted by $\xi_n(\vec{k},E)$ and by looking for the local minima of the function $|E - \xi_n(\vec{k},E)|$ afterwards.

The real parts of the eigenvalues $E_n(\vec{k})$ can be presented in condensed form as a quasi-particle bandstructure. The result for the TDSHF approximation is shown in figure 14 in comparison to the self-consistent Hartree-Fock approximation [15] which we used as a basis for our investigations. In addition, we give in column 1 and 2 of table 9 (conduction bands) and table 10 (valence bands) the energy-eigenvalues at the highly symmetrical points (Γ,X,L). To simplify the comparison, the values are given relative to the respective upper edge of the valence bands.

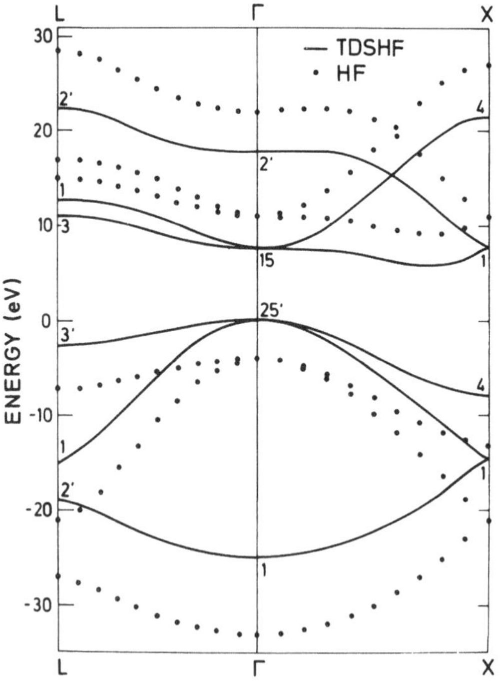

Figure 14. Comparison between the HF and TDSHF bandstructure of diamond.

Table 9. Highly symmetrical points (Γ,X,L) of the conduction bands of diamond in the HF, RPA and TDSHF calculations. The values are given relative to the upper edge of the valence bands.

Brillouin zone point	HF	TDSHF	RPA
Γ_{15}	15.0	7.4	8.25
$\Gamma_{2'}$	26.0	17.6	18.9
X_1	15.0	7.45	8.3
X_4	31.0	21.35	22.8
L_3	19.0	11.0	11.85
L_1	21.0	12.6	13.75
$L_{2'}$	32.5	23.3	24.5

Table 10. As table 9, but for the valence bands of diamond.

Brillouin zone point	HF	TDSHF	RPA
Γ_1	-29.0	-25.2	-26.1
$\Gamma_{25'}$	0.0	0.0	0.0
X_1	-17.0	-14.75	-15.25
X_4	- 9.0	- 7.9	- 8.2
$L_{2'}$	-23.1	-19.7	-20.75
L_1	-17.0	-15.0	- 15.5
$L_{3'}$	- 3.1	- 2.8	- 2.9

On an absolute energy scale the self-energy corrections shift the valence bands to higher energies and the conduction bands to lower energies. This result confirms the trends expected from model-calculations [70] . Through the combined effects of a positive shift of the upper edge of the valence bands by 4.25 eV and a negative shift of the lower edge of the conduction bands by 3.35 eV a drastic reduction of the band gap from 15 eV (HF value) to 7.4 eV in the TDSHF approximation is found. The theoretical value is thus in good agreement with the experimental value [17] of 7.3 eV, obtained in reflectance measurements.

The magnitude of the self-energy corrections is increasing at any given point in the Brillouin zone with increasing distance from the band gap. As a consequence one obtains a shrinking of the band widths. In particular for the case of the valence bands of diamond a reduction from 29.0 eV in the HF calculation to 25.2 eV in the TDSHF calculation is found. Again our results are in quite good agreement with the experimental data of 24.2 $\pm$ 1.0 eV from X-ray photoemission (XPS) experiments [18] . In comparison with the strong energy dependence of the self-energy corrections, the wave-vector dependence is very weak. This clearly points to the short-range property of the self-energy.

In connection with the results for the valence band width two remarks are in order. First, the inaccuracy of the theoretical description of the plasma-resonance in the screening may introduce slight errors for the calculated valence band width, even though the total contribution of the plasmon to the correlation shift is quite small (see below). This inaccuracy (plasma-resonance at too high energies, contribution to sum-rule too small for $\vec{q} = 0$) should lead to a further reduction of the theoretical band-width. Estimates give a maximum reduction to about 23-24 eV. The second comment is connected with new UPS (ultraviolet photoemission spectroscopy) measurements of Himpsel et al [78] for diamond which point to a valence band width of 21 $\pm$ 1 eV. Such a result would strongly confirm LDA results for the bandwith (see table 11) and would apparently be in contrast with the XPS measurements. However, the penetration depth of the photons is much smaller in UPS than in XPS (about 0.5 nm in comparison to 2 nm) [79] . The UPS method has therefore a greater surface sensitivity. Mehta and Fadley [80] recently showed for the d-bands of copper that the band width is reduced by 12 % near the surface. Assuming a similar reduction for diamond, this would account for the difference between XPS and UPS data.

A further feature, which can be compared to experiment, is the indirect bandgap. The TDSHF calculation places the lowest point of the valence band at $\vec{k}_i = (0.75,0,0)\,\frac{2\pi}{a_o}$. The indirect

Table 11. Comparison of the experimental band gap and valence band width of diamond with the results of several calculations. Values are given in eV.

	experiment	LDA (16)	HF (15)	TDSHF	RPA	EH	COHSEX
E_{gap}	7.3 (17)	6.3	15.0	7.4	8.25	7.4	7.2
E_{val}	24.2 (18)	20.4	29.0	25.2	26.1	26.45	28.8

bandgap to the upper edge of the valence band at Γ is found to be 5.7 eV in very good agreement with experiment (5.5 eV - 5.7 eV) [81].

The quasi-particle spectrum has also been calculated with the screened interaction taken in the RPA approximation including local-field effects in the screening. The results for the highly symmetrical points are given in the third column of table 9 and 10. In the RPA the dielectric properties of the crystal (see II.B) are described worse than in the TDSHF, because no vertex corrections of electron-hole type (excitonic effects) are taken into account in the two-particle Green's function G. In the RPA single-particle spectrum this leads to self-energy corrections which are clearly smaller than in the TDSHF. In particular the bandgap (8.25 eV) and the valence band width (26.1 eV) are about 1 eV wider. This result points to the necessity for a treatment of the dielectric function which is as exact as possible. In other words, the single-particle excitations of the system are strongly influenced by the two-particle excitation. The two-particle spectrum on the other hand is directly connected to the one-particle spectrum (see (II.4)). Because of these inter-relations a self-consistent treatment of one- and two-particle excitations is necessary. In our calculations the self-consistency requirement is fulfilled in a pragmatic way and the experiment is taken as a reference to judge the quality of the two-particle excitations.

An interesting additional information which can be obtained from the quasi-particle bandstructure is the absolute value of the upper valence band edge relative to an ideal vacuum-level. In comparison to the Hartree-Fock value of -4 eV this edge is shifted to -0.15 eV in the RPA and to 0.25 eV in the TDSHF calculation. Both results point to a negative electron-affinity of diamond. The real vacuum level is however shifted somewhat relative to our zero of the energy-scale, because of the crystal surface. In order to make a statement about the electron-affinity, one has therefore to include the influence of the surface (model of a dipole-layer). Since typical estimates for surface-potentials of metals give values between 2 eV and 5 eV [82], one is, in spite of the problem of extending a similar assumption to non-metals, tempted to conclude to a negative electron-affinity (vacuum-level in bandgap) of diamond. In fact recent photoemission experimetns [83] clearly show that diamond represents the unusual case of a covalent crystal with negative electron-affinity.

To get a better understanding of the applicability and physical meaning of possible simplifications for the evaluation of the quasi-particle states, we discuss in the following two further calculations (EH, COSEX), we have performed. For this purpose a number of results (theoretical, experimental) for the bandgap (E_{gap}) and the valence band width (E_{val}) are compiled in table 11.

Important information can be extracted from the relative importance of the different types of elementary excitations (electron-hole pairs, plasmons), in the dynamically screened interaction. In order to clarify this question we have performed a calculation which excludes the plasmon-part of W' by restricting the energy integration in (III.19) to an upper value of 32 eV. The results of this investigation which includes only the electron-hole contribution to the dynamical screening are given in the column EH of table 11. We note that the value 7.4 eV for the direct bandgap is unchanged when compared with the full TDSHF. This implies that the quasi-particle properties near the upper edge of the valence bands and the lower edge of the conduction bands are exclusively determined by the coupling to electron-hole pair excitations. This of course is due to the fact that quasi-particles of a given energy couple most effectively to elementary excitations of comparable energy. Since the character of the correlation effects caused by the electron-hole excitations is predominantly of local, short-range nature, a local ansatz for the correlations in the surroundings of the bandgap seems a good approximation. This must not be the LDA in its ground state formulation, since it is rather insufficient for semiconductors and insulators (table 1). However, the energy-dependent extension of the LDA by Sham and Kohn [10] (see (III.9)) might be a step in the right direction, even if it additionally makes use of the homogeneous electron gas model for the functional dependence of the self-energy on density.

The valence band width is found in the EH calculation as 26.45 eV. Thus, also here the dominant part of the correlation corrections is contributed by the electron-hole excitations which reduce the HF result of 29.0 eV by 2.55 eV. The plasmon-like correlation corrections which are predominantly of non-local long-range nature, lead to a further reduction of the valence band width by 1.35 eV. The fact that LDA calculations give clearly too small values for the valence band width (see table 11), must probably again be attributed to the missing energy-dependence of the self-energy corrections. Too small bandgaps and band widths seem to be a general trend in LDA calculations [13,84].

In order to extract a quantitative picture of the role of dynamics in the self-energy we first discuss the results of a calculation which uses a static approximation, consisting simply of the assumption that the screened interaction W' keeps its value at E = 0 over the whole energy-range (0 - 40.5 eV) of the integration in (III.21a) and (III.21b). Such an approximation seems near at hand if one looks at figure 13 and neglects the region near the plasma-resonance which is anyhow weighted by a factor E^{-1}. It can be regarded as a simulation of the analytic COHSEX approximation [9]. As we can see from the column COHSEX in table 11, the bandgap (7.2 eV) of such a calculation is in good agreement with the full TDSHF and experiment. On the other hand, the valence band width (28.8 eV) is practically unrenormalized in comparison to Hartree-Fock. Altogether the results suggest that an energy-independent approximation for the screened interaction W' leads to meaningful results for the quasi-particle energies near the bandgap. As soon as one leaves the bandgap region, the energy-dependence of the screened interaction and therefore the self-energy becomes increasingly important. This corresponds to the increasing weight of the plasma-resonance in the integrals. We expect a similar behaviour for other covalent crystals with a broad valence band (e.g. silicon).

Finally, we want to discuss the results for the widths (life-times) of the quasi-particle states (especially the holes). For this reason, figure 15 shows the imaginary part of the complex quasi-particle energies $E_n(\vec{k})$ as a function of the real part over the energy-range of the valence bands. The corresponding life-times can be extracted from the formula $(2\mathrm{Im}(E_n(\vec{k})))^{-1}$. The mechanism which is responsible for the decay of the holes, is the Coulomb interaction between the particles of the system. The radiationless transition to a final state with more than one hole in the valence band can thus be termed as an intraband Auger process. Figure 15 shows that the decay rates are zero (within numerical error-boundaries) above a threshold-energy. An elementary treatment of the Auger process would, from energy-conservation arguments, give $\Gamma_{25'} - E_{gap} = -7,15$ eV for such a threshold. Since, however, the first maximum in the valence band density of states reaches half of its maximum value at -1.5 eV, the decay rate should start to become noticeably different from zero below -10.5 eV. In fact the TDSHF calculation agrees with the above considerations and moreover displays a second sharp increase at about -21 eV which probably comes from the onset of the decay into 3 holes. The above results are also in qualitative agreement with empirical broadening factors which are used to bring experimental XPS valence band spectra into agreement with

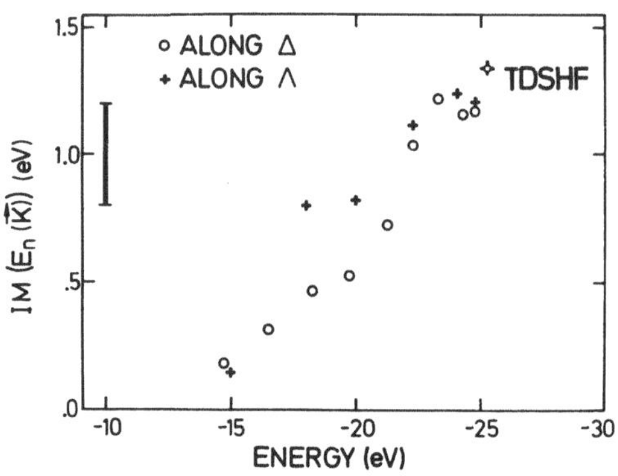

Figure 15. Imaginary part of the hole states in diamond in the TDSHF approximation for two symmetry directions of the Brillouin zone as a function of the corresponding real part. The vertical bar denotes the numerical uncertainty.

theoretical valence band densities of states [79] . One should note that the imaginary part is rather big for quasi-particle states deep in the valence band. This fact raises of course the question, whether a quasi-particle description is still appropriate.

IV. THE INFLUENCE OF MANY-BODY EFFECTS ON THE SCREENING OF IMPURITIES IN PERFECT CRYSTALS : EXAMPLES DIAMOND AND SILICON

The perfect crystal is an ideal case. Real crystals have often a great number of defects. Therefore extensive efforts have been devoted to the aim of getting a quantitative understanding of the corresponding changes in the electronic system [24-26].

The major part of these efforts is concerned with the development and application of efficient methods for the calculation of new, localized quasi-particle states, which appear together with the defects. In this context belong the effective-mass approximation [26] (shallow impurities) and the one-particle Green's function method [25,85,86] (deep impurities). A central quantity in these methods is the effective defect potential, which includes the screening effect of the crystal electrons.

In this chapter we shall investigate the induced potential and the induced charge density of a point-like impurity in covalent crystals (diamond, silicon). The investigations are based on the first-principle treatment of the dielectric function of chapter II and thus include local-field as well as excitonic effects (electron-hole attraction). In addition to the different theoretical approximations ($\overline{\text{RPA}}$, RPA, TDSHF), we are also going to investigate the dependence on the impurity position in the lattice (substitutional or interstitial).

Previous studies of the impurity screening in different covalent materials commonly used model treatments of the wave-vector dependence of the dielectric function, such as the Thomas-Fermi theory [87-89] or the Penn model [46]. There exists also work [91,92], where a pseudo-potential bandstructure is used for the calculation of the dielectric function, but here as well as in the model calculations local-field and many-body effects beyond RPA are neglected. Only recently a model-investigation of the screening of point-like impurities in silicon in the framework of the RPA has been published, which points to the importance of local-field effects [93].

A. Theoretical treatment of the induced potential and the induced charge density

The key quantities in the following investigation are the induced potential $V_{ind}(\vec{r},\vec{u})$ and the induced charge density $P_{ind}(\vec{r},\vec{u})$. u denotes here the position of the impurity relative to a lattice point. In linear-response theory both quantities are directly connected with the static (ω = 0) inverse dielectric function ε^{-1}

$$V_{ind}(\vec{r},\vec{u}) = \int d^3r' \frac{P_{ind}(\vec{r},\vec{u})}{|\vec{r}-\vec{r}'|} \tag{IV.1}$$
$$= \int d^3r' [\varepsilon^{-1}(\vec{r}+\vec{u},\vec{r}'+\vec{u};\omega=0) - \delta(\vec{r}-\vec{r}')] V_{imp}(\vec{r}')$$

In (IV.1) $V_{imp}(\vec{r})$ denotes the unscreened potential of the defect. The inverse dielectric function will be taken from chapter II. This implies a restriction of self-consistent response to the contribution of the valence-electrons which is, as well as the application of a linear theory, certainly justified for shallow impurities. The contribution of the core-electrons will modify the screening mainly near the nucleus. However, it is not our aim to obtain an absolute numerical accuracy in this special region. The aim of the investigation is rather to analyse the importance of local-field effects and excitonic effects in principle. This is, why we also make use of a very simple (point-like) form of the defect potential V_{imp}.

It is useful to employ the wave-vector representation of the inverse dielectric function

$$\varepsilon^{-1}(\vec{r},\vec{r}';\omega=0) = \tag{IV.2}$$
$$= V^{-1} \sum_{\vec{q}}^{BZ} \sum_{\vec{Q},\vec{Q}'} e^{i(\vec{q}+\vec{Q})\vec{r}} \, \varepsilon^{-1}(\vec{q}+\vec{Q},\vec{q}+\vec{Q}';\omega=0) \, e^{-i(\vec{q}+\vec{Q}')\vec{r}'}$$

With the help of the local representation (see II.A) of ε^{-1} one obtains

$$V_{ind}(\vec{r},\vec{u}) = N^{-1} \sum_{\vec{q}}^{BZ} \sum_{\delta\delta'} \bar{F}_{\delta}^{(o)}(\vec{q},\vec{r},\vec{u}) \, G_{\delta\delta'}(\vec{q};\omega=0) \, F_{\delta'}(\vec{q},\vec{u}) \tag{IV.3}$$

for the induced impurity potential. In (IV.3) G is the two-particle Green's function (see (II.9) and (II.14)). The quantities $\bar{F}_{\delta}^{(o)}$ and F are defined by the relations

$$\bar{F}_{\delta}^{(o)}(\vec{q},\vec{r},\vec{u}) = 8\pi\Omega_o^{-1} \sum_{\vec{Q}} e^{i[(\vec{q}+\vec{Q})\vec{r} + \vec{Q}\vec{u}]} \frac{A_{\delta}(\vec{q}+\vec{Q})}{|\vec{q}+\vec{Q}|^2} \tag{IV.4}$$

and

$$F_{\delta}(\vec{q},\vec{u}) = \Omega_o^{-1} \sum_{\vec{Q}} e^{-i\vec{Q}\vec{u}} A_{\delta}^{*}(\vec{q}+\vec{Q}) \, V_{imp}(\vec{q}+\vec{Q}) \tag{IV.5}$$

where $V_{imp}(\vec{q}+\vec{Q})$ is the Fourier representation of the unscreened impurity potential.

The local representation of the induced charge density $P_{ind}(\vec{r},\vec{u})$ can also be cast into the form of eq.(IV.3). The only difference in comparison to the induced potential is, that the term $\overline{F}^{(o)}$ has to be replaced by

$$\overline{F}_{\delta}^{(1)}(\vec{q},\vec{r},\vec{u}) = 2\Omega_o^{-1} \sum_{\vec{Q}} e^{i[(\vec{q}+\vec{Q})\vec{r} + \vec{Q}\vec{u}]} A_{\delta}(\vec{q}+\vec{Q}) \qquad \text{(IV.6)}$$

In eq.(IV.3) the central importance of the elementary excitations for the screening is especially stressed, because the induced potential as well as the induced charge density are written as a sum over form-factors $\overline{F}$ and F multiplied by matrix-elements of the two-particle Green's function. The form factor alone contains the information about the position of the impurity relative to the crystal lattice $(\vec{u})$ and the position $\vec{r}$ (measured from the impurity) at which the induced charge density and the induced potential are to be calculated.

From (IV.3) we obtain the TDSHF approximation, if we consider in the two-particle Green's function (see (II.9)) the electron-hole exchange Z as well as the electron-hole attraction Y. If we neglect the electron-hole attraction, we obtain the RPA.

The $\overline{\text{RPA}}$ approximation cannot be brought into the form of eq.(IV.3). In this case we find for the induced potential the representation

$$V_{ind}(\vec{r}) = N^{-1} \sum_{\vec{q}}^{BZ} \sum_{\vec{Q}} H^{(o)}(\vec{q}+\vec{Q},\vec{r}) \; \chi(\vec{q}+\vec{Q}) \qquad \text{(IV.7)}$$

Between the density-density correlation function χ and the irreducible polarizability $\tilde{\chi}$ one has in the $\overline{\text{RPA}}$ the relation

$$\chi(\vec{q}+\vec{Q}) = \frac{\tilde{\chi}(\vec{q}+\vec{Q})}{1 - v(\vec{q}+\vec{Q}) \; \tilde{\chi}(\vec{q}+\vec{Q})} \qquad \text{(IV.8)}$$

where $v(\vec{q}+\vec{Q}) = 8\Omega_o^{-1}|\vec{q}+\vec{Q}|^{-2}$ is the Fourier representation of the Coulomb potential $V(r) = 2|r|^{-1}$. The irreducible polarizability is in the local representation defined through the relation

$$\tilde{\chi}(\vec{q}+\vec{Q}) = \sum_{\delta\delta'} A_{\delta}(\vec{q}+\vec{Q}) \; G^{o}_{\delta\delta'}(\vec{q};\omega=0) \; A^{*}_{\delta'}(\vec{q}+\vec{Q}) \qquad \text{(IV.9)}$$

The form factors $H^{(o)}$, appearing in (IV.7), are defined by

$$H^{(o)}(\vec{q}+\vec{Q},\vec{r}) = 8\pi\Omega_o^{-1}|\vec{q}+\vec{Q}|^{-2}\, e^{i(\vec{q}+\vec{Q})\vec{r}}\, V_{imp}(\vec{q}+\vec{Q}) \qquad (IV.10)$$

The induced charge density $P_{ind}(\vec{r})$ can be calculated in the $\overline{RPA}$ with an equation, which is formally equal to (IV.7), but where the form-factors $H^{(o)}$ have to be replaced by

$$H^{(1)}(\vec{q}+\vec{Q},\vec{r}) = 2\Omega_o^{-1}\, e^{i(\vec{q}+\vec{Q})\vec{r}}\, V_{imp}(\vec{q}+\vec{Q}) \qquad (IV.11)$$

At this point we want to stress the fact, that the induced potential and the induced charge density in the $\overline{RPA}$ approximation are not dependent on the impurity position relative to the crystal lattice. This fundamental deficiency of the $\overline{RPA}$ is a direct consequence of the neglect of local-field effects.

Since we are mainly interested in the screening properties of the electronic system, a simple model for the unscreened impurity, namely a positive point-charge $V_{imp}(\vec{r}) = \sqrt{2}/|\vec{r}|$, is chosen.

For the evaluation of the sums over the Brillouin zone, which appear in (IV.3) and (IV.7), we use again the special directions method of Banil [53] . As in chapter III the minimum set of special directions (see (III.22)) will be used.

B. Substitutional impurities

The formulas which were introduced in IV.A will now be applied to the calculation of the static screening of substitutional impurities ($\vec{u}$ = 0 in (IV.3) - (IV.6)) in the covalent crystal diamond and silicon. For the unscreened impurity we use the model of a positive point charge. A comparison between the $\overline{RPA}$, RPA and TDSHF calculations will clarify the importance of the corresponding improvements in the theoretical description of the inverse dielectric function for the impurity screening. The determination of the induced charge density in planes through the impurity position gives a 2-d survey over the charge rearrangement near to the impurity.

B1. Diamond

The two-particle Green's function G and the density form factors A are taken from the results of II.B. Along the special line (see (III.22)) 11 equi-distant points are chosen for the integrations over the Brillouin zone.

Figure 16 shows the induced charge density of a positive substitutional point-charge in two specially chosen directions, which are the diagonal ((1,1,1)-direction) and the edge ((1,0,0)-direction) of the cubic unit cell. Arrows on the horizontal axis characterize the positions of neighbouring atoms. As can be seen in figure 16, the results for the different theoretical approximations deviate strongly from each other, in particular in the region of the tetrahedral bonds and near the atomic positions. The drastic corrections, which are found in the RPA (as compared to the $\overline{\text{RPA}}$) as well as in the TDSHF (as compared to the RPA), make clear, that both local-field effects and excitonic

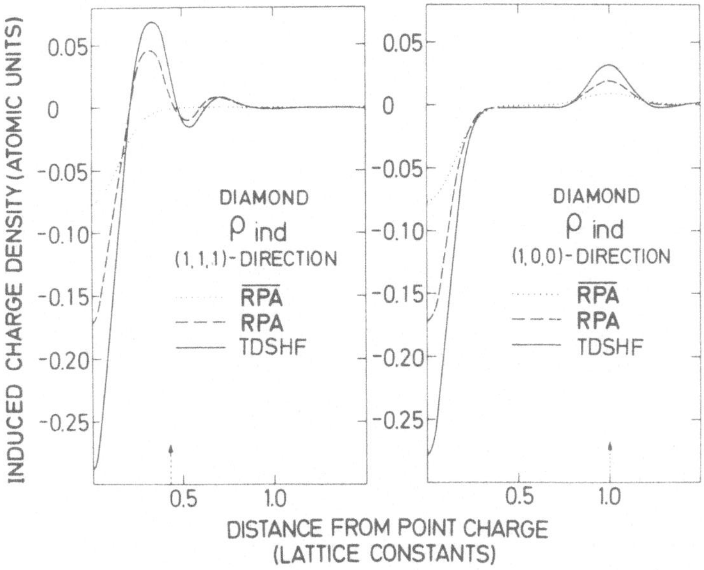

Figure 16. Induced charge density of a positive substitutional point-defect in diamond. Results of various theoretical approximations ($\overline{\text{RPA}}$, RPA, TDSHF) are shown along the (1,1,1)- and (1,0,0)-directions. Arrows on the horizontal axis characterize the positions of neighbouring atoms

effects, which are of comparable magnitude, must be considered in a realistic description of the impurity screening in diamond. Especially interesting is the form of the induced charge-density along the tetrahedral bond to the next-nearest neighbour ((1,1,1)-direction). Here an induced dipole appears which is not present in the $\overline{RPA}$. The inclusion of local-field effects and excitonic effects in the TDSHF is necessary to produce this polarization dipole (see figure 16). One notices that the induced charge density is appreciably different from zero only in the regions of the atomic positions and the tetrahedral bonds, i.e. the high-density regions were also the many-body corrections are dominant.

Figure 17 displays the induced potential along the same two directions as the induced charge density. We compare again the three theoretical approximations. The corrections in the RPA and TDSHF are less pronounced than before, as can be expected from the weighted integration over the induced charge density (see (IV.1)). Nevertheless we see, that local-field and electron-hole effects lead to indispensable corrections for a quantitative calculation of the induced potential. These corrections are not isotropic and increase as one approaches the impurity.

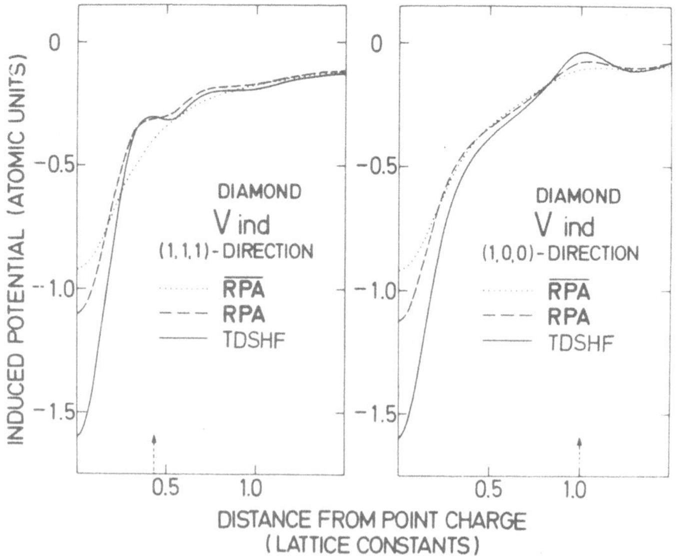

Figure 17. Induced potential of a positive substitutional point-defect in diamond. Otherwise as figure 16.

A two-dimensional plot of the induced charge-density in the surroundings of the impurity is shown in figure 18 for the (1,1,0)-plane through the impurity position, which contains the tetrahedral bonds to the neighbouring atoms. The semi-circle at the left edge gives the impurity position. Crosses denote the locations of neighbouring atoms in the plane. Areas of negative or positive induced charge-density are encircled by continuous or dashed lines, respectively. The left edge is a symmetry line of the (1,0,0)-plane. Figure 18 shows again very clearly the strong anisotropy of the induced charge-density. In the regions, which are far from atomic positions or tetrahedral bonds, the induced charge-density is zero within numerical errors. Along the binding-orbitals, between the atoms (bond charge positions), one finds polarization dipoles, which are ordered in such a way that the impurity is screened as effectively as possible. Furthermore, figure 18 shows, that the polarizability of the electronic charge density of the bonds perpendicular to the bond direction is small. This result is in agreement with a semi-empirical

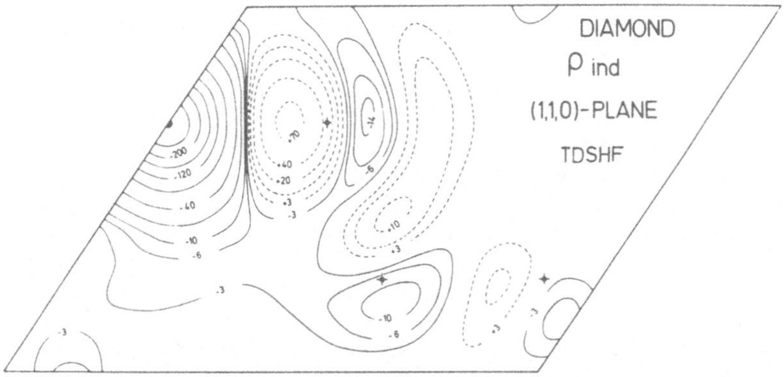

Figure 18. Induced charge-density of a positive substitutional point-defect in diamond for the (1,1,0)-plane. The black semi-circle on the left edge denotes the impurity position. Crosses give the positions of neighbouring atoms in this plane. Continuous or dashed lines encircle regions of negative or positive charge-density, respectively. Numbers multiplied by 10^{-3} give the charge-density in atomic units

analysis of the Raman and infra-red spectra of diamond [93] in the framework of the bond charge model [94], where a negligible polarizability of the bond charge perpendicular to the bond direction is found. The results of figure 18 show furthermore, that the movement of the electrons in the process of screening is practically restricted to the quasi one-dimensional channels of the bond directions. This effective reduction of the translational degrees of freedom for the electrons may also be regarded as an explanation for the unexpected importance of the many-particles effects in the impurity-screening in diamond.

B2. Silicon

The results of II.C are used for the two-particle Green's function G and the density form factors A. For the Brillouin zone integrations 15 equidistant points are chosen along the special line.

Figure 19 displays the induced charge density of a substitutional, positive point-charge along the (1,1,1)- and the (1,0,0)-direction. Between the various theoretical approximations ($\overline{\text{RPA}}$, RPA, TDSHF) we find also in silicon important differences. In quantitative comparison to diamond, however, local-field and electron-hole effects are less strong. In particular the difference between the RPA and TDSHF is clearly smaller in silicon. Remarkable on the other hand is a qualitative difference with respect to diamond. In silicon the minimum of the induced charge density is not at the impurity position, but (in the RPA and TDSHF) about 0.15 lattice constants away. This points to a difference in the polarizability of the charges, contributing to the tetrahedral bonds. To elaborate further on this point, we show in figure 20 the induced charge density in the (1,1,0)-plane through the impurity position. If we make a comparison to the equivalent figure 18 for diamond, the most prominant differences are in the region of the bonding orbital to the nearest neighbour. The impurity is not capable of polarizing the bond charges strongly enough in the bond directions, so as to create a maximum of induced negative charge density at the impurity position. As is clear from figure 20, it is in favor of a greater symmetry in the screening of the impurity advantageous, if the bond charges are also strongly polarized perpendicular to the bond directions. As a result an induced quadrupole in the region of the bond to the first neighbour develops and the polarizabilities of the bond charges parallel and perpendicular to the bond direction are of comparable magnitude in silicon. This result may again be considered as an a-priori confirmation of a semi-empirical analysis of the Raman- and infrared spectra of silicon with the bond charge model [93].

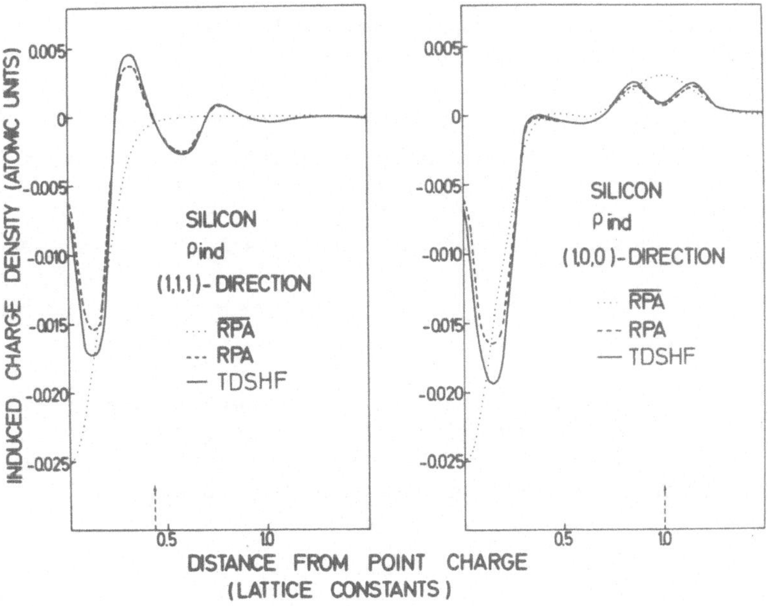

Figure 19. Induced charge-density of a positive subsititutional point-charge in silicon. Otherwise as figure 16.

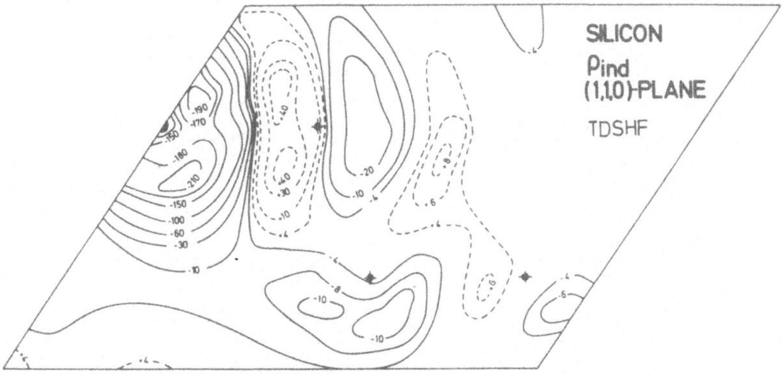

Figure 20. Induced charge-density of a positive substitutional point-defect in silicon for the (1,1,0)-plane. Numbers multiplied by 10^{-4} give the charge-density in atomic units. Otherwise as figure 18.

C. Interstitial impurities

We shall now investigate the static screening of a positive point charge at the interstitial position $\vec{u} = -\frac{a_o}{4}(1,1,1)$ in diamond and silicon. This position is located (as looked upon from a lattice point) in the opposite direction of a tetrahedral bond at next nearest neighbour distance. The calculation of the induced potential and the induced charge density is otherwise analogous to the previous chapter.

C1. Diamond

Figure 21 gives the induced charge density away from the impurity, along the diagonal (1,1,1)-direction and parallel to an edge ((1,0,0)-direction) of the cubic unit cell. The results for the $\overline{\text{RPA}}$, RPA and TDSHF approximations show distinct

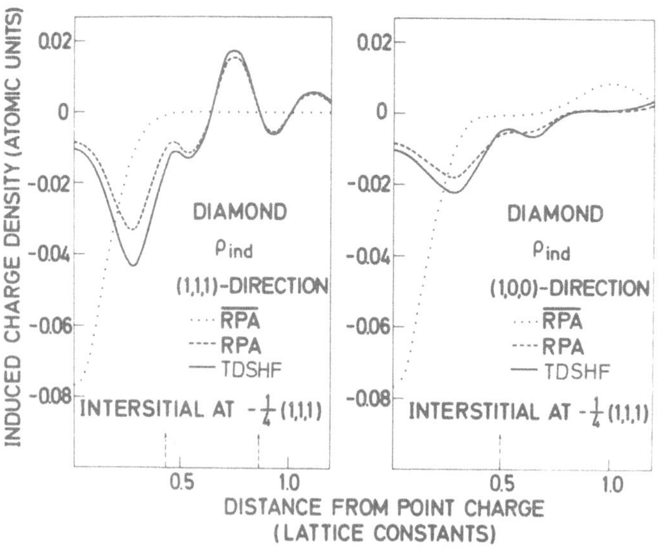

Figure 21. Induced charge-density of a positive interstitial point-charge at $-\frac{a_o}{4}(1,1,1)$ in diamond. Otherwise as figure 16.

differences. Between the $\overline{RPA}$ on one hand and the RPA, TDSHF on the other hand there are qualitative differences in the sense that completely new structures appear. Going from the RPA to the TDSHF leads to more quantitative corrections. Especially striking are the enormous differences induced by the many-body effects in comparison to the induced charge density of a substitutional impurity. While we have still identical results in the $\overline{RPA}$, which is not dependent on the position $\vec{u}$, the inclusion of local-field and excitonic effects leads to corrections in different directions near the impurity. For a substitutional impurity one obtains a large increase, whereas for an interstitial impurity a large decrease results in the induced charge-density. Moreover, the negative charge density of an interstitial impurity has its maximum not at the impurity position, but in both directions at a distance of about 0.275 lattice constants and therefore nearer to neighbouring atoms and bonds of the host-lattice.

These results support the qualitative picture from IV.B1, i.e. the strong coupling of charge response to the quasi one-dimensional channels formed by the ions and the bonds. For the further elucidation of this point, we show in figure 22 the induced charge density in the (1,1,0)-plane through the impurity position. We see that the polarization of the charge, concentrated at the

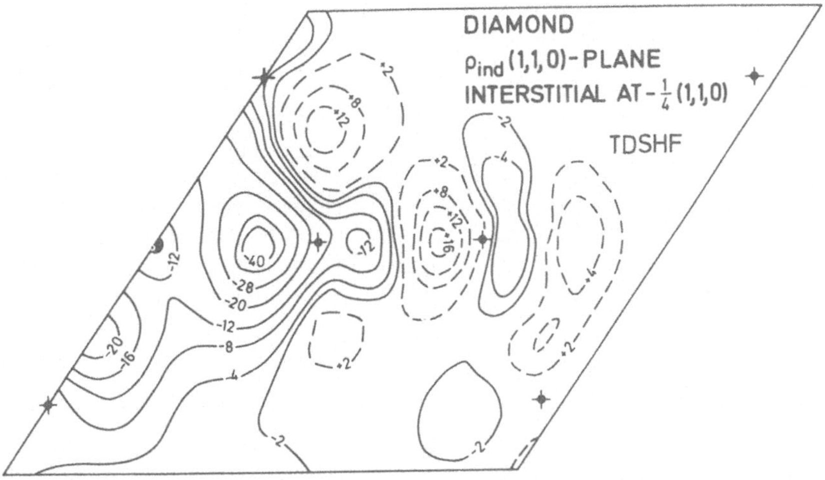

Figure 22. Induced charge-density of a positive, interstitial point-charge at $-\frac{a_o}{4}(1,1,1)$ in diamond for the (1,1,0)-plane. Otherwise as figure 18.

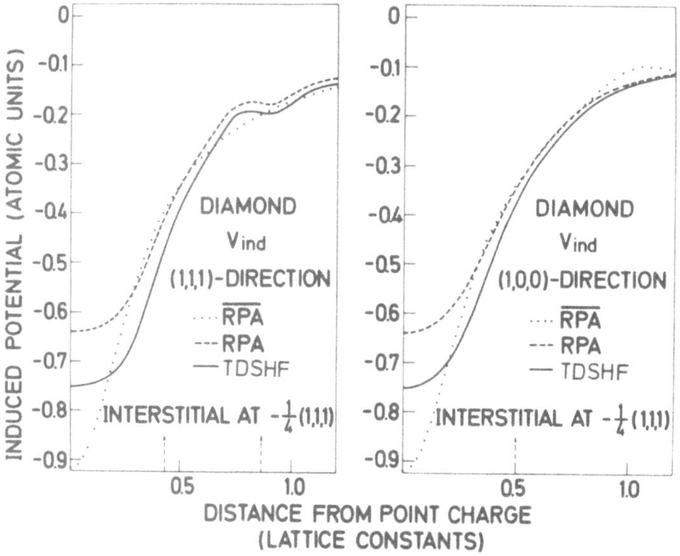

Figure 23. Induced potential of a positive, interstitial point-charge at $-\frac{a_o}{4}$ (1,1,1) in diamond. Otherwise as figure 16.

ions and the bonds, is rather weak. Absolute numbers are much smaller than for a substitutional impurity, and the maxima of the induced negative charge density are near to the neighbouring atoms and bonds.

From the smaller induced charge density a smaller induced potential results. As can be seen from the TDSHF result of figure 23, the induced potential of an interstitial point-charge is reduced by a factor 2 in comparison to the substitutional case. This implies that interstitial impurities are not so effectively screened and are therefore expected to have higher binding energies.

C2. Silicon

Figure 24 gives the induced charge density along the (1,1,1)- and the (1,0,0)-direction. We again notice, that the differences between the RPA and the TDSHF are rather small in Si, while in comparison to the $\overline{\text{RPA}}$ large qualitative and quantitative changes appear. In contrast to diamond the quantitative differences between the induced charge densities of substitutional and interstitial impurities are much smaller. Moreover, the maximum of the

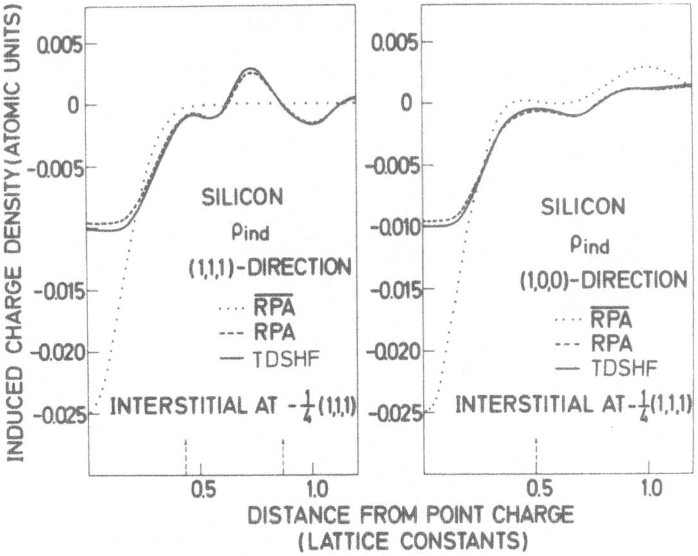

Figure 24. Induced charge-density of a positive, interstitial point-charge at $-\frac{a_o}{4}(1,1,1)$ in silicon. Otherwise as in figure 16.

induced negative charge density is practically at the impurity position and not, as in diamond, a distance of 0.275 lattice constants displaced. These results point to an easier polarizable electronic charge-density in silicon.

Figure 25 gives the induced charge density in the (1,1,0)-plane through the impurity position and reinforces our previous conclusions of the comparatively large polarizability. Charge-density rearrangements over relatively large distances are necessary to have the maximum of the induced negative charge-density at the impurity location.

V. SUMMARY AND DISCUSSION

The main purpose of this article is to present a summary of our recent work on quantitative, first-principles studies of the electronic many-body system in non-metals. The Green's function formalism in combination with a material-adapted local basis description of the single-particle electronic states

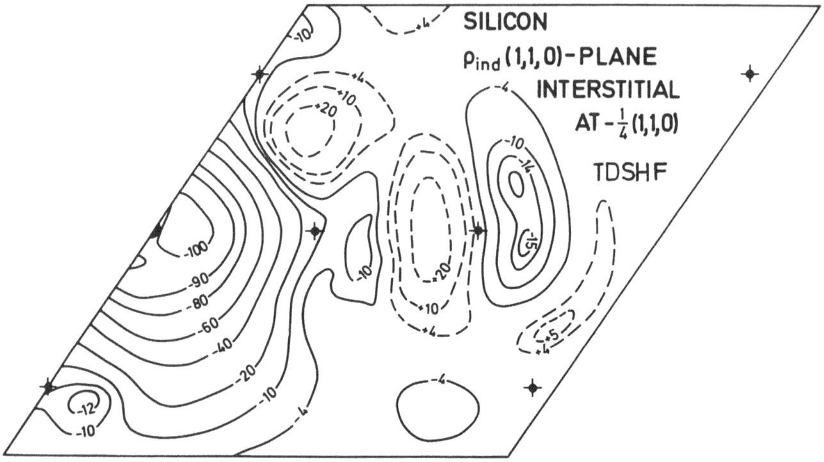

Figure 25. Induced charge-density of a positive, interstitial point-charge at $-\frac{a_o}{4}$ (1,1,1) in silicon for the (1,1,0)-plane. Otherwise as figure 18.

supplies the appropriate tool for this task. In particular, a self-consistent treatment of both the one- and two-particle Green's functions or in other words, the one- and two-particle excitations of the system becomes possible.

As a prototype material for the investigations serves the insulator diamond. In the framework of the Green's function scheme and within Feynman-Dyson perturbation theory, we have calculated the dynamical correlation effect and quasi-particle states of the ideal crystal and the linear response of the electronic system to external perturbations such as electromagnetic fields or defects in the crystal. This then results in a practically complete description of the many-electron properties i.e. single- and two-particle elementary excitations of the system. To the best of our knowledge, such a quantitative description has not been achieved before.

The extension of the concept, which worked successful for diamond, to the semi-conductor silicon leads to an enhancement of the numerical effort because of more spread-out electronic wave functions. A calculation of the quasi particle states is possible also in Si, but was not carried out yet mainly because of the missing self-consistent Hartree-Fock calculation, as a starting point. To circumvent the problem of the Hartree-Fock calculation in the case of silicon it seems also possible to calculate the quasi-particle states by starting from a self-consistent Hartree-calculation or from a LDA calculation.

The work along these lines is presently under way. The basis for the calculation of the quasi-particle states and also the impurity screening is a detailed theoretical description of the frequency- and wave-vector dependent dielectric function of the periodic crystal. This subject is treated in chapter II, where the results of Hanke and Sham [3,4] are summarized and extended in several points. New results are here a significant improvement in the frequency-dependence of the optical response in silicon through the inclusion of higher-order neighbours and orthogonalisation corrections in the local orbitals. For the wave-vector dependence of the dielectric function one gets satisfactory results, with a wave-vector independent approximation for the electron-hole propagator G, stripped of from its matrix elements which in fact give rise to the $\vec{q}$-dependence. This implies, that the wave-vector dependence is mainly carried by the density from factor A.

The Dyson equation for the quasi-particle states of diamond is reformulated in the appropriate local representation and solved with the screened HF (GW) approximation for the self-energy operator Σ. Since a self-consistent Hartree-Fock calculation serves as a basis, the crucial problem is the calculation of the non-local and energy-dependent correlation part Σ' of the self-energy. The matrix elements of Σ' are calculated in chapter III explicitly, while the Hartree-Fock part of the eigenvalue equation is obtained through a Slater-Koster representation of the HF band-structure. The non-hermitian energy matrices yield complex quasi-particle energies, with the imaginary part representing life-times. The relevance of an as accurately as possible determined two-particle Green's function for the single- particle spectrum is especially investigated. It turns out, that it is both sufficient and necessary to include local-field and electron-hole effects. Already the neglect of the electron-hole attraction (RPA) leads to significant deviations (band-widths, gaps) from experiment. Through the study of the relative contributions of the electron-hole excitations and the plasma resonance to the renormalization of the Hartree-Fock quasi-particle energies conclusions to the physical nature (long-range versus short range) of the dynamical correlation can be drawn. Thus, an argument for the possibility of local density approximations for the self-energy can be obtained near the band gap. Decay-rates for the quasi-particles due to intra-band Auger-processes have been calculated for the first time. They also are in agreement with experiment (photoemission).

We expect, that most of the basic conclusions, which we obtained for diamond, apply to a large extent also to other covalent crystals and in particular to the similar covalent semi-conductors silicon and germanium.

In particular we expect the transferability of our result for C, that the energy dependence of the correlation part of the self-energy plays a significantly more important role than the non-localicity, at least near the gap.

The influence of defects in our ideal lattice on the electronic states in diamond and silicon is investigated in chapter IV. We find that also in the screening problem the consideration of local-field and electron-hole effects is a necessary ingredient for a quantitative analysis. Especially important are the local-field effects, because they introduce a complete change in even the qualitative screening pattern. The influence of the lattice structure due to Bragg reflections on the periodic ions then appears. The electron-hole attraction, on the other hand, introduces no new structures in the induced charge-density. However, it is in C a many-body correction which is of similar magnitude to the zeroth-order diagonal screening result. A few important differences in the screening of impurities are found for diamond and silicon. In diamond, the polarizability of the electronic charge-density is tied very much to the quasi one-dimensional channels, given by the tetrahedral bonds. Bond-charges are practically only polarizable parallel to these channels. In silicon the polarizabilities of the bond-charge parallel and perpendicular to the bond-directions are of comparable magnitude. In particular these differences between diamond and silicon are displayed in the comparison of the screening patterns of substitutional and interstitial impurities.

An interesting continuation of the investigations of the impurity screening would be the calculation of the bound states. Rough estimates for the binding energy of impurity states due to the combined local-field and excitonic many-body effects result in 35 % changes in C and 25 % changes in Si.

Finally as a central result of the present work emerges the close relationship between the importance of many-particle effects and the degree of localization of the electronic properties. This is borne out both in the single-particle studies and in the impurity screening work.

REFERENCES

[1] D. Pines, in Elementary Excitations in Solids (Benjamin, New York, 1963).

[2] A.L. Fetter and J.D. Walecka, in Quantum Theory of Many-Particle Systems (McGraw-Hill, New York, 1971).

[3] W. Hanke and L.J. Sham, Phys. Rev. Lett. 33, 582 (1974); Phys. Rev. B12, 4501 (1975).

[4] W. Hanke and L.J. Sham, Phys. Rev. Lett. 43, 387 (1979); Phys. Rev. B21, 4656 (1980).

[5] W. Hanke, in Festkörperprobleme, edited by J. Treusch, (Vieweg, Wiesbaden, 1979), V. 19, p.43; Adv. Phys. 27, 287 (1978).

[6] M.L. Cohen and V. Heine, in Solid State Physics, edited by H. Ehrenreich, F. Seitz and D. Turnbull (Academic, New York (1970), V.24, p.1.

[7] W. Kohn and L.J. Sham, Phys. Rev. 140, A1133 (1965).

[8] L. Hedin and B.I. Lundqvist, J. Phys. C4, 2064 (1971); O. Gunnarson and B.I. Lundqvist, Phys. Rev. B13, 4274 (1976); O. Gunnarson, M. Jonson and B.I. Lundqvist, Phys. Rev. B20, 3136 (1979).

[9] L. Hedin and S. Lundqvist, in Solid State Physics, edited by H. Ehrenreich, F. Seitz and D. Turnbull (Academic, New York, 1969), v.23, p.1

[10] L.J. Sham and W. Kohn, Phys. Rev. 145, 561 (1966).

[11] B.I. Lundqvist, Phys. Konden. Materie 6, 193 (1967).

[12] M. Rasolt and S.H. Vosko, Phys. Rev. B10, 4195 (1974).

[13] A.B. Kunz, Phys. Rev. B6, 606 (1972).

[14] J. Hermanson, Phys. Rev. B6, 2427 (1972).

[15] A. Mauger and M. Lannoo, Phys. Rev. B15, 2324 (1977).

[16] A. Zunger and A.J. Freeman, Phys. Rev. B15, 5049 (1977).

[17] R.A. Roberts and W.C. Walker, Phys. Rev. 161, 730 (1967).

[18] F.R. McFeely, S.P. Kowalczyk, L. Ley, R.G. Cavell, R.A. Pollak and D.A. Shirley, Phys. Rev. B9, 5268 (1974).

[19] D.R. Hamann, Phys. Rev. Lett. 42, 662 (1979).

[20] A. Daunois and D.E. Aspnes, Phys. Rev. B18, 1824 (1978).

[21] L. Ley, S. Kowalczyk, R. Pollak, D.A. Shirley, Phys. Rev. Lett 29, 1088 (1972).

[22] J.C. Slater and G.F. Koster, Phys. Rev. 94, 1498 (1954).

[23] A. Manger and M. Lannoo, Phys. Rev. B15, 2324 (1977).

[24] F. Bassani, G. Iadonisi and B. Preziosi, Rep. Progr. Phys. 37, 1099 (1974).

[25] S.T. Pantelides, Rev. Mod. Phys. 50, 797 (1978).

[26] M. Altarelli and F. Bassani, in Handbook of Semiconductors, edited by W. Paul (North Holland, Amsterdam, in press) v.1.

[27] P. Nozières, in Theory of Interacting Fermi Systems (Benjamin, New York, 1964).

[28] H. Ehrenreich and M.H. Cohen, Phys. Rev. 115, 786 (1959).

[29] H. Fröhlich, in Theory of Dielectrics (Oxford Univ. Press, Oxford, 1949).

[30] V. Ambegaokar and W. Kohn, Phys. Rev. 117, 423 (1960).

[31] L.J. Sham and T.M. Rice, Phys. Rev. 144, 708 (1966).

[32] G.G. Hall, Philos. Mag. 43, 338 (1952).

[33] L. Pauling, J. Am. Chem. Soc. 53, 1367 (1931).

[34] J.L. Whitten, J. Chem. Phys. 44, 359 (1966).

[35] R.N. Euwema, D.L. Wilhite and G.T. Surratt, Phys. Rev. B7, 818 (1973).

[36] I. Shavitt, in Methods in Computational Physics, edited by B. Adler, S. Feshbach and M. Rotenberg (Academic, New York, 1963) p.1.

[37] S.L. Adler, Phys. Rev. 126, 413 (1962).

[38] N. Wiser, Phys. Rev. 129, 62 (1963).

[39] D.L. Johnson, Phys. Rev. B9, 4475 (1974); Phys. Rev. B12, 3428 (1975).

[40] L.J. Sham, in Dynamical Properties of Solids, edited by G.K. Horton and A.A. Maradudin (North-Holland, Amsterdam, 1974) vol.1, chap. 5.

[41] G.S. Painter, D.E. Ellis and A.R. Lubinsky, Phys. Rev. B4, 3610 (1971).

[42] J.C. Slater and G.F. Koster, Phys. Rev. 94, 1498 (1954).

[43] N.V. Cohan, D. Pugh and R.H. Tredgold, Proc. Phys. Soc. London 82, 65 (1963).

[44] R.C. Chaney, C.C. Lin, E.E. Lafon, Phys. Rev. B3, 459 (1971).

[45] P.W. Anderson, Phys. Rev. Lett. 21, 13 (1968).

[46] G. Srinivasan, Phys. Rev. 178, 1244 (1969).

[47] D.R. Penn, Phys. Rev. 128, 2093 (1962).

[48] M. Kastner, Phys. Rev. B7, 5237 (1973).

[49] O. Jepsen and O.K. Andersen, Solid State Commun. 9, 1763 (1971).

[50] G. Lehmann and M. Tant, Phys. Stat. Sol. (b) 54, 469 (1972).

[51] J.A. Van Vechten and R.M. Martin, Phys. Rev. Lett. 28, 446 (1972).

[52] R.F. Egerton and M.J. Whelan, Philos. Mag. 30, 739 (1974).

[53] A. Bansil, Solid State Commun. 16, 884 (1975).

[54] B. Ortenburger and W.E. Rudge, IBM research reports (unpublished).

[55] J.R. Chelikowsky and M.L. Cohen, Phys. Rev. B10, 5095 (1974).

[56] D.J. Chadi, Phys. Rev. B16, 3572 (1977).

[57] E.O. Kane and A.B. Kane, Phys. Rev. B17, 2691 (1978).

[58] P.M. Morse and H. Feshbach, in Methods of Theoretical Physics (McGraw-Hill, New York, 1953).

[59] S.G. Louie, J.R. Chelikowsky and M.L. Cohen, Phys. Rev. Lett. 34, 155 (1975).

[60] H.R. Philipp, J. Appl. Phys. 43, 2835 (1972).

[61] W.A. Harrison, in Electronic Structure and the Properties of Solids (Freeman, San Francisco, 1980).

[62] H.J. Hinz and H. Raether, Thin Solid Films, 58, 281 (1979).

[63] G. Strinati, H.J. Mattausch and W. Hanke, Phys. Rev. Lett. 45, 189 (1980); Phys. Rev. B (to be published).

[64] P. Hohenberg and W. Kohn, Phys. Rev. 136, B 846 (1964).

[65] V.M. Galitskii and A.B. Migdal, Sov. Phys. JEPT 7, 96 (1968).

[66] J.C. Ward, Phys. Rev. 78, 182 (1950); Y. Takahashi, Nuovo Cimento 6, 371 (1957).

[67] T.M. Rice, Ann. Phys. (N.Y.) 31, 100 (1965).

[68] A.W. Overhauser, Phys. Rev. B3, 1888 (1971).

[69] Y. Toyozawa, Progr. Theor. Phys. 12, 421 (1954).

[70] S.T. Pantelides, D.J. Mickish and A.B. Kunz, Phys. Rev. B10, 2602 (1974).

[71] D.J. Mickish, A.B. Kunz and T.C. Collins, Phys. Rev. B9, 4461 (1974);
A.B. Kunz, D.J. Mickish and T.C. Collins, Phys. Rev. Lett. 31, 756 (1973).

[72] W. Brinkman and B. Goodman, Phys. Rev. 149, 597 (1966).

[73] N.O. Lipari and W.B. Fowler, Phys. Rev. B2, 3354 (1970).

[74] N.E. Brener, Phys. Rev. B11, 929 (1975); Phys. Rev. B11, 1600 (1975).

[75] M. Bennett and J.C. Inkson, J. Phys. C10, 987 (1977); J.C. Inkson and M. Bennett, J. Phys. C11, 2017 (1978).

[76] A.J. Layser, Phys. Rev. 129, 897 (1963).

[77] A. Baldereschi, Phys. Rev. B7, 5212 (1973).

[78] F.J. Himpsel, J.F. van der Veen and D.E. Eastman, Phys. Rev. B22, 1967 (1980).

[79] L. Ley, M. Cardona and R.A. Pollak, in Photoemission in Solids, edited by L. Ley and M. Cardona (Springer, Berlin, 1979) p.11.

[80] M. Mehta and C.S. Fadley, Phys. Rev. Lett. 39, 1569 (1977).

[81] W.C. Walker and J. Osantowski, Phys. Rev. 134, A153 (1964); C.D. Clark, P.J. Dean and P.V. Harris, Proc. Roy. Soc. A277, 312 (1964); P.J. Dean, E.C. Lightowlers and D.R. Wight, Phys. Rev. 140, A352 (1965).

[82] J.A. Alonso and M.P. Iniguez, Solid State Commun. 33, 59 (1980).

[83] F.J. Himpsel, J.A. Knapp, J.A. Van Vechten and D.E. Eastman, Phys. Rev. B20, 624 (1979).

[84] S.T. Pantelides, Phys. Rev. B11, 2391 (1975).

[85] G.A. Baraff and M. Schlüter, Phys. Rev. Lett. 41, 892 (1978).

[86] J. Bernhole, N.O. Lipari and S.T. Pantelides, Phys. Rev. Lett. 41, 895 (1978).

[87] R. Resta, Pyys. Rev. B16, 2717 (1977).

[88] F. Cornolti and R. Resta, Phys. Rev. B17, 3239 (1978).

[89] M. Lannoo and G. Allan, Solid State Commun. 33, 293 (1980).

[90] H. Nara, J. Phys. Soc. Japan, 20, 778 (1965).

[91] J.P. Walter and M.L. Cohen, Phys. Rev. B2, 1821 (1970).

[92] R. Car and S. Selloni, Phys. Rev. Lett. 42, 1365 (1979).

[93] S. Go, H. Bilz and M. Cardona, Phys. Rev. Lett. 34, 580 (1975).

[94] J.C. Phillips, Phys. Rev. 166, 832 (1968).

COLLECTIVE PHENOMENA IN NON-UNIFORM SYSTEMS

S. Lundqvist

Chalmers University of Technology
Göteborg, Sweden

1. INTRODUCTION

Collective phenomena have been studied in great detail in the uniform electron liquid, particularly the plasma oscillations of the electron gas. However, the one-electron properties are also very strongly influenced by many-electron effects and there is e.g. a strong coupling between the single particle motion and the plasma modes of the system. The theory of the electron gas applies to simple free-electron like metals such as the alkali metals and Al. For non-uniform electron systems, however, much less is actually known and the theory is still in a stage of development along several different directions.

We have a variety of non-uniform systems having different characteristic properties. For example, the surface of a free-electron like metal introduces an important aspect of the non-homogeneity of the electron liquid, in that the density of the conduction electrons drops to zero within a few Angströms at the surface. In the so called jellium model, the system is still uniform in the two space dimensions parallel to the surface, but the translation symmetry is broken in the direction perpendicular to the surface. In this system we have a new mode of collective excitation, the surface plasmon, and at very long wave length a surface electromagnetic wave, which is usually called a surface polariton. These collective modes are of importance for a number of properties, such as inelastic scattering by charged particles and by electromagnetic radiation, the surface energy, the van der Waals interaction with a physisorbed molecule etc.

Considering e.g. a transition metal we have a rather non-uniform electron distribution on the atomic scale and typically also a strong coupling between plasmons and interband excitations. Such effects will modify the properties of bulk plasmons and surface plasmons compared with the free electron picture. The theoretical framework for collective excitation in periodic systems has been developed in papers by March and Tosi [1], but rather few applications to real metals have yet been made.

An atom has no translational symmetry at all and for a medium heavy atom the electron density changes by several orders of magnitude when we go from the atomic nucleus out to the classical atomic radius. The possibility of collective motion in an atom has been subject for theoretical studies during almost fifty years. However, not until the last two decades has one started to obtain some clearcut experimental evidence that collective effects do exist. Photoabsorption in the far ultra-violet to soft X-ray region has shown the existence of broad resonances in a number of atoms, and electron spectroscopy studies have shown that there are strong many-electron effects in what has up to recent years been believed to be rather ideal one-electron properties. Molecules offer much wider possibilities for exhibiting collective behaviour and experimental data as well as theoretical calculations now provide a steadily increasing body of knowledge about the collective electronic properties. As in atoms the use of synchrotron radiation to study the absorption of light as well as the emission of electrons has provided the experimental possibility for extensive study in the energy regions of interest.

We shall in these lectures be mainly concerned with a discussion of some of the more important theoretical approaches to discuss collective phenomena in non-uniform systems. The published literature is dominated by considerations based on some version of the hydrodynamical model. This method was developed by Bloch [2] in a classical paper from 1933, in which he developed a dynamical extension of the Thomas-Fermi theory, having particularly in mind the application to stopping power. Different modifications of the Bloch theory were later proposed. A formulation similar to Bloch's approach but based on the hydrodynamical extension of the density functional approach has been put forward by Ying et al [3].

In many problems concerning non-uniform systems the interplay between single-particle excitations and many-electron effects plays an important role and in this case a hydrodynamic approach is insufficient. Standard many-body techniques such as the random phase approximation, RPA, has with considerable success been applied by many authors to surface properties and to excitations of heavy atoms and of molecules. However, for some atomic problems and certainly for molecules one has to go beyond the RPA to include dynamical relaxation effects.

The relation between a hydrodynamic model and the many-body approach is not easy to see, and we shall therefore discuss a linearized form of the quantum equations in the RPA in which the classical restoring forces are kept separate from the genuine quantum effects, which depend on the actual excitation spectrum of electron-hole pairs. We shall also present a recently published hybrid form of many-body theory developed by Zangvill and Soven [4]. Their theory has the same form as the time-dependent Hartree theory (which is equivalent with RPA), but in which the local exchange-correlation potential from the density functional theory has been added to the Hartree potential.

It may seem from the last paragraph that one has to do with a wide spectrum of different schemes with rather little in common. However, they all have the same characteristic feature of being based on the idea of a self-consistent field. Just as in a ground state calculation such as Hartree-Fock, the field acting on a given electron has to be the dynamical self-consistent field from all the other electrons. All the formulations we shall discuss are essentially mean field theories, however, in many atomic and molecular problems dynamical relaxation and rearrangement effects are important and one has to go beyond the mean field approximation to include them in the theory.

It should be mentioned that these lectures do not attempt to give anything like a complete review of the progress in this field, which is already too extensive to be summarized in a few lectures. Only a few topics will be taken up for discussion selected among problems with which the lecturer has had some involvment. The emphasis will be on the conceptual basis and intuitive physics and the participants are referred to the original papers to look into the details of the theory.

2. HYDRODYNAMICAL OSCILLATIONS OF AN INHOMOGENEOUS ELECTRON GAS

The hydrodynamical models express the collective motion in terms of the deviations from the equilibrium electron density $\rho_o(\underline{x})$ assuming that all the relevant physical quantities, such as the potential energy, pressure etc., can be expressed in terms of the electron density $\rho(\underline{x})$. The so called density methods will play an important role in our discussion. The earliest density method to describe ground state properties was formulated by Thomas and Fermi and is usually referred to as the Thomas-Fermi-method. The model was improved by Dirac to include exchange and later work incorporated correlation effects, strong density gradients, relativistic correlation etc. Since the Thomas-Fermi model forms the underlying theoretical basis for the development by Bloch [2] of the hydrodynamical model for an inhomogeneous electron gas, we shall briefly remind about the basic ideas. (We shall use atomics units : $e = \hbar = m = 1$ in most of our discussions).

For an ideal uniform degenerate electron gas we have the following relation between the electron density ρ and the Fermi-momentum P_F : $\rho = P_F^3/3\pi^2$. Furthermore the chemical potential μ is simply given by $\mu = P_F^2/2$. If the gas is weakly non-uniform, it is natural to consider the first relation as a local relation :

$$\rho(\underline{x}) = \frac{P_F^3(\underline{x})}{3\pi^2} \tag{2.1}$$

The potential $U(\underline{x})$ due to the other electrons and external fields (e.g. the nuclei) is given by

$$U(\underline{x}) = \int d^3x' \frac{\rho(\underline{x}')}{|\underline{x}-\underline{x}'|} + U_{ext}(\underline{x}) \tag{2.2}$$

Now, the total energy of the fastest electrons E_f is given by the classical equation

$$E_f = \frac{P_F^2(\underline{x})}{2} + U(\underline{x}) \tag{2.3}$$

The maximum energy E_f must be constant in space, because otherwise the electrons would redistribute themselves to lower the energy, and E_f is of course equal to the chemical potential μ of the non-uniform system. Thus we may write

$$P_F^2(\underline{x}) = 2\ (\mu - U(\underline{x})) \tag{2.4}$$

Combining equations (2.1) and (2.4) we obtain the relation

$$\rho(\underline{x}) = \frac{2^{3/2}}{3\pi^2}\ (\mu - U(\underline{x}))^{3/2} \tag{2.5}$$

This is the famous Thomas-Fermi relation between density and potential. It is of course not exact except for a constant potential. However if the variation of $U(\underline{x})$ is small over a de Broglie wave-length for the electrons, it is a good approximation. Equation (2.5) can be used to obtain a self consistent theory by combining it with the Poisson equation :

$$\nabla^2 U(\underline{x}) = -\nabla^2 \frac{P_F^2(\underline{x})}{2} = 4\pi(\rho(\underline{x}) + \rho_{ext}(\underline{x})) \tag{2.6}$$

where $\rho_{ext}(\underline{x})$ is the density of external charges. From equations (2.5) and (2.6) one obtains a nonlinear differential equation for $P_F^2(\underline{x})$ or $U(\underline{x})$ (the Thomas-Fermi equation). This can be solved numerically using the appropriate boundary conditions and in this way it is possible to obtain a complete description of the ground state of the non-uniform system, i.e. the equilibrium condition is characterized by the ground state density $\rho_o(\underline{x})$ and the one electron potential energy $U_o(\underline{x})$.

The non-uniform gas is capable of oscillations about this steady state and we shall briefly review Blochs classical theory. He considered the non-uniform gas characterized by the following quantities : the density $\rho(\underline{x},t)$, the kinetic pressure $p = p(\rho)$ and the hydrodynamical velocity $\underline{v}(\underline{x},t)$, which for these problems can be represented by a velocity potential $\varphi(\underline{x},t)$, such that $\underline{v}(\underline{x},t) = -\nabla\varphi(\underline{x},t)$. The Hamiltonian function for the system can be written as

$$H = \int d^3x \, \frac{1}{2}|\nabla\varphi|^2 \rho(\underline{x},t) + \frac{3}{10}\left(\frac{3}{8\pi}\right)^{2/3} \int d^3x \, \rho^{5/3}(\underline{x},t) + \frac{1}{2}\int d^3x \, d^3x' \, \frac{\rho(\underline{x},t)\rho(\underline{x}',t)}{|\underline{x}-\underline{x}'|} + \int d^3x \, U_{ext}(\underline{x})\rho(\underline{x},t) \tag{2.7}$$

The first term is the hydrodynamic kinetic energy and the second is the internal kinetic energy of the Fermi gas. The latter can alternatively be expressed in terms of an energy density expressed in terms of the kinetic pressure $p(\underline{x})$ through the wellknown formula

$$E_{kin}(\underline{x}) = \frac{3}{2}\, p(\underline{x}) = \frac{3}{10}\left(\frac{3}{8\pi}\right)^{2/3} \rho^{2/3}(\underline{x}) \tag{2.8}$$

The third term is the Coulomb self-interaction and the last term represents the interaction of the electrons with nuclei and external potentials.

The equations of motion are obtained from the variational principle

$$\delta \int_{t_1}^{t_2} L \, dt = 0 \tag{2.9}$$

where the Lagrangian function L is given by

$$L = \int d^3x \frac{\partial\varphi}{\partial t} \rho - H \tag{2.10}$$

Bloch obtained in this way the Euler equations

$$\frac{\partial\rho}{\partial t} - \nabla(\rho\nabla\varphi) = 0$$

$$\frac{\partial\varphi}{\partial t} = \frac{1}{2}|\nabla\varphi|^2 + \frac{1}{2}\left(\frac{3}{8\pi}\right)^{2/3} \rho^{2/3} + \int d^3x' \frac{\rho(\underline{x}'t)}{|\underline{x}-\underline{x}'|} + U_{ext}(\underline{x},t) \tag{2.11}$$

The first is the continuity equation and the second is the Bernoulli equation (which comes from integrating Newton's second law).

The equations are nonlinear and Bloch replaced them by a linear set of equations by expanding for small amplitude motion as follows :

$$\rho(\underline{x},t) = \rho_o(\underline{x}) + \rho_1(\underline{x},t)$$

$$U(\underline{x},t) = U_{ext}(\underline{x},t) + \int d^3x' \frac{\rho(\underline{x}',t)}{|\underline{x}-\underline{x}'|} = U_o(\underline{x}) + U_1(\underline{x},t)$$

$$\varphi(\underline{x},t) = 0 + \varphi_1(\underline{x},t) \tag{2.12}$$

The linearized form of the equations of motion becomes :

$$\frac{\partial\rho_1}{\partial t} + \nabla\cdot(\rho_o\nabla\varphi_1) = 0$$

$$\frac{\partial\varphi_1}{\partial t} = \frac{1}{3}\left(\frac{3}{8\pi}\right)^{2/3} \frac{\rho_1}{\rho_o^{1/3}} + \int d^3x \frac{\rho_1(\underline{x},t)}{|\underline{x}-\underline{x}'|} + U_{ext}(\underline{x},t) \tag{2.13}$$

Here we imply that $U_{ext}(\underline{x},t)$ is a weak external field, small of first order.

We now consider the corresponding expansion of the Hamiltonian and obtain

$$H = H_o + H_1 + H_2 + \ldots \tag{2.14}$$

H_o is the energy of the static unperturbed system. For H_1 and H_2 we obtain

$$H_1 = \int d^3x \, \rho_o \, U_{ext} \tag{2.15}$$

and

$$H_2 = \int d^3x \left\{\frac{\rho_o}{2}|\nabla\varphi_1|^2 + \rho_1 U_{ext} + \frac{1}{6}\left(\frac{3}{8\pi}\right)^{2/3}\frac{\rho_1^2}{\rho_o^{1/3}}\right\} + $$

$$+ \frac{1}{2}\int d^3x\, d^3x' \frac{\rho_1(\underline{x},t)\rho_1(\underline{x}',t)}{|\underline{x}-\underline{x}'|} \quad (2.16)$$

We note two important properties :

1) H_o and H_1 do not contain the unknown functions ρ_1 and φ_1, so that only H_2 is important for the linearized problem. Therefore, the variational problem can be written as

$$\delta \int_{t_1}^{t_2} L_2 dt = 0$$

$$L_2 = \int d^3x \frac{\partial\varphi_1}{\partial t}\rho_1 - H_2 \quad (2.17)$$

where we can vary ρ_1 and φ_1 independent of each other.

2) Except for the term in equation (2.16) which contains the external field $U_{ext}(\underline{x},t)$, the Hamiltonian H_2 is a quadratic form in the two unkonwn functions ρ_1 and φ_1. This makes it possible to introduce a normal mode analysis using the eigen-modes for free oscillations.

In order to calculate the response to an external field U_{ext}, we first look at the linearized equations for free oscillations obtained by putting $U_{ext} = 0$.

$$\frac{\partial\rho_1}{\partial t} + \nabla.(\rho_o\nabla\varphi_1) = 0$$

$$\frac{\partial\varphi_1}{\partial t} = \frac{1}{3}\left(\frac{3}{8\pi}\right)^{2/3}\frac{\rho_1}{\rho_o^{1/3}} + \int d^3x' \frac{\rho_1(\underline{x}',t)}{|\underline{x}-\underline{x}'|} \quad (2.18)$$

i.e. two coupled first order equations for ρ_1 and φ_1. The Hamiltonian is given by

$$H_2 = \int d^3x \, \{\frac{\rho_o}{2} |\nabla\varphi_1|^2 + \frac{1}{6}(\frac{3}{8\pi})^{2/3} \frac{\rho_1^2}{\rho_o^{1/3}}\} +$$

$$+ \frac{1}{2} \int d^3x \, d^3x' \, \frac{\rho_1(\underline{x},t)\rho_1(\underline{x}',t)}{|\underline{x}-\underline{x}'|} = \tag{2.19}$$

$$= \int d^3x \, [\rho_1 \frac{\partial\varphi_1}{\partial t} - \varphi_1 \frac{\partial\rho_1}{\partial t}]$$

where the second form is obtained from the equation of motion after a partial integration. We now look at normal mode solutions of the form

$$\rho_1(\underline{r},t) = -\omega_n \rho_n(\underline{r},t) \sin \omega_n t$$
$$\varphi_1(\underline{r},t) = \varphi_n(\underline{r}) \cos \omega_n t \tag{2.20}$$

Inserting this ansatz in equation (2.19) we obtain

$$H_2 = \frac{1}{2} \omega_n^2 \int d^3x \, \rho_n \, \varphi_n \tag{2.21}$$

It was shown in the original paper by Bloch and it can be easily proven by using the linearized equations of motion (2.18) that the functions ρ_n and φ_n satisfy the orthogonality relation

$\int d^3x \, \rho_n \, \varphi_{n'} = 0$ if $n \neq n'$.

For $n = n'$ we can of course normalize the amplitudes ρ_n and φ_n so that $\int d^3x \, \rho_n \, \varphi_n = 1$.

The functions ρ_n and φ_n therefore form a <u>biorthogonal</u> system, having the property

$$\int d^3x \, \rho_n \, \varphi_{n'} = \delta_{nn'} \tag{2.22}$$

A general solution for free oscillations can now be expanded in terms of the normal mode functions ρ_n and φ_n as follows :

$$\rho_1(\underline{x},t) = \sum_n a_n \, \rho_n(\underline{x}) \sin \omega_n t$$
$$\varphi_1(\underline{x},t) = \sum_n b_n \, \varphi_n(\underline{x}) \cos \omega_n t \tag{2.23}$$

The total energy in a state of free oscillations can be written in the form

$$E_2 = \frac{1}{2} \sum_n \omega_n^2 c_n^2 \tag{2.24}$$

which is independent of time, the coefficients c_n being constants.

We next turn to the question how the system absorbs energy from an monochromatic external radiation field. We shall assume that the wavelength of the radiation is large in comparison with the size of the system. Then the potential energy per electron of a plane wave polarized in the z direction is given by the formula

$$U_{ext}(\underline{x},t) = z\, E_o \sin \omega t \tag{2.25}$$

where E_o is the constant field amplitude. This perturbing potential will only excite modes of the same symmetry,i.e. dipolar modes. It is clear that the perturbing force is irrotational which justifies the use of a velocity potential. The response of the system will be a superposition of normal modes having dipolar symmetry, with time-dependent coefficients.

$$\begin{aligned} \rho_1(\underline{x},t) &= \sum_n a_n(t)\, \rho_n(\underline{x}) \\ \varphi_1(\underline{x},t) &= \sum_n b_n(t)\, \varphi_n(\underline{x}) \end{aligned} \tag{2.26}$$

In addition it is convenient to expand the spatial part of U_{ext} in terms of the normal modes of the velocity potential.

$$U_{ext}(\underline{x},t) = \sin \omega t \sum_n U_n\, \varphi_n(\underline{x}) \tag{2.27}$$

The expansion coefficient U_n is then given by

$$U_n = \int d^3x\, U_{ext}(\underline{x})\, \rho_n(\underline{x}) = E_o \int d^3x\, z\, \rho_n(\underline{x}) \tag{2.28}$$

Substitution of equation (2.26) into the linearized equations (2.13) including the external field gives the following equations of motion for the time-dependent coefficients $a_n(t)$ and $b_n(t)$:

$$\begin{aligned} \dot{a}_n(t) &= -\omega_n^2\, b_n(t) \\ \ddot{a}_n(t) + \omega_n^2\, a_n(t) &= -\omega_n^2\, U_n \sin \omega t \end{aligned} \tag{2.29}$$

Thus, the amplitude of each normal mode responds as a driven harmonic oscillator with frequency ω. The strength of the driving force is given by U_n, which according to equation (2.28) is proportional to the dipole moment of charge in the mode n.

In order to calculate the absorption of energy from the external field, we assume that the external field is switched on at $t = 0$ and then switched off after a time t much larger then ω_n^{-1}. The corresponding solution to equation (2.29) is

$$a_n(t) = U_n \frac{\omega\omega_n \sin \omega_n t - \omega_n^2 \sin \omega t}{\omega_n^2 - \omega^2} \tag{2.30}$$

Afther the field has been switched off the state of the system is a fixed superposition of free oscillations. The amplitude of the mode n at time t and thereafter is obtained by comparing equation (2.24) with equation (2.19) using equation (2.26) and (2.29) with $U_n = 0$. We obtain the result

$$E_n = \omega_n^2 c_n^2 = a_n^2(t) + \dot{a}_n(t)/\omega_n^2 \tag{2.31}$$

which gives the energy of the system in the mode n. For $\omega \neq \omega_n$ this will be an oscillatory function of time according to equation (2.30). A real absorption will be obtained if we consider instead of a discrete frequency ω a band of frequencies around ω_n. Integrating over this frequency interval we obtain an absorption proportional to time, and after time t the absorption of energy in the mode n is given by

$$E_n(t) = \frac{\pi}{4} U_n^2 \omega_n^2 t \tag{2.31}$$

i.e. a constant rate of absorption. The total absorption rate of the system can be written as

$$\frac{dE}{dt} = \sum_n \frac{dE_n}{dt} = \frac{\pi}{4} \sum_n U_n^2 \omega_n^2 \delta(\omega - \omega_n) \tag{2.32}$$

The photoabsorption cross section equals the rate of energy absorption divided by the energy flux of the incident radiation $cE^2/8\pi$ and we obtain

$$\sigma(\omega) = \frac{2\pi^2 e^2}{mc} g(\omega) \tag{2.33}$$

where

$$g(\omega) = \sum_n \omega_n^2 \left| \int d^3x \, z \, \rho_n(\underline{x}) \right|^2 \delta(\omega - \omega_n)$$

is the differential oscillator strength distribution, which contains all the information about the absorption spectrum. To give some orders of magnitude we note that

$$\frac{2\pi^2 e^2}{mc^2} = 8.067\ 10^{-18}\ \text{Ry cm}^2 \tag{2.34}$$

The result of the hydrodynamical theory is similar but not identical in form to the standard result in quantum theory, which is

$$g(\omega) = \sum_n f_n\ \delta(\omega - \omega_n)$$

$$f_n = \frac{2m}{e^2}\,\omega_n\ |\langle n|z|o\rangle|^2 \tag{2.35}$$

is the oscillator strength for the transition $o \to n$.

3. EXTENSIONS AND SIMPLIFICATIONS OF THE HYDRODYNAMICAL MODEL

In the previous section we discussed the hydrodynamical theory of small oscillations around the Thomas-Fermi ground state following essentially the classical paper by Bloch [2]. His theory serves as a model for all the different hydrodynamical models now in use. The use of the Thomas-Fermi model has a particular advantage in that the results for atoms scale with the atomic number Z so that the results have a universal character. On the other hand the Thomas-Fermi ground state of an atom is not a very good approximation, and the equations describing small oscillations are also very approximate. Therefore one has taken many different steps to extend the Thomas-Fermi method to incorporate important physical effects that were not included in the original model. The most important modification was introduced by Dirac [5] by including the effect of exchange and the extended statistical model is usually referred to as the Thomas-Fermi-Dirac-model or TFD-model.

The TFD results are obtained by adding to the TF expression for the total energy, the total exchange energy of the system

$$E_x = -\frac{3}{4}\left(\frac{3}{\pi}\right)^{1/3} \int d^3x\ \rho^{4/3} \tag{3.1}$$

This adds a corresponding term to the hydrodynamical equations of motion and we obtain including exchange the following linearized equations for ρ_1 and φ_1 :

$$\frac{\partial \rho_1}{\partial t} + \nabla.(\rho_o \nabla \varphi_1) = 0$$

$$\frac{\partial \varphi_1}{\partial t} = \frac{1}{3}\left(\frac{3}{8\pi}\right)^{2/3} \rho_o^{-1/3} \rho_1 - \frac{1}{3}\left(\frac{3}{\pi}\right)^{1/3} \rho_o^{-2/3} \rho_1$$

$$+ \int d^3x' \frac{\rho_1(\underline{x}',t)}{|\underline{x} - \underline{x}'|} \qquad (3.2)$$

The inclusion of exchange improves the ground-state density, but the statistical theory with exchange is still quite crude in comparison e.g. with the Hartree or Hartree-Fock self-consistent treatment. Of more general interest is the fact that the statistical treatment of exchange introduced by Dirac is the first "local density" theory of exchange in the literature and gives the same exchange potential as that introduced about 25 years later by Kohn and Sham [6] in the density functional scheme.

Correlation effects have also been added to the statistical model, first by Gombàs [7] who used the interpolation formula proposed by Wigner [8] for the electron gas. Other modifications based on many-body theory have later been proposed. The effect of correlation is usually small.

We should also mention for completeness corrections to the kinetic energy in systems with strong density gradients and relativistic corrections for heavier elements.

In studying collective excitations of non-uniform systems one has usually not considered these extensions of the statistical model. In fact there are very few applications indeed in which even the full contents of the Bloch theory has been used in applications.

Most of the applications of the hydrodynamical model have been based on a simplified version of the statistical model. We mentioned, when we introduced the Thomas-Fermi-model, that we assumed that the density variation was slow enough that the change in the potential will be small over a de Broglie wavelength of the electrons. In the dynamical equations one usually gets one step further and neglects terms containing the gradient of ρ_o. This implies the assumption that we consider the medium locally as a uniform electron gas. This results in the following simple equation for the density oscillation amplitude (which we now write as ρ, letting ρ_o denote the unperturbed ground state density) :

$$\frac{\partial^2 \rho}{\partial t^2} + (\omega_{p\ell}^2(\underline{x}) - \beta^2(\underline{x})\nabla^2)\rho = 0 \tag{3.3}$$

where

$$\omega_{p\ell}^2(\underline{x}) = \frac{4\pi e^2 \rho_o(\underline{x})}{m} \tag{3.4}$$

$\omega_{p\ell}$ is the classical plasma frequency in a medium of density $\rho_o(\underline{x})$.

Looking for normal modes of the form $\rho(\underline{x},t) = \rho(\underline{x})\, e^{i\omega t}$ we obtain the equation

$$(\omega^2 - \omega_{p\ell}^2(\underline{x}) + \beta^2(\underline{x})\nabla^2)\rho(\underline{x}) = 0 \tag{3.5}$$

The coefficient $\beta^2(\underline{x})$ is obtained from the Bloch equations in section 2. We note that the Bloch theory for small oscillations around the Thomas-Fermi ground state is an adiabatic approach, assuming that the local distribution of electrons at any moment is equal to the Fermi distribution corresponding to the instantaneous local density. This is a low frequency approximation in which static screening applies that we have local equilibrium at any instant. The coefficient β is then determined by the static pressure in the Fermi gas and the value of β is given in terms of the Fermi velocity V_F, as

$$\beta^2(\underline{x}) = \frac{1}{3}\, V_F^2(\underline{x}) \tag{3.6}$$

However, when we discuss plasma-like oscillations of small systems we are often in the high-frequency regime; the regime in which the motions are so fast that the collisions cannot establish local equilibrium. A typical example is that of a plasma wave in a degenerate Fermi gas. In this case the Fermi gas will not reach internal equilibrium in one period of oscillation. Only the distribution along the direction of the wave will change with the local change in density, but the perpendicular directions will not be affected. The theory of plasma oscillations in a uniform electron gas in the random phase approximation (RPA) gives the result :

$$\beta^2(\underline{x}) = \frac{3}{5}\, V_F^2(\underline{x}) \tag{3.7}$$

The distinction between these two situations represented by equations (3.6) and (3.7) was discussed already by Bloch (1934).

We note that the use of the high frequency version combining equation (3.5) with the high-frequency form (3.7) for β^2

is equivalent to assume a (local) plasmon dispersion relation

$$\omega^2 = \omega_{p\ell}^2 + \frac{3}{5} V_F^2 q^2 \tag{3.8}$$

where q is the wave number. This is the well known RPA result. However, in many of the applications it tends to give the reader a picture of a modulated wave like motion in the medium, that could locally be interpreted as a plasma wave with a local wave number $q(\underline{x})$. This physical picture is correct for plasmon-like modes where $\omega > \omega_{p\ell}(\underline{x})$. However, we often encounter surface-like modes where ω is smaller than the average plasmon frequency. In this case there is no oscillations throughout the bulk of the system. Instead the amplitude of oscillation is negligibly small except for a small region around the surface. (In a non-local many-body theory this will not necessarily be true).

We started out in these lectures to consider small oscillations around the ground state obtained using the Thomas-Fermi theory. However, we have never made much use of the actual TF theory describing the actual structure of the ground state. In fact we could now easily decouple from the Thomas-Fermi assumptions and consider collective oscillations around the exact ground state density $\rho_o(\underline{x})$ of the system. For example, the equations (3.3) and (3.5) have no reference to the Thomas-Fermi model and we could take $\rho_o(\underline{x})$ from any self-consistent theory like the Hartree-Fock theory.

The density functional theory provides us with a technique to deal with small disturbances from the exact ground state density $\rho_o(\underline{x})$. The density functional method formulated by Hohenberg and Kohn [9] has shown that the ground state energy of an inhomogeneous electron system is a functional $E[\rho]$ of the density ρ. The hydrodynamic theory can now be obtained by considering slow adiabatic perturbations of the ground state. The theory developed by Ying [3] et al implies that we replace the Thomas-Fermi formula for the energy by the more general expression :

$$E = \int d^3x \, \frac{1}{2} \, |\nabla\varphi|^2 \, \rho(\underline{x},t) + \int d^3x \, d^3x' \, \frac{\rho(\underline{x},t)\rho(\underline{x}',t)}{|\underline{x}-\underline{x}'|} + \int d^3x \, U(\underline{x})\rho(\underline{x},t) + \int d^3x \, G[\rho(\underline{x},t)] \tag{3.9}$$

The functional $G[\rho]$ represents the kinetic, exchange and correlation energies.

Using equation (3.9) we can derive the hydrodynamical equations. The continuity equation is of course unchanged but the second of the Euler equations becomes :

$$\frac{\partial\varphi}{\partial t} = \frac{1}{2}\,|\nabla\varphi|^2 + \frac{\delta G}{\delta\rho} + U(\underline{x}) + \int d^3x'\,\frac{\rho(\underline{x}',t)}{|\underline{x}-\underline{x}'|} \tag{3.10}$$

The simplest choice for G is to use a local density description. This gives the kinetic energy contribution as in the Bloch theory, but contains also the contributions from exchange and correlation effects. The term $\frac{\delta G}{\delta\rho}$ represents the effects of the pressure. In Bloch's theory only the kinetic pressure in the Fermi gas is taken into account. Here we also have the contributions from exchange and correlations. Many recent applications have shown the importance of treating exchange and correlation in a satisfactory way. The extension of the Bloch theory, using the density functional approach represents a significant improvement. However, the theory by Ying et al is essentially a theory for adiabatic disturbances around the ground state and cannot obviously be extended to deal with high frequency excitations.

4. APPLICATIONS AND CRITIQUE OF THE HYDRODYNAMICAL MODEL

The hydrodynamical model has been applied to a variety of problems. There is a vast literature in the field, particularly if we go into a discussion of the optical properties studied by considering the hydrodynamic equations together with Maxwell's equations. In these lectures, however, we shall only consider the quasi-static case and not consider the full electromagnetic theory. We shall only discuss a few typical applications. We shall start by considering the model of a small electron gas sphere, which serves as a very crude model for an atom, but also serves as a model for a small metallic particle. This model was the first application of the Bloch theory by Jensen [10] , and has later been treated by many authors (a lucid and simple treatment has been given by Ruppin [11]). Another area of applications has to do with surface problems and we shall briefly discuss the plasma oscillations at a metal surface. The third major application is to the photoabsorption and possible collective resonances in atoms. The applications to surface problems and to atoms also show the weak points in this approach, which we are going to discuss before starting with the quantum formulation.

We consider first the oscillations of a uniformly charged small sphere with radius R. The equilibrium density ρ_0 is assumed to be compensated by a fixed positive background and we assume that ρ_0 is constant up to the surface of the sphere and then changes discontinuously to $\rho_o = 0$, i.e. we neglect the shape of the surface profile.

We first review briefly the classical theory of the oscillations of a conducting sphere of radius R. Assuming complete screening inside the sphere we have that the electrostatic potential $V(\underline{x})$ satisfies the Laplace equation both outside and inside the sphere thus $\nabla^2 V = 0$. For dipolar oscillations we have that

$$V(\underline{x},t) = V(v,t) \cos\theta \tag{4.1}$$

where θ is the polar angle. Separating the time-dependence we obtain

$$V(r,t) = a(t)r \qquad r < R$$

$$V(r,t) = a(t)\frac{R^3}{r^2} \tag{4.2}$$

The discontinuity of the normal derivative of V(r,t) across the surface determines the surface charge and we obtain

$$4\pi\sigma = 3\, a(t) \cos\theta \tag{4.3}$$

where σ is the surface charge density. Equations (4.1) and (4.2) imply that we have an electric field $\underline{E}$ inside the sphere given by

$$\underline{E} = a(t)\, \hat{e}_z \tag{4.4}$$

where $\hat{e}_z$ is the unit vector along the polar axis. This field will accelerate the electrons and the Newton equation is

$$m\dot{\underline{v}} = e\, a(t)\, \hat{e}_z \tag{4.5}$$

The flow of electrons will change the surface charge distribution and we have the continuity equation

$$\rho_o\, e\, \underline{v}\, \hat{e}_n = \dot{\sigma} = \frac{3}{4\pi}\dot{a} \cos\theta \tag{4.6}$$

$\hat{e}_n$ is the unit vector along the normal of the surface ($\hat{e}_z . \hat{e}_n = \cos\theta$). Combining equations(3.5) and (3.6) we obtain the equation

$$\ddot{a}(t) + \frac{4\pi e^2}{3m}\rho_o\, a(t) = 0 \tag{4.7}$$

Thus the sphere will oscillate with a frequency

$$\omega_{\text{sphere}} = \frac{\omega_{p\ell}}{\sqrt{3}} \tag{4.8}$$

in dipolar oscillations.

We note that the oscillating charge density is localized to the spherical surface and has the form

$$\rho(\underline{x},t) = \text{const}\ \delta(v - R) \cos\theta \cos\left(\frac{\omega_{p\ell}}{\sqrt{3}} t\right) \tag{4.9}$$

We now consider the hydrodynamical model and look at the solution of the equation

$$\rho(\underline{x},t) = \text{const}\ \delta(v - R) \cos\theta \cos\left(\frac{\omega_{p\ell}}{\sqrt{3}} t\right) \tag{4.10}$$

$$\frac{\partial^2 \rho}{\partial t^2} + \omega_{p\ell}^2 \rho - \beta^2 \nabla^2 \rho = 0$$

and we use the high frequency result $\beta^2 = \frac{3}{5} V_F^2$. For eigenmodes of the form

$$\rho(\underline{x},t) = \rho(\underline{x})\ e^{i\omega t}$$

we have to solve the equation

$$\nabla^2 \rho(\underline{x}) + K^2 \rho(\underline{x}) = 0$$

with

$$K^2 = \frac{\omega^2 - \omega_o^2}{\omega_o^2} \tag{4.11}$$

The electrostatic potential $V_i(\underline{x})$ inside the sphere satisfies the equation

$$(\nabla^2 + K^2)\nabla^2 V_i(\underline{x}) = 0 \tag{4.12}$$

The corresponding equation outside the sphere is the Laplace equation

$$\nabla^2 V_o(\underline{x}) = 0$$

The boundary conditions on the solution is that the radial component of the velocity V_r vanishes at the surface at all times. The second of the linearized Euler equation gives, using that

$$V_r = -\frac{\partial\varphi}{\partial r}$$

$$\frac{\partial}{\partial t}\frac{\partial\varphi}{\partial r}\bigg|_{v=R} = (\alpha^2\frac{\partial\rho}{\partial r} + \frac{\partial V_i}{\partial r}\bigg|_{r=R} = (-\frac{\alpha^2}{4\pi}\frac{\partial}{\partial r}(\nabla^2 V_i) + \frac{\partial V_i}{\partial r})\bigg|_{r=R} =$$

$$= (-\frac{\beta^2}{\omega_{p\ell}}\frac{\partial}{\partial r}(\nabla^2 V_i) + \frac{\partial V_i}{\partial r})\bigg|_{r=R} = 0 \tag{4.13}$$

The general solution inside the sphere can be written as an expansion of the form :

$$V_i(\underline{x}) = \sum_{\ell m}[A_{\ell m}(\frac{r}{R})^{\ell} + B_{\ell m}\frac{j_\ell(kr)}{j_\ell(kR)}]\, Y_{\ell m}(\theta,\phi) \tag{4.14}$$

where $j_\ell(kr)$ are the spherical Bessel functions
We can now use the boundary condition to eliminate the coefficients $A_{\ell m}$ and we obtain

$$V_i(\underline{x}) = \sum_{\ell m} B_{\ell m}[-\frac{kR}{\ell}\frac{j'_\ell(kR)}{j_\ell(kR)}(1 + \frac{\beta^2 k^2}{\omega_p^2})(\frac{r}{R})^{\ell} +$$

$$+ \frac{j_\ell(kr)}{j_\ell(kR)}]\, Y_{\ell m}(\theta,\phi) \tag{4.15}$$

Outside the sphere the solution has the form

$$V_o(\underline{x}) = \sum_{\ell m} C_{\ell m}(\frac{R}{r})^{\ell+1}\, Y_{\ell m}(\theta,\phi) \tag{4.16}$$

The boundary conditions are now that

$$V_i(R) = V_o(R)$$

$$\frac{\partial V_i}{\partial r}\bigg|_{r=R} = \frac{\partial V_o}{\partial r}\bigg|_{r=R} \tag{4.17}$$

The second condition is because there is now a continuously varying induced charge density near the surface. These boundary conditions give the dispersion formula

$$\frac{2\ell+1}{\ell}\frac{\omega^2}{\omega_p^2} - 1 = \frac{(\ell+1)}{kR}\frac{j_\ell(kR)}{j'_\ell(kR)} \tag{4.18}$$

The classical case of no dispersion is obtained as the limit $kR \gg 1$ in which case the right hand side vanishes and we obtain the corresponding eigenmodes

$$\omega = \omega_{p\ell} \sqrt{\frac{\ell}{2\ell+1}} \tag{4.19}$$

In the hydrodynamical theory we have two distinct types of modes depending on whether $\omega > \omega_{p\ell}$ (k real) or $\omega < \omega_{p\ell}$ (k purely imaginary).

We have two types of solutions for a spherical particle. For $\omega > \omega_{p\ell}$ the wave number k is real and we have modes in analogy with plasmons in bulk matter. The radial amplitudes are spherical Bessel functions, which show an oscillatory behaviour, with amplitudes extending throughout the volume of the sphere.

Because of the finite size, the spectrum of frequencies is discrete rather than continuous as in an infinite medium. For a very small sphere with a radius of the order 10-20 Å there is an appreciable separation between the modes for typical metallic densities, but for a radius of the order of 50-100 Å the separation between neighbouring modes is a small fraction of $\omega_{p\ell}$.

More interesting are the surface modes having $\omega < \omega_{p\ell}$. The interior of the sphere cannot sustain a plasma type oscillation at $\omega < \omega_{p\ell}$ and the density oscillation is now confined to a thin shell around the surface of the sphere. The wave number is now imaginary and the equation for the density oscillations becomes, making the replacement $k^2 = -\alpha^2$

$$(\nabla^2 - \alpha^2)\ \rho(\underline{x}) = 0$$

$$\alpha^2 = \frac{\omega_{p\ell}^2 - \omega^2}{\beta^2} \tag{4.20}$$

Instead of the spherical Bessel functions the radial solution obtained will now be the modified spherical Bessel functions of the first kind,

$$\rho_\ell(\alpha r) = \sqrt{\frac{\pi}{2\alpha r}}\ I_{\ell+1/2}(\alpha r) \tag{4.21}$$

The solution for dipolar modes, $\ell = 1$, will have the form

$$\rho_1(\alpha r) = \frac{\cosh \alpha r}{\alpha r} - \frac{\sinh \alpha r}{(\alpha r)^2} \tag{4.22}$$

The dispersion relation will be of the same form as equation (4.18) but with the proper radial functions :

$$\frac{2\ell+1}{\ell}\frac{\omega^2}{\omega_p^2} = \frac{(\ell+1)}{\alpha R}\frac{\rho_\ell(R)}{\rho'_\ell(R)} \tag{4.23}$$

The solutions give a sequence of surface modes, one for each value of ℓ ($\ell = 1,2,3,\ldots$). Considering a large sphere we have that αR is also large and for $\ell = 1$, we can easily make contact with the classical result discussed earlier. The right-hand side of equation (4.23) tends to zero and we obtain $\omega_s = \omega_{p\ell}/\sqrt{3}$.

We next look into the variation in amplitude of dipolar oscillation close to the surface, using equation (4.22) which gives for large αr

$$\rho_1(\alpha r) = \frac{\cosh \alpha r}{\alpha r} - \frac{\sinh \alpha r}{(\alpha r)^2} \rightarrow \frac{e^{\alpha r}}{2\alpha r} \tag{4.24}$$

Near the surface the variation can be written as

$$\rho_1(\alpha r) = \frac{e^{\alpha R}}{2\alpha R}\, e^{-\alpha(R-r)} \sim e^{-\alpha(R-r)} \tag{4.25}$$

Thus the amplitude decreases exponentially inwards with a decay constant α^{-1}. Now the length corresponding to α^{-1} is of the order of the Fermi wave length, so that the density oscillation is confined to a thin shell with a thickness around a few Ångström. In a jellium model with a surface profile extending over a few Ångström, the result would be qualitatively the same. In a quantum description where one considers the virtual and real excitations of electron-hole pairs the situation would be somewhat different. The density oscillation would then show an oscillatory behaviour extending quite deep into the sphere.

We next turn to a discussion of collective modes at a metal surface, the surface plasmons. There is a recent extensive treatment by Barton [12] including many aspects of the hydrodynamical problem that are often ignored or treated in a non-satisfactory manner. In these notes we shall only make a few brief remarks and make no attempt to summarize the work on surface plasmon modes described in the hydrodynamic approximation.

We first remind about the classical limit. Here we consider the case of a sharp interface between the metal and we describe the metal by a local dielectric function $\varepsilon(\omega) = 1 - \frac{\omega_{p\ell}^2}{\omega^2}$. Furthermore we consider the quasi-static limit where we can neglect the retardation effects. Classical theory gives the resonance condition $\varepsilon(\omega) + 1 = 0$ and therefore the classical surface plasmon frequency is $\omega_s = \omega_{p\ell}/\sqrt{2}$.

We now turn to the dispersion of surface plasmons. We note that irrespective of the surface profile, the long wave length limit gives the classical surface plasmon frequency $\omega_s = \omega_{p\ell}/\sqrt{2}$. Therefore the observation of the limiting frequency gives no information about the surface properties. We should remind that at very long wave length the surface plasma wave couples strongly to the electromagnetic field and we have a strongly dispersive mode called a surface polariton. We are here not considering this coupled mode with the electromagnetic field, and with long waves we mean here a wave length long compared with the lattice distance but considerably shorter than the wave length of light in medium. The dispersion of surface plasmons is given by $\omega_s(q)$, where q is the two-dimensional wave vector parallel to the surface. For small q we can write the dispersion relation as an expansion

$$\omega_s(q) = \omega_s[1 + (A_1 + iA_2)q + (B_1 + iB_2)q^2 + \ldots] \qquad (4.26)$$

There is an important result independent of any model, which was first found by Harris and Griffin [13] using the RPA, and soon thereafter derived in a general form by Garcia-Moliner and Flores [14]. The result is usually written in the form

$$\omega_s^2(q) = \omega_s^2(o)\,[1 + q\,\frac{\int dz z\rho(z)}{\int dz\rho(z)}] \qquad (4.27)$$

where $\rho(z)$ is the eigenfunction for the surface plasmon density oscillation for $q = 0$. The formula is valid for any model in the long-wave limit. It shows explicitly that the linear dispersion coefficient is given by the total dipole moment of the charge density fluctuation. This dipole moment is itself a macroscopic quantity and plays a major role for many surface properties as has recently been discussed by Feibelman [15] and by Apell [16].

Experimental data seem to indicate that the dispersion coefficient is <u>negative</u>. (We have chosen the coordinates so that the metal extends in the positive z-direction. A negative coefficient means that the density fluctuation occurs essentially outside the edge of the positive background). Most model calculations, including the hydrodynamical model, give a <u>positive</u> dispersion coefficient. Feibelman [17] has obtained a negative value by using as a basis the self-consistent potential and charge distribution by Lang and Kohn [18]. The comparison between different models [16] shows that the results are very sensitive to details in the models and that the simplified models do not give a satisfactory agreement. In particular the diffuseness of the density profile is of importance. The influence of the density profile on surface plasmons has been discussed by Equiluz et al [19]. They consider the hydrodynamical model based on the density functional scheme which we mentioned earlier. For a sufficiently diffuse surface they do not only obtain the usual surface plasmon but also higher multipole excitations were obtained. The existence or non-existence of such multipolar modes depends very sensitively on the choice of surface profile, as has recently been discussed by P. Ahlqvist and P. Apell [20].

We note that the basic hydrodynamical model is based on the assumption of a step density profile. Modifications exist where one uses a smooth profile, but the insight obtained from more detailed investigations shows the importance of using a fully self-consistent approach where the study of the density oscillation and the calculation of the ground state profile stand on the same theoretical basis. The coupling to electron-hole excitations seems important to account for e.g. the damping of surface plasmons and such effects are beyond the scope of the hydrodynamic approach.

The third and last application of the hydrodynamical approach is the original one considered by Bloch [2], namely, the excitation of hydrodynamical modes in atoms, which was briefly reviewed in section 2 of these lectures. These ideas were applied by Jensen [10] in the simplified atomic model assuming a constant density up to a sharp boundary, which was the first model we discussed in this section. The Jensen model gave the following formulas for the eigenfrequencies ω_n and the corresponding oscillator strengths f_n

$$\omega_n = k_n Z, \qquad f_n = q_n Z$$

where Z is the number of electrons in the atom and k_n and q_n are numerical coefficients of the order of unity (the frequency is measured in units of the Rydberg and the oscillator strengths fulfill the sumrule $\sum_n f_n = Z$). The eigenmodes correspond to surface modes and plasma modes of a charged liquid sphere. In this simplified theory the frequencies fall in the intermediate energy range between the optical and the X-ray spectra. The applications to the energy loss of a fast charged particle showed good agreement with experimental data from that period.

The interest in the Bloch model was renewed in the forties and fifties and attempts were made to apply more realistic models. To solve the Bloch equations for the Thomas-Fermi model had to wait until adequate computing facilities became available. The problem of calculating the photoabsorption cross section of atoms for frequencies in the far ultraviolet and soft X-ray regions turned out to be a complicated problem. This gave strong impetus to an analysis of the Bloch semiclassical model of hydrodynamical oscillations around the neutral Thomas-Fermi atom. A very complete analysis of this problem has been made by Wheeler and collaborator. The first results were published in a report (probably no longer available) by Wheeler and Fineman (1957). Further results are contained in a thesis by J.A. Ball, 1963, (unpublished). Fortunately the complete work by Wheeler and his collaborators on this problem has been reported in an extensive review [21].

As indicated in the second part of these lectures, Bloch assumed that the collective hydrodynamical modes would form a discrete spectrum, a feature which was of course retained in the applications by Jensen [10]. On the contrary the work by Wheeler et al actually solving the Bloch equations showed that one obtained a continuous absorption spectrum. An indication about this kind of result comes obviously if we consider the hydrodynamical equation for density oscillations if we neglect the explicit effect of the density gradient :

$$[\nabla^2 + K^2(\underline{x})]\,\rho(\underline{x}) = 0$$

with

$$K^2(\underline{x}) = \frac{\omega^2 - \omega^2_{p\ell}(\underline{x})}{\frac{3}{5}\, V_F^2(\underline{x})} \qquad (4.28)$$

Consider oscillations at a fixed ω. When going out into the low-density outer region of the atom the local plasma frequency $\omega_{p\ell}(\underline{x})$ as well as the Fermi velocity $V_F(\underline{x})$ both tend to zero. Consequently the local wave number $K(\underline{x}) \to \infty$ as the unperturbed density $\rho_o(\underline{x}) \to 0$.

Mathematically this means that we have a solution at any given frequency. Physically, the fact that the local wave length $\lambda(\underline{x}) \to 0$ when $\rho_o(\underline{x}) \to 0$ implies that we have oscillations with a wave length small compared to the average distance between electrons which implies a breakdown of the physical foundation of the theory. Nevertheless, being based on the Thomas-Fermi method, the theory gives a universal photoabsorption cross section, which scales with atomic number Z. However, for the reasons just mentioned the absorption cross section will not peak around some characteristic values of ω as suggested by the papers by Bloch and Jensen. Instead the photoabsorption cross section will show a monotonous decrease with frequency for any given atomic number Z. All hydrodynamical theories for small oscillations around this Thomas-Fermi ground state will give a "universal photoabsorption curve" of the form indicated in figure 1.

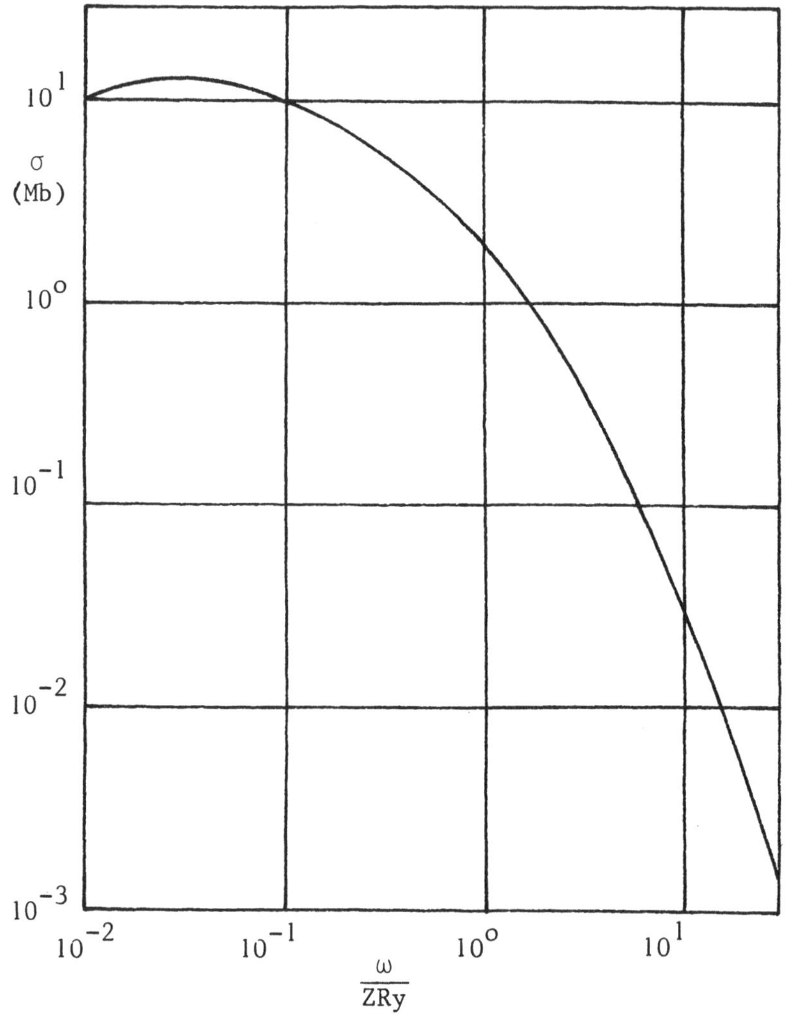

Figure 1. A typical "universal" photoabsorption curve using a hydrodynamical model.

It should be emphasized that the hydrodynamical model by Wheeler et al as well as many similar models do not give good quantitative agreement for any particular atomic species, but gives a reasonable average description average a wide range of energies and atomic numbers. This is indicated in figure 2 in which a typical theoretical curve is compared with some experimental data for a few atoms.

Much of the unphysical nature of the hydrodynamical theory for oscillations around the Thomas-Fermi ground state is because of the ever-decreasing wave-lengths of oscillations in the

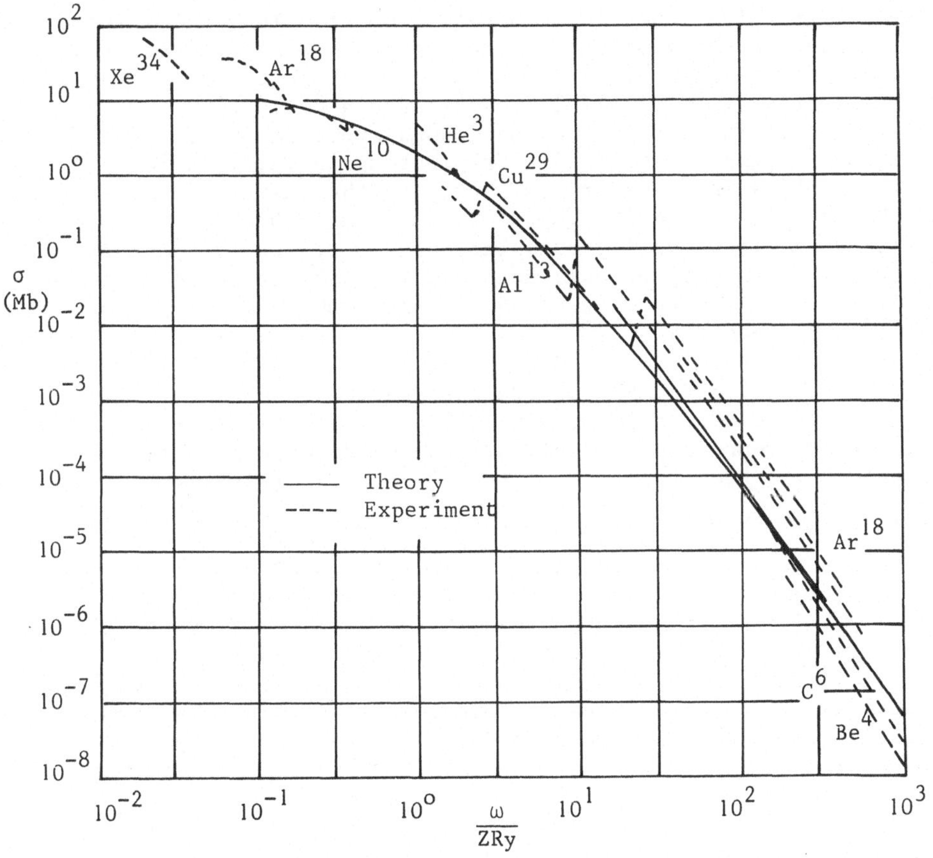

Figure 2. Comparison between a typical "universal" photoabsorption curve and some experimental cross sections. From ref. 24.

low density region. This effect is exaggerated in the Thomas-Fermi model because of the (unphysical) infinite extent of the charge distributions. In an ionized atom the Thomas-Fermi density has a finite radius and consequently we have a discrete spectrum of excitation. This situation has been treated in particular by Walecka [23] .

The statistical model including the theory of collective oscillations has been studied in a series of papers by Kirzhnitz [23] and coworkers (the reference is to a major review paper containing references to the original papers). They discuss at some length the shortcomings of the hydrodynamic model. Clearly an oscillation that continues out into the low density region with ever decreasing wave-length is not what we mean with a well-defined collective mode. If we start an oscillation in the intermediate part of the atom with a frequency $\omega \sim Z$ Ry, we note that the frequency is much higher than the ionization energy of the outer electrons ($\omega \sim 1$ Ry). Therefore such a collective oscillation is in fact in the continuous spectrum of the outer electrons. A charge oscillation originating in the interior of the atom will soon transfer its energy to the outer electrons and thereby cease to exit. Therefore the simple boundary condition that the density oscillation goes to zero at infinite distance will indeed not only describe the collective oscillation itself but also include its decay products, which do have indeed a continuous spectrum.

In order to have a more satisfactory theory we must be able to have solutions where the collective oscillation is confined to a region around a spherical shell with radius R on which the frequency ω is such that $\omega \sim \omega_{p\ell}(R)$. In the far interior the local plasma frequency will satisfy the condition $\omega_{p\ell}(r) \gg \omega_{p\ell}(R)$ and the oscillation will therefore damp <u>inwards</u>. However, the solution must also damp <u>outwards</u> and this can only be achieved by introducing some damping mechanism in the equation of motion. One such mechanism is given by the Landau damping but there is also a damping because of the Coulomb interaction leading to the excitation of electron-hole pairs. Clearly some special dissipative term has to be included and this has been done in an ad hoc fashion by several authors. We note that the form of the damping determines the frequency and width of the collective oscillations, and since the damping means including very important additional microscopic information, we have no longer a simple hydrodynamic picture. We shall return to this question later in these lectures.

5. THE RESPONSE OF A NON-UNIFORM SYSTEM TO AN EXTERNAL FIELD

In the preceding sections we have discussed the application of hydrodynamical models and concluded that in applications to atoms and molecules, we need a microscopic approach to account for important physical effects such as the decay of a collective oscillation into single particle excitations. In developing a microscopic approach it is not always useful to develop a formalism for eigenmodes of free oscillations, since they have often a rather short life-time and a complex line-shape. It is often more fruitful to study the response to an external field, which gives a more direct approach to calculate the full spectral profile. We shall use photoabsorption as the key example and study the linear response to a weak electromagnetic field. The linear response theory for absorption by atoms and similar systems has been published by many authors. Out of the many possible references we refer to a lucid presentation by Roulet and Nozières [25] of the general theory with a good discussion of the atomic problem.

We limit ourselves to the case where the wave length of the external field is very long compared with the dimensions of the system. This is the case of dipolar excitations and the quantity to study is the induced charge density $\rho_1(\underline{x},t)$ induced by a weak external scalar field $V_{ext}(\underline{x},t)$. In linear response theory the average induced charge density is given by

$$\rho_1(\underline{x},t) = \int d^3x' \, dt' \; R(\underline{x},\underline{x}',t-t') \; V_{ext}(\underline{x}',t') \tag{5.1}$$

The <u>response function</u> $R(\underline{x},\underline{x}',t-t')$ is defined as follows :

$$R(\underline{x},\underline{x}',t-t') = -i \; \theta(t-t') < [\rho(\underline{x},t),\rho(\underline{x}',t')] > \tag{5.2}$$

$\rho(\underline{x},t)$ is the density operator, $\theta(t)$ is the Heavside step function $[\ldots,\ldots]$ denotes the commutator and $<\ldots>$ is the average over the ground state at zero temperature, and the usual statistical average at finite temperatures.

The total scalar potential inside the electron system is the sum of the external potential $V_{ext}(\underline{x},t)$ and the polarization potential arising from the induced charges $\rho_1(\underline{x},t)$, thus

$$V_{tot}(\underline{x},t) = V_{ext}(\underline{x},t) + V_{pol}(\underline{x},t) \tag{5.3}$$

where

$$V_{pol}(\underline{x},t) = \int d^3x \; v(\underline{x},\underline{x}') \; \rho_1(\underline{x}',t)$$

and $v(\underline{x},\underline{x}') = e^2/|\underline{x}-\underline{x}'|$ is the Coulomb potential.

$V_{pol}(\underline{x},t)$ is the average potential due to the induced charges and has to be calculated self-consistently. It is the analog to the Hartree potential for ground state properties, however, here it is the average potential induced by an external time-dependent field.

We now introduce a new response function $P(\underline{x},\underline{x}',t-t')$ which describes the linear response to $V_{tot}(\underline{x},t)$, thus

$$\rho_1(\underline{x},t) = \int d^3x' \; dt' \; P(\underline{x},\underline{x}',t-t') \; V_{tot}(\underline{x}',t') \tag{5.4}$$

The relation between the two response functions R and P is straightforward and can be written as

$$R = P + P\,v\,R \tag{5.5}$$

where we have used a shorthand notation and regarded all the functions as continuous matrices in the space and time variables.

The relation between the total scalar potential $V_{tot}(\underline{x},t)$ and the external potential $V_{ext}(\underline{x},t)$ defines the inverse dielectric function through the formula

$$V_{tot}(\underline{x},t) = \int d^3x'dt'\; \varepsilon^{-1}(\underline{x},\underline{x}',t-t')\; V_{ext}(\underline{x}',t') \tag{5.6}$$

Using the matrix language we can write

$$\rho_1 = PV_{tot} = P(V_{ext} + v\rho_1) = \frac{1}{1-Pv}\,PV_{ext} \tag{5.7}$$

and proceed to find that

$$V_{tot} = [1 + v\,\frac{1}{1-Pv}\,P]\,V_{ext} = \frac{1}{1-vP} = \frac{1}{\varepsilon}\,V_{ext} \tag{5.8}$$

The last line shows that $\varepsilon = 1 - vP$, or written out in full :

$$\varepsilon(\underline{x},\underline{x}',t) = \delta(\underline{x}-\underline{x}') - \int d^3x''v(\underline{x}-\underline{x}'')P(\underline{x}'',\underline{x}',t) \tag{5.9}$$

P is usually referred to as the irreducible polarization propagator. In the same way we can write

$$\varepsilon^{-1}(\underline{x},\underline{x}',t) = \delta(\underline{x}-\underline{x}') + \int d^3x''\; v(\underline{x}-\underline{x}'')\; R(\underline{x}'',\underline{x}',t) \tag{5.10}$$

What we have been discussing here are retarded response functions. Many techniques, e.g. diagrammatic methods, use instead the corresponding time-ordered quantities. The time-ordered dielectric function and the retarded dielectric function are closely related an differ only with respect to their different causality properties. The difference is easiest to see if we take the Fourier transform with respect to time and consider them as functions of $(\underline{x},\underline{x}',\omega)$. The two functions are equal for positive frequencies. The time-ordered dielectric function is an even function of the frequency, whereas the retarded function has an even real part but the imaginary part changes sign for negative frequencies. It is often convenient to work in momentum space and use instead of the variables $(\underline{x},\underline{x}',t)$ the set of variables $(\underline{q},\underline{q}',\omega)$. The Fourier transforms are defined as

$$F(\underline{q},\underline{q}',\omega) = \int d^3x\; d^3x'\; dt\; e^{i\omega t}\; e^{-i\underline{q}\cdot\underline{x}}\; e^{i\underline{q}\cdot\underline{x}'}\; F(\underline{x},\underline{x}',t) \tag{5.11}$$

In this way we obtain the dielectric functions as matrices in q-space defined as

$$\varepsilon^{-1}(\omega)_{\underline{q}\underline{q}'} = \delta_{\underline{q}\underline{q}'} + \frac{4\pi e^2}{q^2} R(\underline{q},\underline{q}',\omega) \tag{5.12}$$

and

$$\varepsilon_{\underline{q}\underline{q}'}(\omega) = \delta_{\underline{q}\underline{q}'} - \frac{4\pi e^2}{q^2} P(\underline{q},\underline{q}',\omega) \tag{5.13}$$

When q = q' the response function is directly related to the dynamical form factor $S(\underline{q},\omega)$, defined as

$$S(\underline{q},\omega) = \Pi \sum_n |\langle n|\rho_{\underline{q}}|o\rangle|^2 \, \delta(\omega - \omega_{on}) \tag{5.14}$$

$|n\rangle$ is the exact eigenstate n of the system ρ_q is the Fourier component of the density $\rho(\underline{x})$ and $\omega_{on} = E_n - E_o$ are the exact excitation energies. The relation between the dynamical form factor and the response function R is simply

$$S(\underline{q},\omega) = -\text{Im}\, R(\underline{q},\underline{q},\omega) \tag{5.15}$$

for $\omega > 0$. As is wellknown $S(\underline{q},\omega)$ describes the inelastic scattering of e.g. electrons and photons by the system and for long wave lengths the photoabsorption cross section is given by the formula

$$\sigma(\omega) = \frac{4\pi\omega e^2}{cq^2} S(\underline{q},\omega) \tag{5.16}$$

Alternatively we can express $\sigma(\omega)$ in terms of the oscillator strength function $g(\omega)$ as follows

$$\sigma(\omega) = \frac{2\pi^2 e^2}{mc} g(\omega) \tag{5.17}$$

where

$$g(\omega) = -\frac{\omega}{2\pi^2} \int d^3x \, d^3x' \, \text{Im}\, \varepsilon^{-1}(\underline{x},\underline{x}',\omega) \tag{5.18}$$

The dynamical structure factor $S(q,\omega)$ or, equivalently, the oscillator strength function contain all the information about the density excitation spectrum of the system. Much of the recent development has been focused on ways to calculate these quantities. However, for pedagogic reasons it may be worthwhile keeping the connection with the earlier history and look for normal solutions where the oscillations now are damped, which is generally the case. The linear response relation can be written in the form

$$\int d^3x' \, \varepsilon(\underline{x},\underline{x}',\omega) \, V_{tot}(\underline{x}',\omega) = V_{ext}(\underline{x},\omega) \tag{5.19}$$

Self-oscillations occur in the absence of a driving field and are obtained from the complex eigenvalues ω of the equation

$$\int d^3x' \, \varepsilon(\underline{x},\underline{x}',\omega) \, V_{tot}(\underline{x}',\omega) = 0 \tag{5.20}$$

which is the proper generalization of the condition $\varepsilon(q,\omega) = 0$ in a uniform electron liquid.

In order to calculate the response, many workers have introduced a local density approximation assuming that the non-uniform system responds locally as a uniform electron liquid with the same density $\rho(\underline{x})$. The simplest approximation neglects the spatial non-locality completely and describes the response by the classical formula

$$\varepsilon(\underline{x}) = 1 - \frac{\omega_{p\ell}^2(\underline{x})}{\omega^2} \tag{5.21}$$

This gives the result, using equation (5.18)

$$g(\omega) = \int d^3x \, \rho_o(\underline{x}) \, \delta(\omega_{p\ell}(\underline{x}) - \omega) \tag{5.22}$$

The formula has a simple meaning : the single particle excitations are completely screened out and the atom absorbs at each point at the local plasma frequency. This gives a smooth absorption curve which reflects the charge distribution and shows no resonant behaviour. Calculations based upon various approximate forms for $\varepsilon(q,\omega)$ to introduce the non-local aspects of the response give very similar results. In fact both the simple classical formula (5.22) and the calculation using a non-local $\varepsilon(q,\omega)$ for a uniform electron gas all give universal photoabsorption curves which are qualitatively the same as that obtained by Ball, Wheeler and Fireman [21] by solving the Bloch equations.

It is clear from this discussion that the straightforward extensions of the theory of the uniform electron liquid do not describe in an adequate way the possible collective excitations in a non-uniform system. The approaches we have mentioned describe in an average way the photoabsorption and stopping power of matter but do not describe the actual dynamics of a given non-uniform system. The limitations of the statistical model of matter has been discussed by several authors and we refer to a review by Kirzhnitz et al [26] . They conclude that the conventional free-electron like approach is not valid for these dynamical problems. Instead they introduce a different type of semi-classical theory in which they calculate the trajectories of particles moving in the Thomas-Fermi field. Using such an approach Kirzhnitz and Lozovik [27] and Gadiyak [28] et al calculated the dielectric function $\varepsilon(\underline{x},\underline{x}',\omega)$. Studying the condition for a bounded collective motion they found a collective oscillation which is confined between an outer and an inner radius and is strongly damped inside as well as outside of this shell. They find two eigenfrequencies at $\omega_1 = 13.74$ ZeV, $\omega_2 = 36.04$ ZeV. They also calculated the life-time and oscillator strengths of these modes. The possible experimental verification of these results has not been discussed. The experimental data obtained in the last decade or so rather seem to indicate that collective oscillations in atoms are not a general property of atoms but seem to be a property of special atoms or group of atoms and depend on their shell structure. This implies that a description solely in terms of the density will not be sufficient to describe the frequency-dependent properties of atomic-like systems, but that the theory has also to include aspects due to the shell structure and associated with the one-electron excitations. We have already remarked that the proper inclusion of damping depends crucially on further microscopic information e.g. about how a density oscillation in the intermediate part of an atom decays into electron-hole excitations in the valence shell. In the remaining sections of these lecture notes, we shall discuss how the physics of a hydrodynamical model will be changed by the discrete nature of the one-electron levels.

6. SOME MANY-BODY ASPECTS OF ATOMIC-LIKE SYSTEMS

In the last two decades there has grown up a vast literature on the many-body problem of atoms and molecules. A large variety of methods have been used such as perturbation expansions, diagrammatic methods, Green's function schemes, equation of motion methods, the density functional approach etc. The present state of the art is summarized e.g. in the proceedings of the Nobel Symposium on the Many-body theory of atomic systems [29]. In addition I would like to add a couple of references to reviews by Amusia and Cherepkov [30] and by Wendin [31,32] for atoms and by Cederbaum and Domcke [33] for molecules.

The recent Nobel symposium summarized the present status of the theory for atoms and small molecules both with regards to concepts, methods, techniques and applications. It deals with the ground state properties, excited states, relativistic effects, scattering theory, photoemission, photoabsorption, atoms and molecules bound in solids or at solid surfaces etc. The present situation seems to be that the many-body theory of atoms has now reached a state where most physical effects are not only understood in a qualitative way but can also be calculated quite accurately in most cases. Also for small molecules there has been substantial progress but the molecular problem seems considerably more difficult and progress has been slower, and the methods cannot easily be extended to larger molecules.

Most applications of many-body techniques to dynamical properties of atoms and molecules have been concerned with the description of the entire spectrum in a particular spectral region and have not been particularly focussed on the possible collective nature of the motion. There is an appreciable amount of technology in the many-body description of atoms and molecules and here we are mainly concerned with the collective aspects and need only parts of the technical machinery of many-body theory. We therefore limit ourselves to a brief discussion of the physical basis of the theory.

The theoretical backbone of the many-body theory of extended systems as well as finite systems such as atoms, molecules and nuclei has been some version of the time-dependent Hartree theory or the time-dependent Hartree-Fock theory alias the random phase approximation (RPA) or the random phase approximation with exchange (RPAE). It should be said at once that these theories have been modified and extended in many treatments. For the description of the collective properties of atoms, molecules and surface the RPA and RPAE seem to be fairly adequate. Just as in the case of the electron gas, the RPA method in atomic systems is designed to account for the long range correlations between electrons (longer than the average distance between electrons). Therefore the influence of ordered, collective motion of the electrons on any atomic property will be taken into account, but the short-range correlations between the electrons will be ignored in such a treatment.

We start by writing down the Hartree and Hartree-Fock (HF) equations for a system with a Hamiltonian for N electrons of the form

$$H = \sum_{i=1}^{N} h(\underline{x}_i) + \frac{1}{2} \sum_{i,j}{}' v(\underline{x}_j,\underline{x}_i) + V_{nucl} \qquad (6.1)$$

The non-electron part of the Hamiltonian is written as

$$h(\underline{x}) = -\frac{\hbar^2}{2m}\nabla^2 - \sum_g Z_g\, v(\underline{x},\underline{R}_g) \qquad (6.2)$$

where $\underline{R}_g$ denotes the position of the nucleus g and v is the Coulomb interaction. We let for future use $\underline{x}$ include the spin variable as well and we have for a moment put back the fundamental constants, e, m and ħ in the formulas.

In the Hartree theory the wave function is taken as a

product of N one electron functions u_K, where each one satisfies a one-electron wave equation of the form

$$(h + V_H)u_K = \varepsilon_K u_K \tag{6.3}$$

where the Hartree potential V_H is given by

$$V_H(\underline{x}) = \int d^3x' \, v(\underline{x},\underline{x}') \, \rho(\underline{x}') \tag{6.4}$$

and

$$\rho(\underline{x}) = \sum_{K=1}^{N} |u_K(\underline{x})|^2$$

is the electron density. The Hartree potential replaces the actual Coulomb interaction between pairs of electrons by the average electrostatic potential at $\underline{x}$.

The HF theory is obtained by considering an anti-symmetrized product of one-electron wave functions, usually written as a Slater determinant

$$|\psi\rangle = (N!)^{-1/2} \det \{U_K(\underline{x}_i)\} \tag{6.5}$$

This leads to the HF equations

$$(h + V_{HF})U_K = \sum_{\ell=1}^{N} \lambda_{K\ell} \, U_\ell \tag{6.6}$$

where

$$\begin{aligned} V_{HF} &= V_H + V_x \\ V_x &= -\int v(\underline{x},\underline{x}') \, \rho(\underline{x},\underline{x}') \, d^3x' \\ \rho(\underline{x},\underline{x}') &= \sum_{K=1}^{N} U_K(\underline{x}) \, U_K^*(\underline{x}') \end{aligned} \tag{6.7}$$

V_x is the HF exchange potential, $\rho(\underline{x},\underline{x}')$ is the HF density matrix and $\lambda_{K\ell}$ are Lagrangian multipliers. If we have no restrictions on the form of the solution one can transform the HF equations to diagonal form, thus

$$(h + V_{HF})U_K = \varepsilon_K U_K \tag{6.8}$$

in which case ε_K can be interpreted as approximate one-electron energies.

The Hartree and HF theories both replace the actual Coulomb interaction by an average static field. The important feature is that this field depends itself on the solutions U_K and the problem is therefore non-linear. This leads to the physically important notion of self-consistency : the electrons move in the field set up by their own motion.

These ideas can immediately be carried over to the dynamics of atomic systems. The Hartree-Fock equation (6.8) has the obvious time-dependent generalization

$$[h(\underline{x}) + V_{HF}(\underline{x},t)]\, U_K(\underline{x},t) = i\hbar \frac{\partial}{\partial t} U_K(\underline{x},t) \qquad (6.9)$$

The self-consistent potential $V_{HF}(\underline{x},t)$ depends on t through the time-dependence of the one-electron wave functions $U_K(\underline{x},t)$. These equations can be derived from the variational principle

$$\delta\langle\psi(t)|H - i\hbar \frac{\partial}{\partial t}|\psi(t)\rangle = 0$$

with the restriction that $\psi(t)$ is a single Slater determinant.

A particular solution to the time-dependent equations is

$$U_K(\underline{x},t) = U_K(\underline{x})e^{-i\varepsilon_K t}$$

which gives back the ordinary HF theory. We may also find solutions in which we superimpose a small amplitude time-dependence on the ground state solution. In order to study a possible collective behaviour we consider the case where we superimpose a periodic modulation of the ground state orbitals with the same frequency ω. We consider for simplicity a system with a closed shell structure, so that we have a clear and simple distinction between occupied one-electron state i and unoccupied states m. We now consider a solution of the form

$$U_i(\underline{x},t) = U_K(\underline{x})e^{-i\varepsilon_i t} + \sum_m x_{mi} U_m(\underline{x})e^{-i\omega t - i\varepsilon_m t} + \sum_m y^*_{mi} U_m(\underline{x})e^{i\omega t - i\varepsilon_m t} \qquad (6.10)$$

By imposing the same additional frequency dependence for all orbital states it is clear that we consider an oscillation of frequency ω for all electrons, i.e. oscillations of the entire system at a frequency ω above the ground state.

The solution of the Hartree or HF equations using this ansatz will give the eigenfrequencies of free or damped oscillations of the system. For simplicity we consider only the Hartree approximation, i.e., we neglect the exchange potential V_x. We consider also only small amplitude motion and retain only terms linear in the amplitudes x_{mi} and y_{mi}. In this way we obtain the <u>linearized time-dependent Hartree equations</u>

$$(\varepsilon_m - \varepsilon_i - \omega)x_{mi} + \sum_{n,j} \langle jm|v|ni\rangle x_{nj} + \sum_{n,j} \langle mn|v|ij\rangle y_{nj} = 0$$

$$(\varepsilon_m - \varepsilon_i + \omega)y^*_{mi} + \sum_{n,j} \langle mn|v|ij\rangle x^*_{nj} + \sum_{n,j} \langle jm|v|ni\rangle y^*_{nj} = 0 \tag{6.11}$$

The matrix elements of the Coulomb potential are defined as

$$\langle ij|v|k\ell\rangle = \int d^3x\, d^3x'\; U^*_i(\underline{x})\; U^*_j(\underline{x}')\; v(\underline{x},\underline{x}')\; U_k(\underline{x})\; U_\ell(\underline{x}') \tag{6.12}$$

In the same way one obtains the linearized HF equations which have a similar structure, including also the exchange matrix elements. The solutions to these equations describe the excited levels as the eigenmodes of free oscillations around the ground state. In a strict single-particle model the excitation frequencies would just be given by the difference in single-particle energies $\omega = \varepsilon_m - \varepsilon_i$. The Coulomb interaction with other excitations nj couples the excitations and produces an energy shift. In many cases the shift is small and the excited states are just slightly modified (screened) electron-hole excitations. In other cases many single-particle excitations will act coherently to produce a large shift and we may obtain a collective excitation.

We could equally well have studied the forced oscillations when applying a time-dependent external field of frequency ω. We would still look for solutions of the form given in equation (6.10). The equations for the amplitudes will be of the form given by equation (6.11), except that we have to add the driving term due to the external field in the right-hand side. The physical interpretation is again simple. The electrons move in the external field plus the induced field from all the other electrons. In a mean field theory such as the RPA or RPAE the induced field is the average field which follows the frequency of the external field and we neglect the fluctuations around the average.

We could equally express the RPA approach in the space representation, using the dielectric approach formulated in the preceding section. The RPA is in fact equivalent to calculate the irreducible polarization propagator using the non-interacting ground state and excitations. The RPA theory in the dielectric formulation is condensed in the formulas

$$\varepsilon(\underline{x},\underline{x}',\omega) = \delta(\underline{x}-\underline{x}') + \int d^3x'' \frac{P_o(\underline{x}'',\underline{x}',\omega)}{|\underline{x}-\underline{x}''|}$$

with

$$P_o(\underline{x},\underline{x}',\omega) = \sum_n \left[\frac{\phi_n(\underline{x})\,\phi_n^*(\underline{x}')}{\omega + i\varepsilon - \omega_n} - \frac{\phi_n(\underline{x}')\,\phi_n^*(\underline{x})}{\omega + i\varepsilon + \omega_n}\right] \qquad (6.13)$$

Here ω_n is the particle-hole excitation energy and the corresponding amplitudes for the density fluctuations are $\phi_n(\underline{x}) = U_p^*(\underline{x})\;U_h(\underline{x})$ where $U_p(\underline{x})$ and $U_h(\underline{x})$ are the one-electron wave functions for the particle and hole states, respectively.

In order to look into the nature of the solutions, we shall consider a simplified model which is exactly soluble. We consider the dipolar oscillations of a single atomic shell. Assuming that the quantum numbers are high, the matrix elements factorize and can be written as

$$\langle ij|v|k\ell\rangle \approx \lambda P_{ik} P_{j\ell} \qquad (6.14)$$

where P_{ik} and $P_{j\ell}$ are the matrix elements for the dipolar transitions $i \to k$ and $j \to \ell$ respectively. $\lambda = R^{-3}$ where R is the mean radius of the shell. This is a different form of the semi-classical limit than the hydrodynamical theory. Here, we express the interaction as the dipolar coupling between orbital excitations, which in the limit of high quantum numbers approach the Bohr-Sommerfeld orbits in the atomic self-consistent field.

Using this model we can solve the equation (6.11). We obtain

$$x_{mi} = \text{const}\,\frac{P_{mi}}{\varepsilon_m - \varepsilon_i - \omega}\,, \qquad y_{mi} = \text{const}\,\frac{P_{mi}}{\varepsilon_m - \varepsilon_i + \omega} \qquad (6.15)$$

The energy eigenvalues are obtained from the dispersion relation

$$1 + \lambda \sum_{mi} \frac{2\,P_{mi}^2(\varepsilon_m - \varepsilon_i)}{(\varepsilon_m - \varepsilon_i)^2 - \omega^2} = 1 + \lambda\alpha_o(\omega) = 0 \qquad (6.16)$$

where

$$\alpha_o(\omega) = \sum_{mi} \frac{f_{mi}}{(\varepsilon_m - \varepsilon_i)^2 - \omega^2} \qquad (6.17)$$

is the polarizability of the atomic shell in one-electron theory, f_{mi} being the oscillator strength for the transition $i \rightarrow m$ ($f_{mi} = 2(\varepsilon_m - \varepsilon_i)P^2_{mi}$ in atomic units)

The contents of this formula is easiest to see graphically, as is done in figure 3.

We have used a simplified picture with only a discrete electron-hole spectrum with a finite number of levels. We see that we obtain one solution between any pair of neighbouring electron-hole excitations. These levels correspond to screened particle hole excitations. Above the highest particle-hole excitation we obtain a new solution which splits off from the other levels with a shift that will be large if the coupling constant λ is large. This level will be the "most" collective of the excitations in the spectrum. The form of the theory indicates the importance of the strength and positions of the bare electron-hole levels which couple through the Coulomb interaction to give a collective shift. Clearly all of the classical concepts of the hydrodynamical approach have been lost in the kind of approach just described.

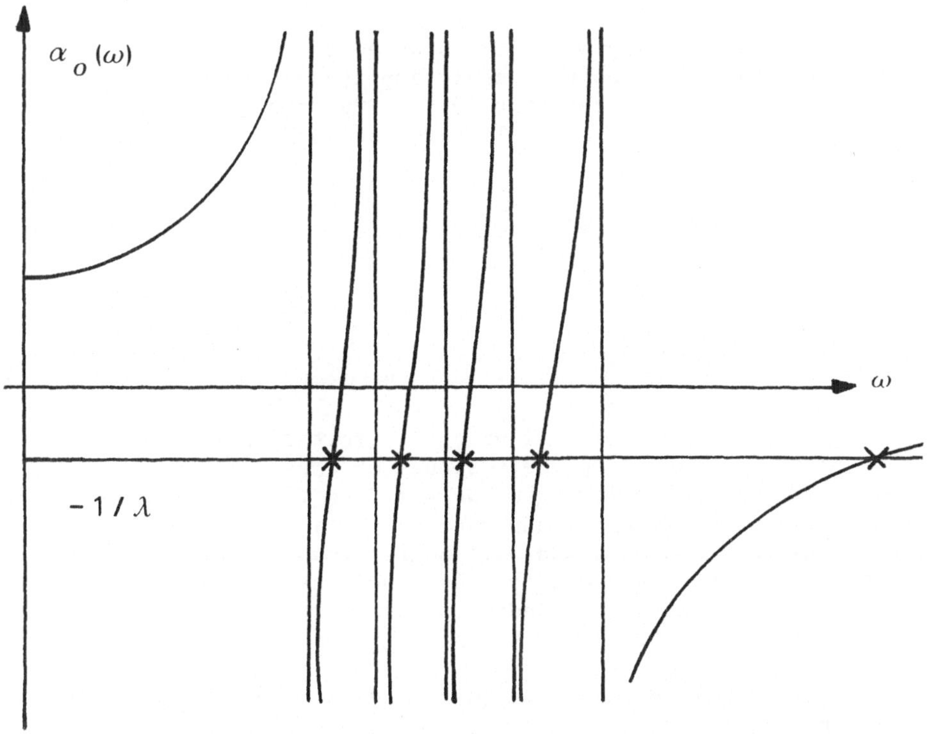

Figure 3. Graphical solution of equation 6.15.

A particularly simple situation occurs if all the single particle transitions are degenerate so that $\varepsilon_m - \varepsilon_i = \omega_o$ for all transitions. This is a model of the case where one has a strong dynamical interaction between two neighbouring atomic shells, e.g. the d and an unfilled f shell in heavier atoms. In this case we obtain only one solution to the equations, given by

$$\omega^2 = \omega_o^2 + \lambda F \tag{6.18}$$

where

$$F = 2\omega_o \sum_{mi} P_{mi}^2$$

is the total oscillator strength associated with the shell to which i belongs, which is of the same order as the number of electrons in that shell. In this case, therefore, the single-particle transition ω_o is completely quenched and all the oscillator strength goes into the coupled mode (6.17). A particular case of this is the long wave length limit of plasma oscillations in an electron gas. In this case all the particle hole excitation frequencies tend to zero so that $\omega_o \to 0$ and $\lambda F \to \omega_{p\ell}^2$.

We could equally well have studied the linear response to an external field by adding a small amplitude external field on the right-hand side of equation (6.11). One can then solve for the amplitudes x_{nj} and y_{nx} and calculate the total induced dipole moment

$$p(\omega) = \alpha(\omega)\, E_{ext}(\omega)$$

and obtain

$$\alpha(\omega) = \frac{\alpha_o(\omega)}{1 + \lambda\alpha_o(\omega)} \tag{6.19}$$

The poles of $\alpha(\omega)$ will give the excitation frequencies, which we just discussed, and the residues of these poles will determine the oscillator strength. We see from (6.19) that the poles of $\alpha_o(\omega)$ which give the unperturbed particle-hole excitations will be completely quenched and will be replaced by the poles of $\alpha(\omega)$. We see that the vanishing of the denominator in equation (6.19) is identical with the condition for free oscillations given by equation (6.15).

The simplified model we just discussed assumed that we had only one atomic shell. In a real atom the excitations of a given shell will be modified due to presence of other shells. The inner shells will typically have excitation frequencies considerably higher than the range we consider and they will couple only weakly. The shells outside the one we consider have frequency below and in the range we consider. Typically, they are in the continuous spectrum of the atom and the coupling to the continuum will change the nature of the spectrum drastically. In some cases such as the $4d^{10}$ spectrum of Xe, Ba, La and other elements will show a strong, broad resonance with a rather complex non-Lorentzian profile. In other cases such as the spectra of Ne, Ar, Kr, there is a strong enhancement of the photoabsorption cross section. The experimental cross section is 3-5 times higher than one would obtain from one-electron theory but there is no resonance structure in the spectrum [26] .

The discussion just given has just meant to give some qualitative insight in the physics of the self-consistent field approach as applied to collective dynamics of electrons. In actual applications a straightforward application of the theory would usually not be sufficient. Many refined formulations of the theory tend to preserve some of the formal structure we have indicated by extensive redefinitions and renormalizations of the single-particle energies and wavefunctions, the effective interaction, the structure of the ground state etc. For example, the energy levels of the excited electron is in general always calculated in the presence of the hole in an approximate way, usually so that the static Coulomb attraction between the electron and the hole is taken into account. This gives a zerothorder approximation to the electron-hole spectrum which is not too far from the actual spectrum. It is often also an advantage to renormalize the energies and wave functions for localized hole states in order to account for the relaxation of the electron cloud which takes place around the hole. This will take into account the relaxation shift of the core levels which is of importance to get the correct positions e.g. for ionization thresholds.

The approach outlined here has been applied successfully to the photoabsorption of the rare gas atoms, Ba and several other heavy atoms by Amusia et al, Wendin and many others. We refer to ref. 29-31 for extensive review and references to the original papers.

Photo electron spectroscopy also shows many interesting examples of collective dynamics and strong deviations from the one-electron picture. This field has recently been summarized by Wendin [32] .

The case of collective motion in molecules is more complex. In contrast to atoms a self-consistent approach of the RPA type is usually not sufficient and in many cases one obtains imaginary excitation frequencies, which shows that the form of the ground state which has been assumed is unstable against small oscillations. It is likely that the partially two-dimensional and even one-dimensional structure of many molecules requires in general a more accurate treatment of correlations both in the ground state and for the dynamical properties. We refer to the recent review papers in ref.29 and to the extensive review by Cederbaum and Domcke [33] for discussions of the molecular problem.

We discussed earlier the dispersion of surface plasmons in the long wave length limit, where the linear dispersion coefficient only depends on the density eigenfunctions for the surface plasmon with $\underline{k} = 0$. In a hydrodynamical model the plasmon wave function is a real smooth function, being non-zero essentially only in the region of the surface profile. In a RPA type of description the surface plasmon frequency is degenerate with electron-hole excitations. As a consequence the surface plasmons will be damped for all values of the two-dimensional wave vector $\underline{k}$. Furthermore the density eigenfunction will be complex and show an oscillatory behaviour inwards from the surface as a consequence of the coupling of the plasmon to the electron-hole pairs. The first application of the RPA theory to the dispersion of surface plasmons over the entire range of moments was done by Inglesfield and Wikborg [34].

7. LINEARIZED QUANTUM EQUATIONS FOR DENSITY OSCILLATIONS

In the first part of these lectures we discussed the hydrodynamical approach to collective motion in bounded systems. The theory was based on the Thomas-Fermi idea to treat the system locally as an electron gas characterized by its density. We concluded that an input of microscopic information is needed to account for important physical effects such as the decay of a collective motion by electron-hole excitations etc. In the preceding section we gave an introductory discussion of the many-body approach to collective oscillations. However, in this approach the contact with the hydrodynamical approach was lost and although we certainly saw the dynamical self-consistency in action, the fact that every thing was expressed in terms of single-particle energies, wave functions and matrix elements tended to obscure the possibility to establish the relation to the semi-classical theory. However, it is straightforward to write the quantum mechanical equations of motion in a form where one can separate the purely classical contributions from the genuine quantum dynamics of the system. The method is to derive directly an approximate equation of motion for the second time derivative of the average density oscillation, which will have the classical terms appearing explicitly but still include all the many-body interactions of the system. Such equations have been discussed by many authors and we refer to ref. 1 for a discussion adapted to the problems discussed here. In these lectures we use an approach based on the linear response equations discussed in section 5 [35].

The major point is to rewrite the theory in a form such that the classical part of the response, which arises from the instantaneous Coulomb forces will appear explicitly and is separated to the retarded part, which depends on the dynamics of the electrons. In order to achieve this separation we differentiate the linear response formula for the induced charge density twice with respect to time and obtain on the right-hand side two contributions. One arises from the equal time contribution t' = t, which enters because of the unit step function in the density-density response function. The equal time commutator can be worked out from the commutation rules of the operators and does not depend on the dynamics. The other contribution comes from the retarded response and describes the effect of the actual dynamics of the electrons. In order to show the structure of the theory in its simplest form we put and look for the free oscillations, which are described by the equation

$$\{\frac{\partial^2}{\partial t^2} + \omega^2_{p\ell}(\underline{x}) - \nabla\rho_o.\nabla v\}\rho = -\int_{-\infty}^{t} dt'\ \ddot{\tilde{P}}(t-t')\ v\rho(t') \quad (7.1)$$

where

$$\ddot{\tilde{P}}(t-t') = -i\,\frac{d^2}{dt^2} <[\rho(\underline{x},t);\rho(\underline{x}',t')]>_{irr} \quad (7.2)$$

The subscript irr means that we have considered the response to the total acting field defined by the response function $P(\underline{x},\underline{x}',t)$.

The eigenvalue equation for oscillations at a complex frequency ω takes the form

$$\{\omega^2 - \omega^2_{p\ell}(\underline{x}) + \nabla\rho_o.\nabla v\}\rho = \ddot{\tilde{P}}(\omega)v\rho \quad (7.3)$$

The classical terms appear on the left-hand side. The classical restoring force is given by the local plasma frequency $\omega_{p\ell}(\underline{x}) = (4\pi\rho_o(\underline{x}))^{1/2}$. The term containing $\nabla\rho_o$ is characteristic of a non-uniform system. In classical theory with discontinuous boundaries, this term is replaced by the usual boundary conditions. In a hydrodynamical approach, the right-hand side would be of the form $\alpha\nabla^2\rho$ where the density dependent coefficient depends on whether we consider the low frequency limit of oscillations of the order of the local frequency $\omega_{p\ell}(\underline{x})$. The formulation given is quite general and P can be calculated from any approximation chosen for the irreducible polarization propagator. In the RPA we obtain the explicit formula

$$\ddot{\tilde{P}}(\omega) = - \sum_n \omega_n^2 \left[\frac{\phi_n(\underline{x})\, \phi_n^*(\underline{x}')}{\omega + i\varepsilon - \omega_n} - \frac{\phi_n(\underline{x}')\, \phi_n^*(\underline{x})}{\omega + i\varepsilon + \omega_n} \right] \tag{7.4}$$

with the notations as used in the preceding section.

We wish to point some general properties of this formulation.

(1) Core contributions cancel out and give no restoring force for oscillations of the system. At any frequency ω we can define the core as those parts of the system for which $\omega_{core} \gg \omega$. The core contributions in the right-hand side of equation (7.3) cancel out against the core contributions to the classical terms on the left-hand side. This implies that at frequencies $\omega \ll \omega_{core}$ we can replace $\rho_o(\underline{x})$, $\nabla\rho_o(\underline{x})$ and $\omega_{p\ell}(\underline{x})$ with the corresponding contributions from the outer shells only. This feature shows that the major response of the system comes from parts which have excitation frequencies of the same order as the applied frequency ω.

(2) The response function is a complex function and the imaginary part describes the decay of the collective motion into particle-hole excitations.

(3) The detailed nature of the particle-hole spectrum is reflected in the response function and gives rise to structure in the spectrum, which may partly be a screened particle-hole spectrum with more or less pronounced collective resonances as was discussed in the preceding section.

(4) The detailed nature of the particle-hole spectrum is reflected in the response function and gives rise to structure in the spectrum, which may partly be a screened particle-hole spectrum with more or less pronounced collective resonances as was discussed in the preceding section.

The advantage of this formulation will be in applications where one expects that the classical theory should be a good first approximation. We note that the evaluation of the equal time contributions requires the use of closure properties of the one-electron states, exact relations involving the operators etc. In an approximate application of many-body theory only part of these classical terms will be included.

We should mention that this formulation makes it easy to make contact with different versions of the hydrodynamic theory. In the low-frequency limit we would retrieve in essence the Bloch theory. As remarked earlier, however, we are rather interested in high frequencies. In the limit where ω is considerqbly higher than the typical electron-hole energies one can expand the response function in inverse powers of ω, with coefficients which can be calculated exactly (at least in principle). Such treatments have been published by several authors and we refer to ref. 36 for a recent discussion. In application to surface plasmons and similar problems, the frequency does not always seem high enough to ensure that the high-frequency expansion will converge, which leads to some uncertainty about the applicability of the high frequency expansion.

We shall give a few simple illustrations considering only self-oscillations in the simplest case when we neglect the electron-hole spectrum, i.e. we use only the classical theory.

a. Surface plasmons

We consider the jellium model with the edge at $z = 0$ and let $\underline{r}$ denote the two-dimensional vector parallel to the surface. In the classical approximation we obtain

$$(\omega^2 - \omega_{p\ell}^2(z))\, \rho_{sp}(\underline{r},z,\omega) = = -\frac{e^2}{m}\frac{d\rho_o}{dz}\frac{d}{dz}\int \frac{d^2r\, dz'\, \rho_{sp}(\underline{r}',z',\omega)}{\sqrt{(\underline{r}-\underline{r}')^2 + (z-z')^2}} \tag{7.5}$$

Introducing the Fourier transform in the xy-plane we obtain

$$(\omega^2 - \omega_{p\ell}^2(z))\, \rho_{sp}(k,z,\omega) = = \frac{2\pi e^2}{m}\frac{d\rho_o}{dz}\int_{-\infty}^{+\infty} dz'\, \exp[-k|z-z'|]\, \mathrm{sgn}(z-z')\, \rho_{sp}(k,z',\omega) \tag{7.6}$$

In general this equation can be solved only if we specify the form of $\rho_o(z)$. The exception is the case $k = 0$, when we obtain

$$\rho_{sp}(0,z,\omega) = A\frac{d}{dz}\left[\frac{1}{\omega^2 - \omega_{p\ell}^2(z)}\right] \tag{7.7}$$

and the corresponding eigenvalue is the surface plasmon frequency $\omega_s = \omega_{p\ell}/\sqrt{2}$.

b. Spherical systems

We consider here the self-oscillations in spherical systems, systems where $\rho_o(\underline{x}) = \rho_o(r)$. Our integral equation becomes

$$\rho_{sp}(\underline{x},\omega) = \frac{e^2}{m} \frac{1}{\omega^2 - \omega^2_{p\ell}(\underline{x})} \frac{d\rho_o}{dr} \frac{d}{dr} \int d^3x \frac{\rho_{p\ell}(\underline{x}',\omega)}{|\underline{x} - \underline{x}'|} \quad (7.8)$$

Expanding $\rho_\ell\ (\underline{x},\omega)$ in spherical harmonics

$$\rho_{sp}(\underline{x},\omega) = \sum_{\ell m} \rho_\ell(v,\omega)\ Y_{\ell m}(\theta,\varphi)$$

we obtain

$$\rho_\ell(r,\omega) = -\frac{4\pi e^2}{m(2\ell+1)} \frac{1}{\omega^2 - \omega^2_{p\ell}(v)} \frac{d\rho_o}{dr} \int dr'\ \rho_e(r',\omega)\ \times$$

$$\times \left[\ \ell\theta(r'-r)\left(\frac{r}{r'}\right)^{\ell-1} - (\ell+1)\ \theta(r-r')\left(\frac{r'}{r}\right)^{\ell+2}\right] \quad (7.9)$$

As a typical example, we write

$$\rho_o(r) = \rho_{in}\ \theta(R-r) + \rho_{out}\ \theta(r-R) \quad (7.10)$$

where ρ_{in} is the electron density inside a sphere of radius R and ρ_{out} is the density outside the sphere. Such a form of $\rho_o(r)$ describes a sphere filled with electrons of constant density in an infinite electron gas. The cases ρ_{in} or ρ_{out} respectively represent a void in an electron gas or a sphere in vacuum.

Since the density is constant except the discontinuity at the surface of the sphere, the solutions are confined to the spherical surface and we write

$$\rho_\ell(r,\omega) = A_e\ \delta(r-R)$$

From equation (7.9) we then obtain the well-known classical results for the plasmon modes

$$\omega^2_\ell = \omega^2_{out}\ \left(\frac{\ell+1}{2\ell+1}\right) + \omega^2_{in}\left(\frac{\ell}{2\ell+1}\right) \quad (7.11)$$

where

$$\omega^2_{in/out} = \frac{4\pi e^2}{m}\ \rho_{in/out}$$

It is worthwhile to point out that these well-defined modes are a consequence of the discontinuity of the density at the surface of the sphere. For a smooth profile there would be a distribution of frequencies around the classical values. Only when R is very large compared with the region over which $\rho_o(r)$ varies, we can usually neglect the width of the spherical plasmons.

c. A spherical shell

As a final example we consider the case of a spherical shell for which

$$\rho_o(r) = \rho_o[\theta(r - R + d/2) - \theta(r - R - d/2)] \qquad (7.12)$$

where d is the width and R is the average radius of the shell.

Again the oscillations will be confined to the surfaces and we can write

$$\rho_\ell(r,\omega) = A_\ell\ \delta(r - R + d/2) + B_\ell\ \delta(r - R - d/2) \qquad (7.13)$$

The plasmon modes, determined by equation (7.9) are then found to be

$$\omega^2_{\ell\pm 1} = \omega^2_{sp}[1 \pm \frac{1}{2\ell + 1}\{1 + 4\ell(\ell+1)\ x^{2\ell+1}\}^{1/2}] \qquad (7.14)$$

with

$$x = \frac{R - d/2}{R + d/2} .$$

To see the meaning of these two modes we consider the limiting case of a large sphere with a small hole, which corresponds to the limit $d/2 \to R$ and $x \to 0$. We find then by comparison with equation (7.11) that the $\omega_{\ell+}$ modes correspond to the oscillations on the inner surface and the $\omega_{\ell-}$ modes will correspond to the oscillations localized on the outer surface.

In the limit $x \to 0$ these two modes are uncoupled but for $x \neq 0$ the oscillations on the inner and outer surfaces will couple and the frequencies will shift as given by equation (7.14). The situation is therefore analog to the coupling of surface plasmons on the two sides of a thin film.

The oscillations of a spherical shell may serve as a model for the oscillations of a well confined atomic shell in a heavy atom, by replacing the actual density distribution by a constant density between an inner and an outer radius. The sharp density change at these radii will give the well defined modes just discussed whereas a solution of the integral equation for the actual density distribution would give a width to these modes caused by the inhomogeneity. An application of the model to the $4d^{10}$ shell in Xe has recently been reported [37] . The frequencies for the two dipolar modes were found to be $\omega_- = 95$ eV and $\omega_+ = 160$ eV. About 80 % of the oscillator strength resides in the lower mode. It is interesting to note that the peak of the photoabsorption cross section occurs around 100 eV. Too much significance should not be attached to the close agreement between the dipolar frequency of a classical shell and the strong maximum in the photoabsorption spectrum. However, it is significant to note that the classical forces give a restoring force having a frequency of the right order of magnitude. More significant is the qualitative fact that the collective oscillations in a medium with strong density gradients differs in character from the bulk plasmon type of oscillation in an almost uniform system. It is not the average bulk plasma frequency that determines the frequency of the collective oscillation. Indeed the frequency and therefore the oscillations will occur near the inner and outer boundaries. We see from equation (7.14) that the dipolar frequency ω_- is smaller than the collective frequency $\omega_{p\ell}/\sqrt{3}$ of a uniform sphere and that $\omega_- \to 0$ when the thickness of the shell $d \to 0$.

8. DENSITY FUNCTIONAL THEORY APPLIED TO PHOTOABSORPTION

The density functional scheme provides an efficient and accurate method to calculate ground state properties. Small perturbations of the ground state density due to a weak external field can also be handled comparatively easy. Excited states are in general beyond the power of the theory, since in principle density functional theory does not provide any information about the excited states produced e.g. in photoabsorption. Nevertheless it seems possible to use the method to calculate the response to a time-dependent external field, and the applications of the method have been very successful.

The essence in the calculations using the density functional technique is that the exact ground state particle density can be obtained by solving a set of on-electron wave equations

$$[-\frac{1}{2}\nabla^2 + V_n + V_H + V_{xc}]\varphi_i = \varepsilon_i\varphi_i \tag{8.1}$$

V_n represents the nuclear potential, V_H is the self-consistent Hartree potential, and V_{xc} is the exchange-correlation potential.

Kohn and Sham [6] have shown that the exchange-correlation energy of the system is a universal functional of the density $\rho(\underline{x})$ and the exchange-correlation potential is the functional derivative

$$V_{xc}(\underline{x}) = \frac{\delta G}{\delta\rho(\underline{x})} \tag{8.2}$$

The ground state density $\rho_o(\underline{x})$ is given by

$$\rho_o(\underline{x}) = \sum_i^{occ} |\varphi_i(\underline{x})|^2 \tag{8.3}$$

The orbitals are occupied according to the Pauli principle following the ordering of the corresponding eigenvalues ε_i. Although the orbital functions φ_i and the eigenvalues ε_i do not have a rigorous meaning, we shall follow the general practice and treat φ_i and ε_i as one-electron wave functions and energies respectively.

It is straightforward to calculate the perturbation of the ground state density caused by a weak static external field $V_{ext}(\underline{x})$. The density will now change to $\rho(\underline{x}) = \rho_o(\underline{x}) + \rho_1(\underline{x})$ and we can calculate $\rho_1(\underline{x})$ from the first order change in the orbital functions $\varphi_i(\underline{x})$. The perturbing field $V_1(\underline{x})$ will not only be the external field, since the induced density $\rho_1(\underline{x})$ will give a change in the Hartree potential, $\delta V_H(\underline{x})$ as well as a change in the exchange-correlation potential, $\delta V_{xc}(\underline{x})$, given by the formulas

$$\delta V_H(\underline{x}) = \int d^3x' \frac{\rho_1(\underline{x}')}{|\underline{x} - \underline{x}'|} \tag{8.4}$$

and

$$\delta V_{xc}(\underline{x}) = \left.\frac{\partial V_{xc}}{\delta\rho}\right|_{\rho=\rho_o(\underline{x})} \rho_1(\underline{x}) \tag{8.5}$$

The first order change in the density is then given by the self-consistent relation

$$\rho_1(\underline{x}) = \int d^3x' \; P_o(\underline{x},\underline{x}',0)\,[V_{ext}(\underline{x}') + \delta V_H(\underline{x}') + \delta V_{xc}(\underline{x}')] \tag{8.6}$$

In equation (8.6) $P_o(\underline{x},\underline{x}',0)$ is the static response function in first order perturbation theory, and depends only on the one-electron orbitals and energies φ_i and ε_i. This approach has been applied successfully to the calculation of atomic polarizabilities by Stott and Zaremba [38] and by Mahan [39] and others.

The idea to use a self-consistent approach based on the density functional theory also for time-dependent phenomena was first put forward by Zangwill and Soven [4] and has been applied with considerable success to atomic xenon, barium and cerium [40].

Considering now a time-dependent external field of wave length long compared with the size of the atom having frequency ω, we obtain by the same arguments as those leading to equation (8.6) the self-consistent formula

$$\rho_1(\underline{x},\omega) =$$

$$= \int d^3x' \; P_o(\underline{x},\underline{x}',\omega)\,[V_{ext}(\underline{x}',\omega) + \delta V_H(\underline{x}',\omega) + \delta V_{xc}(\underline{x}',\omega)] \tag{8.7}$$

The response function $P_o(\underline{x},\underline{x}',\omega)$ takes the form as for independent particles, thus

$$P_o(\underline{x},\underline{x}',\omega) = \sum_n \left[\frac{\phi_n(\underline{x})\,\phi_n^*(\underline{x}')}{\omega + i\varepsilon - \omega_n} - \frac{\phi_n(\underline{x}')\,\phi_n^*(\underline{x})}{\omega + i\varepsilon + \omega_n}\right] \tag{8.8}$$

where the energies and amplitudes are the ones calculated for the ground state density $\rho_o(\underline{x})$.

We notice that the theory has the same general structure as the random phase approximation (RPA) discussed in section 6. The procedure corresponds to self-consistent first order perturbation theory. The perturbation is the sum of the external and induced fields, however the induced fields now include the contribution from exchange and correlation. The scheme is similar to an RPAE calculation but with the HF orbitals, energies and exchange matrix elements replaced by the orbitals, energies and linearized potentials in the density functional approach. The calculations are greatly simplified by the fact that all potentials are local functions of position. As already mentioned the applications to rare gas atoms, barium and cerium have shown very good agreement and compare very well with the many-body calculations by Amusia, Wendin and others. This suggests that this approach may be useful in applications to small and medium size molecules.

We should finally note that Zangwill and Soven have recently applied the method to photoabsorption and photoemission of copper [41]. Similar calculations have also been reported by Zaremba (private communication).

REFERENCES

[1] N.H. March and M.P. Tosi, Proc. Roy. Soc., A330, 373 (1972).

[2] F. Bloch, Z. Phys. 81, 363 (1933).

[3] S.C. Ying, J.R. Smith, and W. Kohn, J. Vac. Sci. and Tech. 9, 575 (1972).

[4] A. Zangwill and P. Soven, Phys. Rev. Lett. 45, 204 (1980).

[5] P.A.M. Dirac, Proc. Cambridge Phil. Soc. 26, 376 (1930).

[6] W. Kohn and L. Sham, Phys. Rev. 140, A1133 (1965).

[7] P. Gombas, Z. Physik 121, 523 (1943).

[8] E.P. Wigner, Phys. Rev. 46, 1002 (1934).

[9] P. Hohenberg and W. Kohn, Phys. Rev. B136, 864 (1964).

[10] H. Jensen, Z. Physik, 106, 620 (1937).

[11] R. Ruppin, J. Phys. Chem. Solids, 39, 233 (1978).

[12] G. Barton,

[13] J. Harris and A. Griffin, Phys. Lett. 34A 51 (1971).

[14] F. Garcia-Moliner and F. Flores, Introduction to the Theory of Solid Surfaces, Cambridge University Press 1979. The book contains references to the original papers.

[15] P.J. Feibelman, Phys. Rev. B14, 762 (1976).

[16] P. Apell, Physica Scripta (in press).

[17] P.J. Feibelman, Phys. Rev. B9, 5077 (1974).

[18] N.D. Lang and W. Kohn, Phys. Rev. B7, 3541 (1973).

[19] A. Equiluz, S.C. Ying and J.J. Quinn, Phys. Rev. B11, 2118 (1975).

[20] P. Ahlqvist and P. Apell, Physica Scripta (in press).

[21] J.A. Ball, J.A. Wheeler and E.L. Fireman, Rev. Mod. Phys. 45, 333 (1973).

[22] J.D. Walecka, Phys. Lett. 58A, 81 (1976).

[23] D.A. Kirzhnitz, Yu. E. Lozovik and G.V. Shpatakovskaya, Sovjet Physica-Uspekhi, 18, 649 (1975).

[24] W. Brandt, L. Eder and S. Lundqvist, J. Quant. Spectrosc. Radiat. Transfer, 7, 185 (1967).

[25] P.B. Roulet and P. Nozières, J. Physique 29, 167 (1968).

[26] G. Wendin, J. Phys. B, 3, 467 (1970).

[27] D.A. Kirzhnitz and Yu. E. Lozovik, Sovjet Physica Uspekhi, 9,430 (1966).

[28] G.V. Gadiyak, D.A. Kirzhnitz and Yu. E. Lozovik, JETP Lett., 21, 61 (1975).

[29] I. Lindgren and S. Lundqvist (ed.), Many-body theory of atomic systems, Physica Scripta 21, No 3/4 (1980).

[30] M.Ya. Amusia and N.A. Cherepkov, Case studies in Atomic Physics 5, 47 (1976).

[31] G. Wendin, in Photoionization and other probes of many-electron interactions (ed. F. Wuilleumier) NATO Advanced Study Institute Series, Plenum Press 1976.

[32] G. Wendin, Structure and bonding, 45, 1 (1981).

[33] L.S. Cederbaum and W. Domcke, Adv. Chem. Phys. 36, 205 (1977).

[34] J.E. Inglesfield and E. Wikborg, Solid State Comm. 14, 661 (1974).

[35] G. Mukhopadhyay and S. Lundqvist, Nuovo Cimento, 27B, 1 (1975).

[36] G. Mukhopadhyay, Physica Scripta (in press).

[37] G. Mukhopadhyay and S. Lundqvist, J. Phys. B, 12, 1297 (1979).

[38] M.J. Stott and E. Zaremba, Phys. Rev. 21A, 12 (1980).

[39] G.D. Mahan, Phys. Rev. A, 22, 1780 (1980).

[40] A. Zangwill and P. Soven, Phys. Rev. (in press).

[41] A. Zangwill and P. Soven, Phys. Rev. (in press).

DYNAMICAL STRUCTURE FACTOR OF AN ELECTRON LIQUID

H. Yasuhara

College of Arts and Sciences
Tohoku University
Sendai, Japan

I. INTRODUCTION

I would like to mention our recent theoretical study on dynamical aspects of electron correlations at metallic densities [Awa, Yasuhara and Asahi, 1981]

The plasmon excitation in metals has been widely investigated since the fifties [Pines, 1963] Recently experimental observations by inelastic electron and X-ray scattering have been extended to the wave-number region beyond the cut-off wave-number q_c [Zacharias,1975; Batson, Chen and Silcox, 1976; Gibbons, Schanatterly, Ritsko and Fields, 1976; Platzman and Eisenberger, 1974; Priftis, Boviatsis and Vradis, 1978] It has been commonly observed for several metals that just before the cut-off wave-number q_c the plasmon dispersion curve bends appreciably from the one theoretically predicted by the RPA and that the plasmon peak persists in the one-pair excitation region though its strength is much weaker.

Particularly inelastic X-ray scattering experiments carried out by Platzman and Eisenberger [1974] have first revealed that the dynamical structure factor $S(q,\omega)$ in the intermediate wave number and intermediate frequency region has an interesting structures which cannot be explained by the RPA. That is, it consists of two peaks or one peak accompanied by a well-developed shoulder. One is a sharper plasmon-like peak and shows almost no dispersion. In some cases it even shows a negative dispersion. The other is a broad peak or a shoulder and shows a considerable dispersion.
Such an interesting structure has also been observed by Priftis, Boviatsis and Vradis [1978] It is generally believed that these experimental features of $S(q,\omega)$ are ascribed to correlation effects

of an electron liquid and that the band structure is of little importance in this case.

A theoretical interpretation of the two-peak structure in $S(q,\omega)$ has been the subject of controversy in the last several years. Mukhopadhyay, Kalia and Singwi [1974] have tried to tackle this problem in a somewhat adhoc way. They have added the imaginary part of the self-energy to the one electron energy occurring in Vashishta-Singwi's dielectric function. As a consequence they have obtained excitation spectra which have some resemblence to the observed ones. However, their calculated structure in $S(q,\omega)$ could be washed out when convoluted with an experimental resolution function of width 5 eV. Their work has also been open to question chiefly because it violates the continuity equation.

According to the memory function approach, the continuity equation, the compressibility sum rule and the third frequency moment sum rule can easily be satisfied. Several authors [Mukhopadhyay and Sjoelander, 1978; Yoshida, Takeno and Yasuhara, 1980] have therefore tried to tackle this problem through the memory function approach but no one seems to have obtained much success.
We may say that a satisfactory solution has not yet been given to this problem. Under these circumstances we think it necessary to reconsider the problem from a fundamental point of view.
Diagrammatic considerations are, though complicated, very helpful for us to understand dynamical aspects of electron correlations.
We have pursued our studies in this direction.

The RPA gives an adequate description of long-range correlations such as plasmon excitations and screening. Its validity, however, is restricted to the high density region where such long-range correlations plays a dominant role. It has been recognized from an early stage that short-range correlations become increasingly important as the electron density is lowered to metallic levels.

A successful treatment of short-range correlations at metallic densities was first devised by Singwi, Tosi, Land and Sjoelander [1968] Later on, a better understanding of short-range correlations has been achieved by the use of diagrammatic method [Yasuhara, 1972, 1974; Hede and Carbotte, 1972; Lowy and Brown, 1975; Awa and Asahi, 1980] It has been found that the particle-particle ladder interactions are indespensable for the description of short-range correlations. In the metallic density region where the magnitude of the potential energy is comparable to that of the kinetic energy, a strongly repulsive part of the Coulomb potential has a considerable influence on electron correlations. One must therefore sum a particle-particle ladder interactions up to infinite order for the correct description of short-range correlations. Intuitively speaking, the short-range correlation implies that those interactions whose momentum transfers are larger than the order of the Fermi momentum are essentially reduced, compared with the lowest order estimation in the RPA.

The short-range correlation has frequently been regarded as a local field correction to the RPA which is essentially the Hartree approximation in the treatment of the effective electric field. Using diagrammatic technique Hubbard [1957] has considered the local field correction arising from the exchange effect. Many authors have used the Hubbard type of dielectric functions with variously modified forms of the static local field correction G(q), with the intention of including the correlation correction as well. These expressions, however, are constructed only from one-pair excitations and do not explicitly include multi-pair excitations which should exist originally. Such a treatment may be tolerable for the description of the pair distribution function g(r) or the static structure factor S(q) which is the integration of $S(q,\omega)$ over all frequencies. It is very probable that higher order excitations such as two-pair excitations and one-pair-one-plasmon excitations makes a significant contribution to the spectral shape of $S(q,\omega)$ in the metallic density region. So far as q,ω values are far from the one-pair excitation region simple perturbation expressions for these excitations are certainly meaningful but they are divergent in the very one-pair region. This is because there is a strong coupling between usual one-pair excitations and higher order excitations in the one-pair region and its neighboring area. We must therefore use the renormalized form of Green functions when calculating $S(q,\omega)$ in the intermediate frequency and intermediate wave number region. The problem of $S(q,\omega)$ thus requires a more sophisticated treatment of exchange and correlation corrections.

II. FORMULATION

The dynamical structure factor $S(q,\omega)$ is directly related to the imaginary part of the inverse dielectric function as,

$$S(q,\omega)= -h/v(q)\pi\cdot \mathrm{Im}\{1/\varepsilon(q,\omega)\} , \tag{2.1}$$

$$\varepsilon(q,\omega)= 1+v(q)\pi(q,\omega) , \tag{2.2}$$

where v(q) is the Coulomb interaction and $\pi(q,\omega)$ is the proper polarization function. A faithful calculation of $S(q,\omega)$ or the proper polarization function by the use of the renormalized form of Green functions is exceedingly difficult. Instead, we shall persue our studies resorting to some physical considerations. The formal expression for the proper polarization function $\pi(q,\omega)$ is given as,

$$\pi(q,\omega) = -2\int\frac{d\varepsilon}{(2\pi i)}\int\frac{dp}{(2\pi)^3}\, G(p,\varepsilon)G(p+q,\varepsilon+\omega)\Lambda(p,\varepsilon;q,\omega) \qquad (2.3)$$

$$= -2\int\frac{d\varepsilon}{(2\pi l)}\int\frac{dp}{(2\pi)^3}\, G(p,\varepsilon)G(p+q,\varepsilon+\omega)\cdot\{\ 1 +$$

$$\int\frac{d\varepsilon'}{(2\pi i)}\int\frac{dp'}{(2\pi)^3}\, I(p,\varepsilon;\ p',\varepsilon',q,\omega)G(p',\varepsilon')G(p'+q,\ \varepsilon+\omega) + \ldots\ \} \qquad (2.4$$

Here, $G(p,\varepsilon)$ denotes the renormalized one-particle Green function. The function $\Lambda(p,\varepsilon;q,\omega)$ and $I(p,\varepsilon;p',\varepsilon\ ;q,\omega)$ are the proper vertex function and the irreducible particle-hole interaction, respectively. Let us first split the renormalized Green function into two parts: one is a coherent part which originates from the contribution of a quasi-particle's pole ($\varepsilon = E(p)$) on the analytically continued plane The other is an incoherent part from configurations of several elementary excitations.

$$G(p,\varepsilon) = \frac{z_p}{\varepsilon - \varepsilon(p) - \Sigma(p,E(p))} + G_{incoh}(p,\varepsilon), \qquad (2.5)$$

where z_p is the renormalization constant of a quasi-particle. For the moment let us put $\Lambda = 1$ in Eq. (2.3). A sharp frequency dependence of $\pi(q,\omega)$ does arise from a convolution integral of two coherent parts of Green functions. On the other hand, the remaining integrals including the incoherent part are not expected to give a significant frequency dependence; they appear as a broad background. At first sight one may suppose that the magnitude of the coherent contribution is considerably reduced owing to the appearance of two renormalization constants. It must however be remembered that there is a large amount of cancellation between self-energy and vertex corrections. In a related problem of many-body effects on the optical absorption strength in the interband region for alkali metals, it has been indicated that the above renormalization effect is overcome by the inclusion of the vertex function $\Lambda(p,\varepsilon;q,\omega)$; the vertex correction is partially cancelled by the effect of z_p . The net effect on the strength of one-pair absorption amounts roughly to the vertex correction calculated in the statically screened interaction[Beeferman and Ehrenreich, 1970; Watabe and Yasuhara, 1971]. Assuming that the above conclusion is also valid for the present problem, we shall obtain an approximate form of $\pi(q,\omega)$. Considering, in advance, the cancellation between vertex and self-energy corrections, we put $z_p = 1$ and neglect the dynamically interacting part of the irreducible particle-hole interaction. Thatis, the many-body corrections to $\pi(q,\omega)$ are effectively

represented by an energy width of the quasi-particle coming from the imaginary part of the self-energy correction and those parts of the vertex correction which can substantially be regarded as static interactions. The energy shift of the quasi-particle coming from the real part of the self-energy is omitted, since it is almost independent of the magnitude of the quasi-particle's wave-number [Hedin and Lundqvist, 1969]
The resulting expression for $\pi(q,\omega)$ is given as,

$$\pi(q,\omega) = \frac{\pi^{(0)}(q,\omega)}{1+I(q)\pi^{(0)}(q,\omega)} , \qquad (2.6)$$

where $\pi^{(0)}(q,\omega)$ is the free polarization function with the modification of the energy width. The function $I(q)$ in Eq. (2.6) is a statically approximated form of the irreducible particle-hole interaction and represents the local field correction to the Hartree field.
The main features of excitation spectra of $S(q,\omega)$ are expected to come from such a coherent part of higher order excitations as makes up the energy width of the quasi-particle. We include the effect of multi-pair excitations of $S(q,\omega)$ only through the energy width of quasi-particle. The remaining part of two-pair and one-pair-one-plasmon excitations may constitute a broad background.
The above approximation may be termed the quasi-one-pair excitation approximation. The problem is then to evaluate the energy width of quasi-particle and the statically approximated form of particle-hole interaction. Taking into account the short-range correlation important at metallic densities, we shall approximate $I(q)$ as,

$$I(q) = -G(q)v(q) = 1/2 \langle I(p,p';q)-v(q) \rangle_{pp'}$$

$$+ 1/2 \langle I(p,p';q)-v(q)-I(p,p';p-p'+q) \rangle_{pp'} , \qquad (2.7)$$

where $G(q)$ is the spin-averaged local field factor and $I(p,p';q)$ is the particle-particle ladder vertex which is the solution of the following integral equation:

$$I(p,p';q) = v(q) + \int \frac{dk}{(2\pi)^3} v(q-k) \cdot$$

$$\frac{(1-f(p+k))(1-f(p'-k))}{\varepsilon(p)-\varepsilon(p+k)+\varepsilon(p')-\varepsilon(p'-k)}\, I(p,p';k)\ . \tag{2.8}$$

Here, $\langle \cdots \rangle_{pp'}$ denotes an averaged value over p and p' within Fermi spheres. The functions $\varepsilon(p)$ and $f(p)$ are the free electron energy and the Fermi distribution function at zero temperature, respectively. The first term on the right hand side of Eq. (2.7) is the local field correction due to spin-antiparallel correlation which is adequately represented by the ladder vertex; note that the first order term in the Coulomb interaction is excluded there since we treat the proper polarization function. An approximate expression of the averaged value $I(p,p';q)$ obtained by the author [1974] is used for $q>p_f$ where the local field correction due to the strong Coulomb repulsion plays an essential role.
The second term on the right hand side of Eq. (2.7) is the local field correction due to spin-parallel correlation; the direct process is accompanied with the exchange counterpart. We here estimate it to first order in the Coulomb interaction, because the higher order effects on spin-parallel correlation probably are little significance. For $q>p_f$ an averaged value of $v(p-p'+q)$ over p and p' is chosen to be $4\pi e^2/(q^2+p_f^2)$ as Hubbard first did [1957]
For $q<p_f$ we use an extrapolated form such as $G(q)=\alpha q^2+\beta q^3+\gamma q^4$, where β and γ are determined so as to reproduce the value of $G(q)$ at $q=p_f$ and its derivative; $\alpha=1/4$ giving the compressibility in the Hartree-Fock approximation.

We shall approximate the energy width $\Gamma(p)$ using the imaginary part of the self-energy estimated at the unperturbed pole.
The energy width $\Gamma(p)$ is proportional to $(p-p_f)^2$ in the immediate neighborhood of p_f and increases suddenly at the immediate vicinity of p_f+q_c where a damping channel due to the plasmon emission opens Hedin and Lundqvist [1969]. For the quantitative estimation of $\Gamma(p)$ at metallic densities it is necessary to include the local field correction arising from the Coulomb repulsion as well as the exchange effect, which is not allowed for in the RPA calculation. The resulting expression for $\Gamma(p)$ is written as,

$$\Gamma(p)=\int\frac{dq}{(2\pi)^3}\, v(q)[1-G(q)][1-C(q)]\ \cdot$$

$$\mathrm{Im}\ \{1/\ \varepsilon(q,\varepsilon(p)-\varepsilon(p-q))\}\cdot\{\theta(\varepsilon(p)-\varepsilon(p-q))$$

$$- \theta(\varepsilon_f - \varepsilon(p-q))\} , \tag{2.9}$$

where ε_f is the Fermi energy and $\theta(x)= 1$ for $x >0$, zero otherwise. The dielectric function in Eq. (2.9) is the Hubbard form with the local field factor $G(q)$ of Eq. (2.7). In Eq. (2.9) the local field factor $C(q)$ due to spin-antiparallel correlation appears, which is defined as follows,

$$- C(q)v(q) = \langle I(p,p';q)\rangle_{pp'} - v(q) \quad . \tag{2.10}$$

III. NUMERICAL RESULT

NUMERICAL RESULT

Numerical calculation of $S(q,\omega)$ has been carefully performed for the electron density $r_s=2.0$ appropriate for Al and roughly for Be ($r_s= 1.88$), over the wave-number range $0 < q/p_f < 2.6$. Calculated spectra for various wave-number ratios, 1.2,1.6,1.8,2.0, and 2.4 are shown in Fig. 1. Not only gross features but also the fine details of the experimental $S(q,\omega)$ for Be can be reproduced by the present calculation(see Fig. 2)

Numerical results are summarized as follows:

(1) for $q/p_f< 0.8$ the plasmon peak has finite line width and shows dispersion smaller than the RPA's one. The cut-off wave-number q_c calculated from the dielectric function with the local field factor $G(q)$ included is $0.68p_f$, while q_c estimated by the RPA is $0.73p_f$. The magnitude of q_c is thus reduced by the local field correction. This reduction effect is more pronounced as r_s increases.

(2) About at $q/p_f =0.8$ there occurs a crossing over of the plasmon-like peak and the individual excitation peak. For $0.9 < q/p_f < 1.5$ a considerably sharp plasmon-like peak accompanied with a well-developed shoulder can be seen.

(3) For $1.5 < q/p_f < 2.0$ a broad peak appears on the high energy side of the sharp plasmon-like peak. As q increases the strength of the broad peak gradually develops relative to that of the sharper one.

(4) Abour at $q/p_f =2.0$ a switching over of the two strengths occurs. For $q/p_f >2.0$ the strength of the plasmon-like peak becomes weaker and fades away as q increases. However, even at $q/p_f=2.5$ a weak peak can be discerned.

(5) As for dispersion of the plasmon and plasmon-like peak its position, starting from the plasmon frequency $\omega_{pl}/\varepsilon_f=1.33$ at $q =0$, reaches its maximum value 1.9 at about $q/p_f= 0.8$.

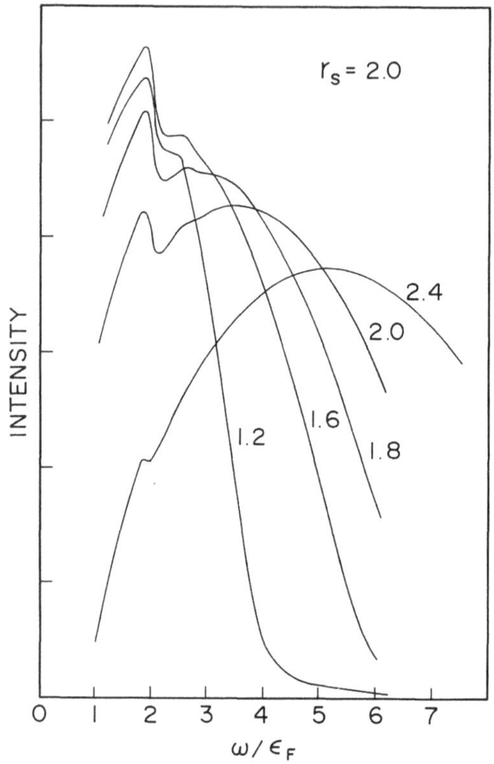

Figure 1. Calculated spectra of $S(q,\omega)$ for wave-number ratios 1.2, 1.6, 1.8, 2.0 and 2.4 for $r_s = 2.0$.

For $0.9 < q/p_f < 1.5$ negative dispersion can be seen. For $1.5 < q/p_f$ it shows almost no dispersion, located at 1.7. (6) There appears a high energy tail outside the usual one-pair region. Calculated spectra of $S(q,\omega)$ as a whole are shifted by a considerable amount to the low energy side, compared with the RPA's spectra, in the intermediated wave-number region. This is for the most part due to the local field correction. The inclusion of the energy width furthermore shifts the spectra to the low energy side. The distinct two peaks and the well-developed shoulder shown in Fig. 1 could not possibly be washed out even if convoluted with an experimental resolution function of width 5 eV.

Electron scattering experiments [Zacharias, 1975; Batson, Chen and Silcox, 1976] for Al have been extended well beyond q_c (see Fig. 3.) Calculated dispersion curves of the plasmon and plasmon-like peak and the broad peak are also shown in Fig. 3. An excellent agreement

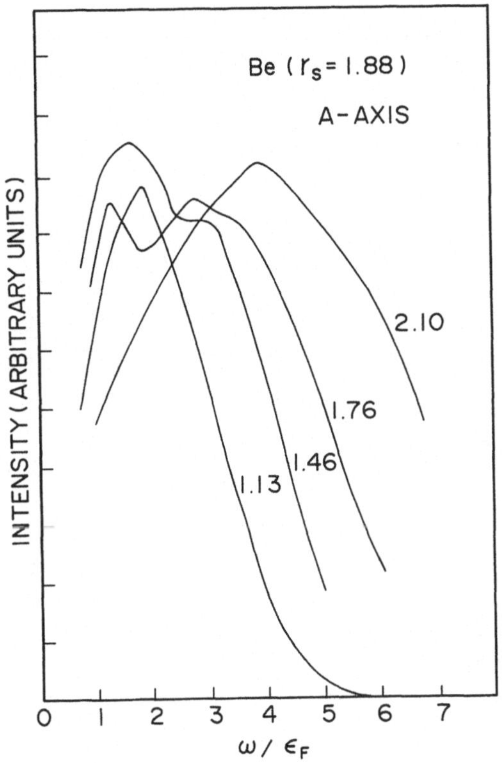

Figure 2. Experimental S(q,ω) by P. M. Platzman and P. Eisenberger [1974] for Be for wave-number ratios, 1.13, 1.46, 1.8, 2.0 and 2.4.

with the experimental plasmon dispersion around q_c has been obtained. For comparison we show in Fig. 4 the experimental dispersion curve for Li (r_s= 3.2) by Priftis, Boviatsis and Vradis [1978] .

In conclusion the existence of the two peaks in the intermediate wave-number region as well as an anomalous dispersion of the plasmon around q_c is ascribed to the striking damping effect of one-electron states originating from the virtual plasmon emission under the influence of strong short-range correlations at metallic densities.

Using their own forms of the local field factor and the energy width, Mukhopadhyay, Kalia and Singwi have first obtained the spectral shape which has some resemblence to the observed one.
Now we may say that though its reasoning seems to be somewhat obscure their theory has a significant implication. In a sense we might say that we have justified its underlying idea from a diagrammatic point of view.

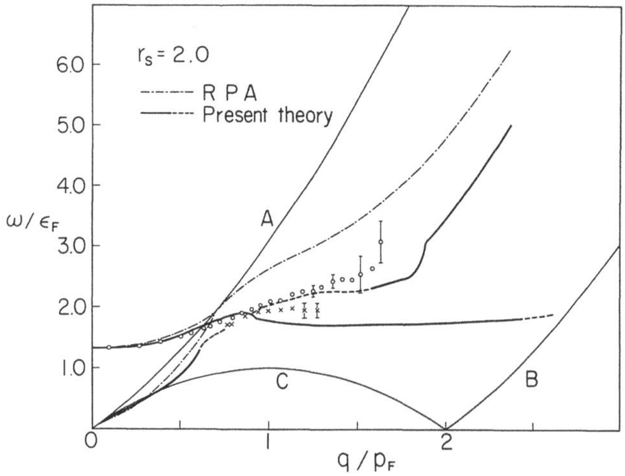

Figure 3. Calculated dispersion curves of the plasmon and plasmon-like peaks and the broad peak. The open circles denotes the experimental results for Al by P. E. Batson, C. H. Chen and J. Silcox [1976] and the crosses those by P. Zacharias [1975]

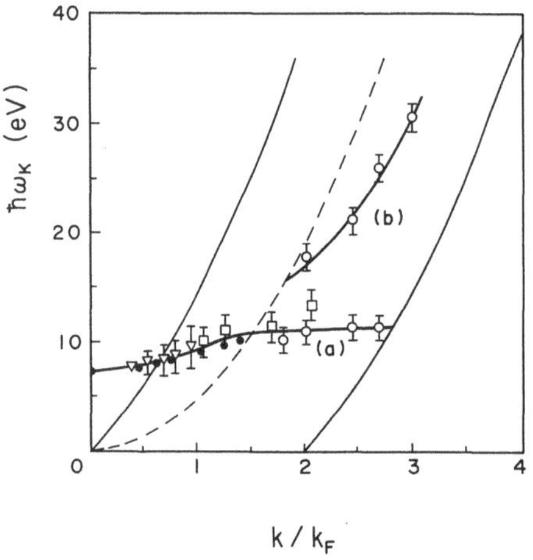

Figure 4. Experimental dispersion curves of the plasmon and plasmon-like peaks and the broad peak for Li (r_s = 3.2) The open circles denotes the experimental results by G. D. Priftis, J. Boviatsis and A. Vradis [1978]

REFERENCES

Awa , K., H. Yasuhara and T. Asahi, 1981, Solid St. Commun. 38, 1285.

Awa, K. and T. Asahi, 1980, J. Phys. Soc. Jpn 48, 757.

Batson, P. E., C. H. Chen and J. Silcox, 1976, Phys. Rev. B13, 937.

Beeferman, L. W.and H. Ehrenreich, 1970, Phys. Rev. B2, 364.

Gibbons, P. C., S. E. Schnatterly, J. J. Ritsko and J. R. Field, 1976, Phys. Rev. B13, 2451.

Hede, B. B. and J. P. Carbotte, 1972, Can. J. Phys. 50, 4512.

Hedin, L. and S. Lundqvist, 1969, in Solid State Physics vol 23, edited by F. Seitz, D. Turnbull and H. Ehrenreich, Academic Press, New York.

Hubbard, J., 1957, Proc. Roy. Soc.(London) A240,539; A243, 336.

Lowy, D. N. and G. E. Brown, 1975, Phys. Rev. B12, 2138.

Mukhopadhyay, G., R. K. Kalia and K. S. Singwi, 1974, Phys. Rev. Letters 34, 950.

Mukhopadhyay, G. and A. Sjoelander, 1978, Phys. Rev. B17, 3589.

Pines, D., 1963, in Elementary Excitations in Solids, Benjamin, New York.

Platzman, P. M. and P. Eisenberger, 1974, Phys. Rev. Letters 33, 152.

Priftis, G.D., J. Boviatsis and A. Vradis, 1978, Phys. Letters 68A, 482.

Singwi, K. S., M. P. Tosi, R. H. Land and A. Sjoelander, 1968, Phys. Rev. 176, 589.

Watabe, M. and H. Yasuhara, 1971, Phys. Letters 34A, 295.

Yasuhara, H., 1972, Solid St. Commun. 11, 1481.

Yasuhara, H., 1974, J. Phys. Soc. Jpn. 36, 361.

Yoshida, F., S. Takeno and H. Yasuhara, 1980, Prog. Theor. Phys. 64, 40.

Zacharias, P., 1975, J. Phys. F5, 645.

AUTHOR INDEX

SUBJECT INDEX